AF803533

Corundum

Butterworths Gem Books
Edited by Peter G. Read

Amber
Helen Fraquet

Beryl
John Sinkankas and Peter G. Read

Corundum
Richard W. Hughes

Garnet
John D. Rouse

Jet
Helen Muller

Pearls: natural, cultured and imitation
Alexander E. Farn

Quartz
Michael O'Donoghue

In preparation

Diamond
Roy V. Huddlestone

Topaz
D. B. Hoover

Butterworths Gem Books

Corundum

Richard W. Hughes
Executive Vice President, Asian Institute of Gemological Sciences, Bangkok, Thailand

Butterworth–Heinemann
London Boston Singapore Sydney Toronto Wellington

PART OF REED INTERNATIONAL P.L.C.

First published 1990

British Library Cataloguing in Publication Data

Hughes, Richard, W.
Corundum.
1. Corundum
I. Title
553.8′4

ISBN 0-7506-1006-9

Library of Congress Cataloging in Publication Data

Hughes, Richard, W.
Corundum/Richard W. Hughes.
p. cm.—(Butterworths gem books)
Bibliography: p.
Includes index
ISBN 0-7506-1006-9:
1. Corundum. I. Title. II. Series.
QE391.C9H84 1990
553.8—dc20 89-15712
CIP

Photoset by Genesis Typesetting, Laser Quay, Rochester, Kent
Printed and bound in England by Courier International Ltd., Tiptree, Essex

To my parents, for their patience, to my brother, for his boundless creativity, and to the night, from which this work has come.

'The price of wisdom is beyond rubies.' (Job 28:18)

'They brought me rubies from the mine,
   And held them to the sun,
I said, "They're drops of frozen wine . . .".' (Emerson)

Profile: summary of constants and characteristics

Chemical composition:	Al_2O_3 (aluminium oxide)
Crystal system:	Hexagonal (trigonal)
Crystal class:	32/*m*
Crystal habits:	Hexagonal bipyramid, tabular hexagonal prism
Varieties:	Red = ruby Blue = sapphire All other colours are termed sapphire with the colour prefix (e.g. yellow sapphire)
Phenomena:	Asterism (six and 12 rays), change of colour, chatoyancy (rare)
Transparency:	Transparent to opaque
Hardness (Mohs):	9
Fracture:	Conchoidal
Cleavage:	Poor, rhombohedral and prismatic
Parting:	Rhombohedral or basal
Streak:	Colourless
Blowpipe:	Infusible
Acids:	Insoluble (soluble in borax)
Lustre:	Vitreous
Specific gravity:	4.00 ± 0.02
Optic character and sign:	Uniaxial negative

Refractive index: 1.762 − 1.770 + 0.008
− 0.006
o-ray = 1.1770
e-ray = 1.762

Birefringence: 0.008 ± 0.001

Pleochroism: Dichroic (weak to strong), o-ray usually more intensely coloured than e-ray
Ruby: o-ray = purplish red; e-ray = orangy red
Blue sapphire: o-ray = violetish blue; e-ray = greenish blue

Absorption spectrum: Ruby (Cr) = fluorescent doublet at 694.2 and 692.8, narrow lines at 668.0 and 659.5, broad band centred at 550, narrow lines at 476.5, 475.0 and 468.5 nm, and general absorption of the violet
Blue sapphire (Fe + Ti) = Three lines at 451.5 (strongest), 460 (weaker), and 470 nm (weakest)

Ultraviolet fluorescence: Ruby = inert to strong red (long wave and short wave)
Blue sapphire = various reactions (see text)

Natural inclusions: Solid crystals of various types, 'silk' of rutile and haematite, boehmite needles, negative crystals, growth twins, repeated twinning (rhombohedral), primary and secondary liquid inclusions, straight and angular (hexagonal) colour zoning, 'treacle' (colour swirls), etc.

Major sources: Burma, Thailand, Cambodia, Sri Lanka, India, Australia, Kenya, Tanzania, Montana and North Carolina (USA), Nigeria, Afghanistan, Pakistan, Laos, China, Colombia, Malawi, Nepal, Zimbabwe, Madagascar, etc.

Synthetic corundum: Verneuil (flame fusion) = all colours, star, colour change
Flux grown = mainly red, blue and orange/yellow
Floating zone = red, blue and orange
Hydrothermal (experimental only) = red and blue

Synthetic inclusions:

1. Verneuil = curved growth lines, gas bubbles, 'Plato' lines (prismatic), rhombohedral repeated twinning (rare), boehmite needles (rare), induced fingerprints or feathers (rare)
2. Czochralski (Inamori) = curved growth lines, gas bubbles, 'rain'-like particles
3. Floating zone (Seiko) = gas bubbles, colour swirls, rectilinear parting
4. Flux grown
   (a) Chatham = primary and secondary flux inclusions, zoned white particle clouds, platinum plates and needles, platinum-rich growth planes, rounded

low-relief crystals, growth (contact) twinning and repeated twinning, straight and angular colour zoning, seed crystal (Verneuil or flux grown)

(b) Kashan = primary and secondary flux inclusions, 'rain'-like flux particles, repeated twinning (rhombohedral), straight and angular colour zoning. The larger flux inclusions often show a two-phase appearance and surface crazing

(c) Knischka = primary and secondary flux inclusions, zoned white particle clouds, two-phase negative crystals (surrounded by bluish clouds at times), platinum plates, growth (contact) twins, straight and angular colour zoning, seed crystal (Verneuil)

(d) Ramaura = primary flux inclusions (flux-filled negative crystals) of an orange-yellow colour, whitish secondary flux inclusions, extremely narrow, straight and angular colour zoning (straight striae), unusual colour swirls and growth disturbances at the junctions of the colour zoning

(e) Lechleitner = thin flux overgrowth (various colours) on a seed (various colours) of natural (rare) or Verneuil synthetic (common) corundum. Secondary 'wispy' flux inclusions which penetrate well into the seed, division between seed often very hard to locate, straight and angular colour zoning in the flux overgrowth, repeated twinning (with natural seed), curved growth lines and gas bubbles in the Verneuil seed

Introduction

Gemstones. The mere mention of the word conjures up images of untold wealth and splendour – from remote mines in distant Eastern lands to the exclusive jewellery salons of the world's major capitals. Gemstones. Since earliest times they have captivated mankind, and their story is the story of man himself.

This fascination with gems is evidenced by the large body of literature that has grown up around them. Thus we have books on gem deposits, the lapidary arts, carved gems, gem treatments, gem identification, history and lore, even books on references to gems in Shakespeare! Individual gem species are also covered in detail. Dozens of monographs may be found which cover the diamond and the pearl; many others exist on emerald, amber, ivory, opal, turquoise, jade, and even the lowly garnet. When it comes to ruby and sapphire, however, there is virtually nothing. Although 1986 and 1987 saw the publishing of a book written in German on identifying ruby (K. Schmetzer, 1986), and a book in English on the Thai ruby (in which the present author played a role) (Hoskin, 1987), there does not exist a single work in English which gives detailed coverage of the gemmological aspects of both ruby and sapphire. Considering the stature of the corundum gems in the precious stone pantheon this is all the more amazing, for together the ruby and the sapphire account for over 50% of the world trade in coloured gemstones. Certainly it is not for lack of interest. The ruby and sapphire have long been considered among the most highly sought after and most precious of gems. No, we must look elsewhere for answers, if indeed they can be found at all.

Corundum, unlike the diamond, the pearl and others, is, by and large, an Asian gem. From the fabled Mogok Stone Tract in Upper Burma to the gem gravels of Sri Lanka, we find the historical occurrences of rubies and sapphires largely restricted to Asia – lands as different from the West as night is from day. Rudyard Kipling summed up these differences best when he made the (oft-quoted) remark: 'East is East and West is West and never the twain shall meet.'

Travellers from the West have been venturing to the Orient continuously since Roman times and before; much of the region was later colonized by the European powers, and many of the nations of the East now display outward signs of Westernization. Even so, today the East remains an alien region for Western travellers, more foreign, in fact, than almost any other point on the globe. Perhaps it is these cultural barriers, then, that have prevented the free flow of information regarding these most Asian of gems, the ruby and the sapphire.

In my many years of residence in the East, I have developed an abiding love and respect for these peoples and their cultures. It is certainly presumptuous even to suggest that such residence has conferred upon me a true and total understanding of these lands. However, it has provided a certain insight into things Eastern in origin, including the ruby and sapphire. And origin is the issue here.

It is a well-known fact that truth is quickly distorted as it travels from one pair of lips to another, diminishing rapidly as the distance from the source grows. Thus seekers of the truth have among their goals a quest for the source, so as to be better able to separate fact from fable. Appreciation of any gemstone is almost always greatest at the source, and this certainly holds true for the ruby and the sapphire. Asians have long prized the ruby and the sapphire, but such has not always been the case in the West (thus it was that Tavernier, in the seventeenth century, was able to take Asian rubies from Europe and sell them at a profit in India); widespread appreciation of corundums in the West is largely a twentieth-century phenomenon.

In this book I have attempted to convey some of this Eastern appreciation of the ruby and the sapphire, while at the same time providing details of the more technical aspects, which have come largely from the West. It is hoped that this, in turn, will foster a greater interest in, and appreciation of, these magnificent gems' for in the world of gems they truly have no peer.

Richard W. Hughes
Bangkok

Acknowledgements

A large number of individuals and organizations have assisted me during the writing of this book. I would like to extend my gratitude to all of them for their cooperation.

Special thanks go to W. K. Ho and Henry Ho of the Asian Institute of Gemological Sciences for their tremendous support and understanding. This book would not have been possible without their help. Others to whom thanks are due include John Rouse, Robert Weiser, the Djevahirdjian Co., Mr and Mrs John Sinkankas, the Gemological Institute of America, Gerald Rogers, Tom Chatham of Chatham Created Gems, Judy Osmer of Ramaura, the Smithsonian Institution, The American Museum of Natural History in New York, the Royal Ontario Museum in Toronto, Barrie O'Leary, Peter Read, Leslie Punchihewa, Paul Kessler, Robert Stevenson, William Spengler, Valaya Rangsit of Christie's in Geneva, Mom Christine Rangsit, Pierre Bariand, Urs Zwyssig, Donna Dirlam, John Koivula, Judith Kinmond, and Rita Pascolini. A special thanks is due the staff of my publisher, Butterworths, for their understanding and support during the many delays and missed deadlines. To them I am truly grateful.

The above is only a partial listing of the many people who gave assistance. To the many others who helped, I would like to extend my heartfelt appreciation. Without such help this work would not have been possible. A special thanks goes to the staff of the Asian Institute of Gemological Sciences in Bangkok. This book, in reality, is their book.

Contents

Chapter 1

History of corundum

In the mountains of Serendib (Ceylon) precious stones are found, of various colours, red, green and yellow, most of which are washed from caverns or crevices by rains and torrents. . . . In some places these are dug out of veins like the ores of metals and the rock has often to be broken to come at the precious stones which it contains (Abu Zeid al Hasan, AD 850 (Ball, 1922))

Introduction

Next to diamond, corundum is the most important of all gem materials. The species name, 'corundum', however, is relatively unknown to laypeople. Much better known are the varietal names of ruby (red) and sapphire (blue).

The name corundum is believed to be derived from the Hindu word *kurand* (Streeter, 1892) or *kuruvinda* (Tagore, 1879). This was the term used in India to describe an impure form of corundum. 'Ruby' means red, and is derived from the Latin *ruber* (red) through the late form *rubinus* (Smith, 1972). 'Sapphire' is a word meaning blue and, when first used, described lapis lazuli. Its exact origin is unknown, but it may have come from Sanskrit originally. The word sapphire came to the West through the Latin form, *sapphirus*, of a Greek word (Smith, 1972). A similar word is found in both Hebrew and Persian.

Today corundum is used solely to describe the mineral species consisting of Al_2O_3 crystallizing in the rhombohedral division of the hexagonal system. In its pure state corundum is colourless; small amounts of impurities, however, are normally present and these give rise to the varieties such as ruby (red = chromium) and sapphire (blue = iron + titanium). The term sapphire alone denotes a blue corundum, while for other colours sapphire is prefixed by the colour – yellow sapphire, green sapphire, etc.

We also know today that the rich red of ruby and the azure blue of sapphire are nothing but different colour varieties of the same mineral – corundum. Such was not always thought to be the case, however. Old Sanskrit texts translated by S. M. Tagore (1879) indicate that in India, at least, this fact may have been known since early times; Westerners did not realize this, however, until the late eighteenth century. Prior to this, in Europe most gems were classified by colour alone. Ruby, for example, was found under the heading of 'carbunculus' (carbuncle), along with many other red stones, such as garnet and spinel ('ballas ruby'). The sapphire was listed under 'hyacinthus' with stones of a yellow and blue colour, such as topaz, quartz, zircon and others.

History

Among the oldest Western texts on precious stones is the *Peri Lithon* (*On Stones*) of the Greek scholar, Theophrastus, which was written about 315 BC. In this famous text gems were grouped primarily according to colour, with different subgroupings separated by properties. Among these it is probable that the true ruby was found under the group carbunculus. Speculations, however, on the exact meaning of the various subgroups to Theophrastus, are, in the opinion of the present author, extremely tenuous as so much time has passed since the work was written. One example of such incorrect deductions made by a modern writer is that of E. H. Warmington (1928) in his book *The Commerce Between the Roman Empire and India.* On page 248 he states that 'corundums were sent from India. . . . I take Pliny's Bactrian stones to be Kashmirian sapphires. . .'. There is one small problem here. The Kashmirian sapphire mines of India were not even discovered until 1881, quite a bit later than Pliny's time.

From the above example it should be very clear that much of what has previously been written about ancient gem sources and gemstones by ancient writers is suspect, with one of the major problems being that the writers themselves were, in general, unfamiliar with gemstones. Thus, to this author, these writings must be taken with a grain of salt.

Much of Theophrastus' writings on carbuncle are concerned with the sources of the stones. These can often be dismissed, as traders, then as now, often try to hide the true origin of the stones. Such places as Turkey and Ethiopia are mentioned but seem improbable in the light of today's knowledge. What we do know is that both rubies and sapphires first made their appearance in Europe in Graeco-Roman times. The only sure way of knowing the origin is to perform a detailed analysis of both inclusions and trace element content on such pieces as exist in museums and collections. Until this is done all such speculations will remain mere conjecture.

Early Western knowledge of precious stones is based largely on the writings of Theophrastus and Pliny. However, because rubies and sapphires were apparently brought from far away, their knowledge of these stones is somewhat suspect. Corundum has always been, and remains today, largely an Asian gemstone. This author has found, during his many years' residence in Asia, that truth tends to be diluted more and more the farther away from the source one goes. This is true *despite* the wonders of modern communications; just imagine what it must have been like during Pliny's time. For this reason the author will not provide here a

Figure 1.1 Dr Sourindro Mohun Tagore, the author of *Mani-Mala*

detailed discussion of what the *anthraces* of Theophrastus or the *lychnis* of Pliny might possibly have been. Entire volumes by Greek and Roman scholars already exist on these subjects (Ball, 1950; etc.). Instead, we will go to the source, India, where the people had close contacts with the corundum gems.

Ruby in ancient India

Few translations of ancient Indian texts on gems into Western tongues exist. Probably the best is the *Mani-Mala* of S. M. Tagore. Written in 1879, *Mani-Mala* consists of translations of facts from several unspecified Indian sources (believed to be the 'Puranas') into three different Indian dialects, as well as English. It has provided us with an unparalleled look at the use and knowledge of precious stones in India. A large section is devoted to ruby, with a somewhat smaller part to sapphire. The highlights of the section on ruby are reprinted below.

152. According to some authorities, the ruby is found in Ceylon and the lands bordering upon the Indus and the 'Ravana Ganga.'

153–154. According to others, ruby quarries are found in Ceylon, Kalpur and Tumbura. The rubies of Ceylon are beautifully red; those of Kalpur are yellow, and are called 'Kuruvinda' (rubicelle); and those of Tumbura are furnished with a blue shade and are hence called 'nilagandhi' (violet ruby). That ruby which shoots a gleam green as the leaf of the asoka goes by the name of 'saugandhike' (spinel).

155. The rubies of Ceylon are good par excellence, while those of Tumbura are at the bottom of the scale, the rubies found in Kalpur occupying the middle position. Rubies vary in value in accordance with the place where they are respectively found.

156–159. Of rubies, some are like the bandhuka flower, some are like the gunja berry, some like the cochineal, some like the China rose, some like blood, some like the seeds of the pomegranate, some like the flower of the butea frondosa, some like red lead, some like the red lotus, some like saffron, some like alakta, some like deep blood, some like deep alakta, some like the kusuma flower, some like alakta with a shade of blue, some like the flower of *Asteracantha longifolia* and kantakari, some like vermilion, some like the eyes of the Greek partridge, *Cuculus indicus*, or the Indian crane, and some like the interior of the half-blown red water lilly.

160. The 'sangandhika' resembles the crystal in brightness, weight and hardness; and its color is like that of the red-lotus dashed with blue.

161. Of 'Kuruvinda' rubies, that which is reddish, owes its origin to crystal; still its inner lustre is such as even to surpass the brightest 'padmaraga'.

162. The 'Kuruvindas' which are produced in the Ravanaganga, are red like the 'padmaraga,' but are not so bright.

163. Rubies have different appellations according to their colors. That which is like the bandhuli flower, is called 'bandhujibi,' that resembling the gunja berry, is called 'sikhandika,' that which is like the cochineal, is called 'indragopi,' that which resembles the China rose, is called 'odrapushpaka,' that which is like blood, is called 'raktakhya,' that which is like the seed of the pomegranate, is called 'kuttima,' that which is like the kinsuka flower, is called 'parna,' and that which is like vermilion, is called 'simantaka.'

164. That ruby the color of which is like mountain slime, is called 'gairikakhya,' that which is like saffron, is called 'surjasanga,' that which is like lac, is called 'drumamaya,' that which is like the kusuma flower, is called 'maharajanagandhi,' that which is like asteracantha longifolia, is called 'gokshura' and that which is like the solanum jaquini, is called 'kantakarika' (Almandine ruby).

165. The ruby the color of which is vermilion, is called 'maniraga,' that which resembles the eye of the Greek partridge, is called 'chakaraksha,' that which is like the eye of the *Cuculus indicus*, is called 'kokilaksha,' that which is like the egg of the Indian crane, is called 'sarasakhya,' and finally that which is like the red-lotus, is called 'kokanada.' The earth produces a splendid variety of the kinds enumerated above.

166. The Andhra country (The modern Telinagana. Its people are called Andarae by Pliny) produces rubies resembling the 'kuruvinda' of Ravanaganga. These have the same value as kuruvindas.

167. The 'padmaragas' of Tumbura, which resemble in color crystalline 'padmaragas,' are of small worth.

168. Heaviness, coolness, evenness, lustre, thickness, deepness of color and auspiciousness are the principal good qualities of gems.

169. The four-fold division into castes obtains also among rubies. The characteristics of each class I shall now describe.

170. The 'padmaraga' (Oriental Ruby) is reckoned a Vipra (Brahmin), the 'kuruvinda' (Rubicelle) a Khetriya, the 'Saugandhika' (Spinel) a Vaisya, and the 'Mansa-khanda' (Ballas Ruby) a Sudra.

171. Some say that the 'Vipra' rubies emit a shade red like the color of the red-lotus or the red-hot charcoal of the Anacia Catechu.

172. The 'kshetriya' ruby casts forward a shade resembling in hue a gunja, vermilion, or the flower of the bandhuka, the orange, or the pomegranate.

173. From the Vaisya saugandhika comes out a shade like vermilion, the flower of the asoka, the China rose, or alakta; or a red lustre tinctured with yellow.

174. A Sudra 'mansakhanda' emits a reddish, unbeautiful, and glossy shade resembling a piece of flesh. The ruby which is red-white is a Brahmin, that which is deep red, is a Kshetriya, that which is red-yellow, is a Vaisya, and finally that which is red-blue, is a Sudra.

175. The authorities mention sixteen kinds of shade, four good and eight bad properties of rubies.

176–177. The rubies of Ceylon emit a gleam like the red-lotus, the glow-worm, fire, or the eye of the pigeon, the *Cuculus indicus*, the Indian crane or the Greek partridge. All these shades are auspicious.

178. The 'Kuruvinda' shoots a shade resembling red lead, the gunja berry, or the flower of *Symplocos racemosa* or of the *Butea frondosa*.

179. The saugandhika gem scatters a gleam either pale-red or like milk or the water colored by the crocus Sativus; – of these, the milky luster blights wealth.

180. The 'inilangandi' shines in a halo like the blue lotus or iron.

181. When the ruby is placed over a mirror and is then exposed to the morning sun, the glass will flash with a lustre composed of all the above-mentioned shades.

182. That ruby which, when exposed to the sun 'new risen,' radiates red rays or paints the whole house with crimson, is surpassingly fair.

183. That 'padmaraga' which, on being cast into a quantity of milk a hundred times its bulk, makes the white mass one entire sheet of red, or sends out a red flame, is nonpareil.

184. That 'padmaraga' which, like the sun, can destroy even Egyptian darkness, is supremely excellent.

185. That 'padmaraga' which, when placed on the lotus-bud, anon makes it blossom, is Perfection's self – and is hardly attained even by the Gods.

186. The 'padmaraga' which is perceptible from a distance like a flame, is entitled 'vansakanti,' – it leads to wealth.

187. All those 'padmaragas' that, on being thrown into a quantity of milk or of water pure or blue two hundred times their bulk, crimson the liquid, conduce to prosperity though they differ among themselves as regards worth.

188. I shall now describe the eight bad properties of rubies, which have been enumerated by the authorities.

189. 'Dwi-chchhayatwa,' 'virupatya,' 'samveda,' 'karkara,' 'asovana,' 'kokila,' 'jara,' and 'dhumra,' are the defects commonly found in rubies.

190. The 'padmaraga' which emits a two-fold shade is called a 'dwi-chchhaya.' A 'dwi-chchhaya' ruby is the death of friends. That ruby which bears a mark like a bird's foot, is called a 'virupa.' A 'virupa' gem always brings humiliation to its master.

191. A 'samveda' ruby is one which shows a mark like a rift. This gem renders its owner liable to blows from a weapon. A 'karkara' ruby bears in its bosom a

gritty fragment. It causes the death of friends and is also fatal to domestic beasts.

192. That 'padmaraga' which naturally appears as if it were immersed in milk, goes by the name of 'asovana'; this gem is the source of a multiplicity of evils.

193. The mark in a ruby which looks like a drop of honey, is called a 'kokila.' A 'kokila' ruby is inimical to life, wealth and fame, and must be shunned.

194. The quality of being discolored is called 'jara.' A 'jara padmaraga' causes loss of wealth. The color of a 'padmaraga' which looks like smoke, is called 'dhumra' (literally smoke). A 'dhumra-colored padmaraga' renders one liable to lightning-stroke.

195. The above-mentioned rubies are worthless and pernicious, and should never be treasured up, even were they to be had for the gathering.

196. Those rubies which send out a two-fold shade, bring about loss; and those which are marked with a bird's foot in two places, are the cause of defeat.

197. Those which bear gritty fragments and are indented, dirty, lack-lustre, rough and discolored, are certainly unlucky, although they may be otherwise good and may belong to a superior caste.

198. If through ignorance, a person happens to wear a defective ruby, loss of friends, destruction of wealth, et cetera, are the inevitable consequences.

199. The man who keeps in his house a sterling 'padmaraga,' can with perfect safety, constantly live in the midst of enemies, and is impervious to the attacks of adverse fortune.

200. He who wears a 'padmaraga' of matchless virtues, need not apprehend any mischief from the foul touch of bad 'padmaragas' without number.

201. According to some, as the contact of a single 'chandala' (a generic name for a man of the lowest and most despised of the mixed tribes, born from a Sudra father and a Brahmin mother) suffices to contaminate ever so many Brahmins, even so one impure gem serves to pollute pure gems ad infinitum.

202. The colors of defective rubies are never like those of good rubies, yet to a certain extent they differ one from the other in color in accordance with their respective mines.

203. Bad rubies should never be worn, although set along with the best rubies – nay, should they happen to be in company with Kaustuva (The name of a celebrated jewel obtained with thirteen other precious things at the churning of the ocean. It was suspended on the breast of Krishna) itself.

204. 'Padmaragas' which resemble in shape a gunja berry, may weigh three, seven or ten gunjas, and the heaviest is the most precious.

205. 'Padmaragas' resembling the fruit of Kroshtu (Terminalia Catappa), may weigh two, eight or twelve gunjas.

206. That 'padmaraga' which is plum-shaped, weighs nine, ten or twelve mashas (a unit of weight for gold), that resembling the fruit of the Emblic Myrobalam in shape, weighs sixteen, twenty or thirty mashas.

207. A rudrakshya-shaped (the berry of a tree, *Elaeocarpus ganitrus*) 'padmaraga' weighs one, two or three mashas; a tambuli-shaped (a type of berry) 'padmaraga' weighs from one to four tolas (an Indian weight unit).

208. That 'padmaraga' which is like a vimba (the bright red gourd of the *Momardica monadelpha* plant), weighs six, eight or ten tolas – this is the

maximum of weight. If any man by good luck happens to obtain a heavier padmaraga, he will succeed in whatever he undertakes.

209. Some vicious rubies are as beautiful as good ones; but the authorities advise us to shun them carefully.

210. That 'padmaraga' which is lack-lustre, or which loses its former brightness on being rubbed by cloth, and which betrays a dark shade on its sides on being held between two fingers, is a false and vicious gem, and should never be worn even when it is selling for a song.

211. Should other tests fail in dispelling doubt, the reality or otherwise of a 'padmaraga' will be indubitably ascertained by rubbing it upon a whetstone or with a 'padmaraga' of tested genuineness. A false 'padmaraga' will receive scratches from the process.

212. That 'padmaraga' which shines out the more, the more it is rubbed, and which is moreover not light, is a pure gem; the others are impure.

213. Excepting 'kuruvinda' and diamond, no gem can make scratches upon either the 'padmaraga' or the 'indranila.'

214. A genuine 'kuruvinda' is worth as much as a 'padmaraga' furnished with every quality.

215. The treasuring up of a 'padmaraga' furnished with every perfection is as meritorious as the celebration of the Aswamedha jajna (an ancient ceremony going back to the Vedic period), and leads to wealth, success and long life.

The following quotations are from the same source (Tagore, 1879) but cover sapphire.

397. The sapphire is found in the country of Padmakara near the Ravana Ganga in Ceylon.

398. All sapphires are not of the same color. Those found in Ceylon resemble one or other of the following objects:– The blue lotus, a cloth dyed in blue, a polished sword, the Bhramara (the black bee), Sri Krishna (a Hindu God represented as beautiful gray), Mahadeva's throat (the throat of this Hindu God is blue), the eye in a peacock's tail, the black aparajita (a name applied to several plants, Clitoria Ternatea, Marsilia Quadrifolia, Sesbania Aegyptiaca).

399–400. The 'dark blue ocean,' the peacock's throat, the bubbles of blue water and the throat of the mad Kokila (Vide nate). These kinds of the Sapphire are generally found.

401–402. A blue Sapphire with a white shade is a Brahmin; a blue one with a red shade is a Kshetrya: a Sapphire of the same color with a yellow lustre is a Vaisya; and finally a blue Sapphire shining in dark lustre is a Sudra.

403. As Rubies are good, middling and bad, – so it is also with Sapphires.

404. That Sapphire whose interior gleams with the iris, is a rare gem of priceless worth.

405. That Sapphire which being thrown into a quantity of milk a hundred times its bulk, paints the whole mass with its native blue, is entitled a 'mahanila.'

406–407. Of gems, those which are blue, are Sapphires; those which are red, are Rubies; those which are green, are Emeralds; those which are ash-colored with a yellow shade, are 'Pushparagas'; those of the last which are red, are called 'Kaurandakas.'

408. Weight, coolness, depth of color, the quality of being 'parsvavarti' (see 411.) and beauty, invest a Sapphire with matchless worth.

409. That Sapphire which combines great weight with small bulk is called 'Guru' (literally 'heavy'): such a gem conduces to increase of family. That which always gives out moisture is called 'Snigdha' (vide ante). A 'Snigdha' Sapphire brings on accession of wealth.

410. That Sapphire which, on being exposed to the morning sun, flames in a blue effulgence, is termed 'Varnadhya.' It brings prosperity.

411. That Sapphire through which any lustrous substance such as crystal, silver, gold, is seen, is entitled 'parsvavarti.' A 'parsvavarti' Sapphire brings on fame.

412. That Sapphire which, when placed in a pot, darkens it all through, is called 'Ranjaka.' This gem increases wealth and is conducive to fame and increase of family.

Defects of the Sapphire

413. 'Avraka,' 'Trasa,' 'C'itraka,' 'Mridgarva,' 'Asmagarva,' and 'Raukshya' are the six defects of Sapphires.

414. A Sapphire the surface of which wears a mica-like sheen, goes by the name of 'Avraka.' Such a gem brings about loss of wealth and life.

415. That mark in a Sapphire which at first sight looks like a rift, is called 'Trasa.' A 'Trasa' renders one liable to bites. That Sapphire which is parti-colored, is called 'C'itraka': it causes loss of family dignity.

416. That Sapphire which contains dirt, has the sobriquet of 'Mridgarva.' This gem produces a variety of skin-diseases like itching.

417. That which contains gritty fragments, is called 'Asmagarva.' This defect is destructive. That which is rough, is called 'Raukshya.' Banishment is the consequence of wearing this jewel.

418. A flawless, sterling Sapphire brings its wearer, strength, fame and length of days.

419. The man who wears a Sapphire of spotless chastity, finds favor with 'Narayana' (the preserver of the Hindu Triad) and acquires longevity, family dignity, fame, understanding and wealth.

420. Those good and bad consequences which follow the wearing of a 'Padmaraga' combining good and bad qualities, are also produced by the wearing of a Sapphire of a like nature.

421. A Sapphire is distinguishable from glass by weight and hardness only.

422. A shining gem which is found in Sapphire quarries, resembling the 'Vramara,' and furnished with a shade, is called 'Vramara.'

423. That Sapphire which shows a dash of red, is called 'Tittiva' (amethyst). No sooner a woman who is in her family way, wears a 'Tittiva' than she is delivered of her child, with perfect ease.

424. The price of the best Sapphire is equal to that of the best 'Padmaraga.'

Although Tagore (1879) does not specifically say so, it is believed that the above is translated from the ancient Sanskrit texts known as 'Puranas' (Sanskrit: 'Ancient lore'). These were Hindu sacred literature, popular encyclopaedic collections of myth, legend and genealogy, varying greatly as to date and origin (*Encyclopaedia Britannica*, 1980).

Traditionally a Purana treats five subjects: primary creation of the universe, secondary creation after periodical annihilation, genealogy of gods and saints, grand epochs, and history of the royal dynasties. Puranas are connected in subject with the *Mahabharata* (*Great Epic of the Bharata Dynasty*) and have some relationship to the lawbooks (*Dharma-sastras*). Around this central core has amalgamated much other material of religious concern during the period from about AD 400 to AD 1000. Puranas were written almost entirely in narrative couplets in much the same easy flowing style as the Greek epic poems (*Encyclopaedia Britannica*, 1980).

It is most interesting to compare the writings of Theophrastus and Pliny with those of the Puranas on ruby. Theophrastus and Pliny spend most of their time discussing slight variations in colour and sources of origin, with methods of testing described only briefly. Pliny (Ball, 1950) states the following about carbunculus:

> In resume, there is nothing more difficult than to attempt to distinguish these various kinds of carbunculi from one another. Further, they are easy to counterfeit and falsify by the art and skill of lapidaries and goldsmiths who put a foil beneath them to make them brilliant and glitter like fire. Some say that the Aethiopians steep their dusky and dark carbunculi in vinegar. As a result, in fourteen days, they become pure and lively and remain so for fourteen months. Carbunculi are imitated by glass and such imitations at first sight are excellent: but by grinding on a mill, the fraud is immediately discovered as is true with any other artificial or false stone: for the substance of the latter is softer and more brittle than that of the true gem. Further, false carbunculi are detected by the lack of hardness of their powder and by their weight: for glass imitations are the lighter of the two. Further, one sees in false carbunculi certain small inclusions, that is blisters and vesicules, which look like silver.

A number of similarities are apparent between Pliny and the Puranas. Both mention hardness as a potential test, with Pliny even noticing the hardness of crushed powders made from the stone. The Puranas go so far as to state that only diamond and 'kuruvinda' (a type of ruby) can scratch the ruby.

The question of density or specific gravity is also brought up by each source. Pliny says only that the carbunculus is heavy, while the Puranas compare the weight of a 'padmaraga' (ruby) to that of a seed of similar size. Here we have the 'seed' of the idea of specific gravity, literally.

Inclusions are also mentioned by both sources, although the Puranas discuss them more in terms of defects, than as identifying features. Pliny, though, does go so far as to state that glass may be identified by reference to included gas bubbles, an extremely astute observation considering that he was writing almost 2000 years ago (Pliny, AD 23–79).

As to the soaking of stones in vinegar to improve their colour, the Puranas also mention a similar process (not reprinted above) but state that the gems must be cooked over a fire. We might immediately read into this an early discovery of heat treatment. We also have a mention in the Puranas of the coolness of sapphires. This may perhaps be an allusion to the warmth of glass to the touch, when compared with crystals.

The most important question, however, is that of ruby and sapphire: were the ancients aware that they were simply different colour varieties of the same mineral? Unfortunately there is no specific mention of this fact in any early writings. From the writings of Theophrastus and Pliny we can be relatively certain that they were not aware of this. With the Puranas it is less certain. Considering what ancient man did know about gems in general, it seems at first hard to believe that this fact was *not* known at the mines. Certainly in Europe, so far from the producing country (probably Sri Lanka), it is amazing that Pliny and Theophrastus got as much right as they did. However, at the mines it is another story entirely. In Sri Lanka the same hole or digging may produce both ruby and sapphire. Not only do the Sri Lankan rubies and sapphires have the same crystal habit (not necessarily true elsewhere for corundum), but in some instances both ruby and sapphire will be found *in the same crystal*. However, in a world where belief in magic and superstition was the norm rather than the exception, it cannot always be assumed that logic and deduction would rise to the top. Even today there is a belief in the mining areas of Sri Lanka, Burma and Thailand that sapphires are nothing more than unripened rubies; if left in the ground for a sufficient length of time it is thought that they will eventually become rubies. The link between the two gems was undoubtedly known. However, this author has not been able to uncover any firm evidence to suggest that ruby and sapphire were regarded as the same material. More likely, it seems that sapphire was thought to *change into ruby*, not that sapphire and ruby were the same substance (except in terms of colour).

After the fall of the Roman Empire, learning in Europe generally halted under the oppressive rule of the Christian Church. To the Christian of the Middle Ages, all knowledge came from God. As God's representatives on earth, church leaders felt that only they should direct learning. The result was a persecution of anyone outside the Church (and a good many within it, as well) with new ideas. Learning, as a result, consisted almost exclusively of repetition of work done by the Greeks and Romans. Thus, we find little of interest on gems from this period. What little there is was derived mainly from Arab sources (who, in turn, got much of their information from the Greeks and Romans). The Puranas are interesting here because they are believed to have been written mainly during this period when much of Europe was intellectually asleep.

The reawakening of Europe came slowly at first, but the pace soon quickened, fuelled by the exploration of far-off lands, to the point where even the Church could not hold it back. Much of this effort was directed towards the East. Early (AD 1200–1500) European travellers to Asia found fabulous lands of untold wealth and prosperity, far in advance of the countries from which they had come. (It is interesting that while Europe began to move forward, throwing off the shackles of the Church, Asia slowly began to regress at the hands of despotic rulers.) Despite attempts by the Church to retain its power (such as the Inquisition), true separation of church and state was eventually effected. The European Renaissance was born,

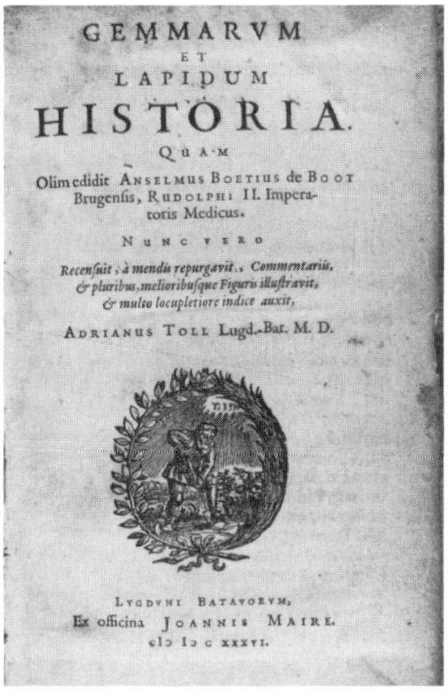

Figure 1.2 The title page of the 1636 edition of Anselm Boetius de Boodt's famous lapidary. It is considered to be the most important lapidary of the seventeenth century

and with it began a flowering of all intellectual pursuits, the likes of which can still be seen continuing today. With the rebirth of democracy came the principle of freedom of thought; this, perhaps more than anything else, produced the modern world in which we live today.

Among the most famous of all early lapidaries is that written by Anselm Boetius de Boodt (1636). Published in Hanover, it is considered by many to be the most important work on gems of the seventeenth century.

De Boodt described the ruby under the heading of 'Carbuncolo' (carbuncle), which literally means 'glowing embers.' He described how these stones were held in high esteem in India and Arabia, and that many kings or other nobles would actually glow at night because they wore so many of these jewels. De Boodt himself did not actually admit to having seen this. He did say, however, that one could separate a true carbuncle from other red stones by its glow at night. Certainly this does not sound like he was referring to a garnet here. Instead, he was most probably referring to rubies or spinels, both of which may fluoresce red to visible light. This concept of ruby emitting a red glow is extremely important and more will be said about it later.

Engraved rubies and sapphires

Despite its great hardness, examples of engraved corundums have come down to us from ancient times. Engraved gems were formerly of much greater importance than at present. Numerous books devoted to them have been written, particularly in the eighteenth and nineteenth centuries. Undoubtedly the most prolific of the authors on this subject was the Rev. C. W. King.

King, writing in the late nineteenth century, authored a number of books on the Glyptic arts. In his *The Natural History, Ancient and Modern, of Precious Stones and Gems, and of the Precious Metals* (1865), he describes several examples of both ruby and sapphire which have been engraved, mostly in Roman times. According to King (1865),

> the experienced Lessing (A. Br. lxxix.), and later the Count de Clarac (Cat. des Artistes Gr. et Rom.), altogether deny the existence of any really antique intagli in these harder gems; . . . . Nevertheless, a few works in Ruby of apparently indisputable antiquity have been observed by me amongst the thousands of other gems examined. First, on account of the quality – a large oval slightly convex stone of, of the true "pigeon's blood" tint, and weighing apparently about 3 carats – is one in the Devonshire Parure (No. 17 in the Bandeau), engraved with a Venus Victrix – a but poor intaglio in the latest Roman manner.

King goes on to mention several other engravings done in spinel, but the general impression is that truly ancient intaglios done in ruby were decidedly scarce.

Early in 1985, a most valuable discovery was made in England. Hidden away amongst old documents at the back of the muniment room in Burghley House, near Stamford, Lincolnshire, was an engraved rock crystal ewer featuring on the front the only known cameo of Queen Elizabeth I – an engraving done on a large ruby. It was discovered by chance when two men from the Royal Commission on Historical Manuscripts came to help sort out the family documents of Lord Exeter.

Four years earlier a schedule written in 1690 of the Countess of Exeter's gems was found. In it the ewer was described as 'A Christall Ewer Garnisht with Diamonds and Rubies one of the Rubies being Queen Elizabeths head'. English photographer Michael Freeman, who has examined the ewer and ruby cameo, described it as being well executed, with the cameo representing an excellent likeness of Elizabeth I. The engraved rock crystal ewer is believed to have been made in the Miseroni workshop in Milan around 1600. It was probably mounted in England around 1660 with an enamelled gold foliage studded with diamonds and rubies (Norman, 1985).

Ancient sapphire ('hyacinthus') engravings were only slightly less scarce. As with the ruby, on account of its extreme hardness the ancients for the most part employed sapphire as a mere ornamental stone for setting in their jewellery, unengraved and unshaped. Apparently they were content to give a light polish to the natural surface of the pebbles.

King (1865) describes

> amongst the Rutupine antiquities preserved in the library of Trinity College, Cambridge, a portion of a necklace of small rough Sapphires drilled at each end and linked together with gold wire, the exact ornament referred to by the poet Naumachius.

According to King, prior to the Imperial epoch engravings in sapphire were extremely rare. He did, however, mention a small Etruscan scarab engraved in poor-quality sapphire, as well as a magnificent head of Jupiter with Greek inscription, executed in the purest Greek style. The latter had been discovered as an ornament on the pommel of a Turkish dagger, the intaglio turned downwards, and the back of the stone crudely faceted by the Oriental lapidary into whose hands this precious monument had fallen, further proof of its antiquity. The stone was a massive 1 inch in diameter. Superior to this as a work of art, and belonging to the same school, was the nearly full face of the Medusa's Head, described by King as one of the chief glories of the famous Marlborough Collection. The carving was said to have been considerably enhanced by the fine quality of the material. Most famous of all, according to King, though, was the Signet of Constantius II (then in the Rinuccini Collection) on a perfect stone weighing 53 carats. The Emperor is represented as spearing a monstrous wild boar before a reclining female figure personifying 'Caesarea of Cappadocia', the scene of the exploit. The inscription CONSTANTIVS AVG in the field suggests that this costly stone had been engraved for the actual signet of the imperial hunter. Also mentioned by King was a sapphire engraving of considerable merit, then in London, of the well-known design of Hebe feeding the Eagle; the stone, heart shaped and of fine colour, measured 1½ × 1¼ inches. It apparently belonged to the time of Hadrian.

The Lapidarium of Marbodus

Among the more well-known lapidaries of the Middle Ages is that of Marbodus, abbot and master of the Cathedral School of Anjou from 1067 to 1081 when he was made Bishop of Rennes. According to C. W. King (1866), much of the poem known today as the Lapidarium of Marbodus was taken from Pliny, but chiefly from Solinus, of whom he paraphrases entire sentences. He also borrows heavily from the so-called Orpheus, a work believed to have been composed in the third century. Later authors, such as Camillo Leonardo (*The Mirror of Precious Stones*), have borrowed largely from Marbodus (King, 1866).

As with the writings of Pliny and Theophrastus, it is often quite difficult to ascertain from the writings of Marbodus exactly which gem he speaks of in each passage. Of the corundum gems, we find several possible mentions in Marbodus. Under 'carbuncle' we may assume that both ruby and spinel were included (King, 1866):

The Carbuncle eclipses by its blaze
All shining gems and casts its fiery rays
Like to the burning coal; whence comes its name,
Among the Greeks as Anthrax known to fame.
Not e'en by darkness quenched its vigour tires;
Still at the gazer's eye it darts its fires;
A numerous race, within the Lybian ground
Twelve kinds by mining Troglodytes are found.

Although Libya is mentioned as a source by Marbodus, today we know of no ruby or spinel coming from this area. Once again this indicates the difficulty in determining exactly to what gem he was referring. Possibly Libya was a trading locality only.

Sapphire is specifically mentioned by Marbodus, but in the time of Theophrastus and Pliny this term designated lapis lazuli. It is more likely that we should find our sapphire under 'hyacinths' (King, 1866):

Three various kinds the skilled as Hyacinths name,
Varying in colour, and unlike in fame:
One, like pomegranate flowers a fiery blaze;
And one, the yellow citron's hue displays.
One charms with pale blue the gazer's eye
Like the mild tint that decks the northern sky:
A strength'ning power the several kinds convey
And grief and vain suspicions drive away.

The above passage contains several interesting points. First of all, we have the apparent knowledge that the hyacinthus may vary in colour – one material, but several varieties. This fits corundum very nicely. Secondly, we see one variety compared to the colour of the pomegranate flowers, a reference that appears throughout history with ruby (the Thai word for ruby, *Thab Tim*, means literally 'pomegranate'). Thirdly, what little we do know of ancient corundum sources suggests that Sri Lanka was probably the major, if not the only, source of corundum in early times. Marbodus mentions three colours under the hyacinthus: red (orange?), yellow and sky blue. It is precisely these three colours that are most common in the corundums of Sri Lanka. Sri Lanka is also perhaps the only major source of light-blue (sky-blue?) sapphires. Furthermore, Sri Lanka is unusual in that all three colours of corundum occur in exactly the same hexagonal bipyramidal habit. In hindsight, it would appear most unusual if the miners in Sri Lanka, at the very least, did *not* notice the resemblance between these gems. Thus, from the above evidence, a very strong case can be made for assuming that, at least among some people, the link between ruby and sapphire was known at an early date. Why then, we may ask, the separate carbuncle category? Possibly this could suggest that most carbuncles were spinels, a gem of quite a different shape from the Sri Lankan corundums. Or perhaps they were spinels plus rubies from another locality (Burma?). The shape of rubies from other sources is almost always distinctly different from that of sapphire. Once polished, the carbuncles might be confused with the red hyacinths, further muddying the proverbial pond. Ah, but this author now finds himself guilty of precisely that which he has accused others – endless speculation without sufficient hard evidence. Tsk, tsk! In any case, it would make a fascinating project for some gemmologist(s) with knowledge of corundum inclusions by locality to study these ancient rubies and sapphires in museum collections, such as those at the Victoria and Albert Museum. Then we might answer once and for all the question of just where the corundums of the ancient world originated from.

Unification of ruby and sapphire

Although the gemmological literature from early times to the Middle Ages does contain hints that at least some people were aware of the relationship between ruby and sapphire, this fact was not proven beyond a shadow of a doubt until the late eighteenth century. The event occurred in France and credit for the discovery goes to the Count de Bournon and Charles Greville, who in 1798 and 1802 published two landmark papers in the *Philosophical Transactions of the Royal Society of London* (Greville, 1798; de Bournon, 1798, 1802).

Prior to this date, the term 'oriental' was used to indicate a superior hardness in the precious stones which we now know as corundum. Thus our blue sapphire was termed 'oriental sapphire', our ruby 'oriental ruby', our yellow sapphire 'oriental topaz', and our green sapphire 'oriental emerald'. The word 'oriental' was derived from the fact that the Orient (India, Sri Lanka, Burma, Thailand, China, etc.) was the major source of these gems. Count de Bournon (1802) described it thus:

> Although the epithet oriental has been for a long time used by the lapidaries, to express, in gems or precious stones, a degree of hardness superior to that of other stones, (the diamond excepted,), which made them capable of taking a more brilliant polish; and although, following the example of the lapidaries, naturalists had employed the same term by way of distinguishing them, there still remained a great uncertainty, respecting the nature of the analogy which really existed between the various stones to which the above epithet was applied.

This nomenclature was not due to any mineralogical knowledge, with the result that a number of different species, related only in hardness and colour, were united together. Slowly, however, mineralogy progressed to the point where the true nature of these substances could be ascertained.

In the eighteenth century, as European colonial empires expanded eastwards, mineral specimens from the new colonies were sent back to Europe. Among these were samples of a material from southern India termed 'convindum' or 'corivindum':

> Nella Corivindum is found in fields where the rice grows: it is commonly thrown up by field rats, and used, as we do emery, to polish iron. (Greville, 1798)
>
> Tella Convindum tis a talky spar, grey, with a cast of green: it is used to polish rubies and diamonds. (Greville, 1798)

The mineral to which British gentlemen stationed in Madras referred was an impure variety of our corundum. As mineralogists began to study these specimens, as well as others of ruby and sapphire, clues were found in their crystal symmetry to link them together. Pioneering work in this regard was performed by Romé de Lisle and the Abbé Hauy, in France, but even these famous mineralogists failed to realize the relationship, due to the small number of perfect rough specimens in the mineral cabinets of Europe. Thus, it was left to Greville and de Bournon to fit the final piece of the puzzle.

> It is, in fact, among those gems or stones, now known by the names of sapphire, oriental ruby, &c. that corundum ought to be placed; but the progress by which we have arrived at this degree of knowledge was necessarily very slow, and was impeded by continual obstacles: for the scarcity and smallness of the crystals of corundum, and the impression naturally made upon our minds by the

various appearances it exhibited to us, were by no means likely to lead us to form a true judgment respecting it. (de Bournon, 1802)

De Bournon was able to unite all of these dissimilar substances under the name of corundum by a combination of keen observation and shrewd application of the then new gemmological tests such as specific gravity, hardness and chemical analysis:

> The substance treated here of, has hitherto presented itself to our notice under two appearances, which differ so much from each other, in the greater number of those characters which most forcibly affect out senses, particularly those which concern the organ of sight, that we cannot be much surprised to find that mineralogists feel some reluctance, at the idea of uniting together substances which appear so very dissimilar.
>
> Under one of these appearances, in which it is known by the name of corundum, this substance presents itself either in fragments, or in crystals of a pretty large size; sometimes, indeed, of a very considerable one. Their colour, which is most commonly rather dull, is a whitish, greenish, and sometimes yellowish gray. Specimens of a purplish red, or of a blue color, have always been extremely rare; indeed, a short time since, no such specimens were known, excepting a very few, preserved in the collection of Mr. Greville, and some small fragments he had given away; but the specimens which have been lately sent from the district of Ellow (Elahera, Sri Lanka?), have contributed to increase their number.
>
> Under the other appearance, (in which this substance is known by the names of sapphire, ruby, &c.) it offers itself, on the contrary, in crystals which are generally of a very small size, and have a smooth and brilliant surface. Their transparency is often very great; and it seldom happens that they are not semitransparent, in a greater or less degree. They are more difficult to break in the direction of their crystalline laminae; and this difficulty increases, in proportion to their purity and their brilliancy. Their colours are much more beautiful, more variegated, and more lively.
>
> With respect to the name of this substance, as, in its most common state, it is known in India (its native country) by the name of corundum, and as that name has been generally adopted in Europe, I have thought proper to continue it, and shall distinguish, by the terms perfect and imperfect, the two different states in which it presents itself to our observation. (de Bournon, 1802)

Thus it was, in the year 1802, that the ruby and sapphire were formally united under the heading of 'corundum.'

Bibliography

Ball, S. H. (1922) The geologic and geographic occurrence of precious stones. *Economic Geology,* **17**, 575–601

Ball, S. H. (1931) Historical notes on gem mining. *Economic Geology,* **26**, 681–738

Ball, S. H. (1935) A historical study of precious stone valuation. *Economic Geology,* **30**, 630–642

Ball, S. H. (1950) *A Roman Book on Precious Stones*, Gemological Institute of America, Los Angeles

Ball, V. and Crooke, W. (1925) *Tavernier's Travels in India*, Oxford University Press, London

de Bood, Anselmus Boetius (1636) *Gemmarum et Lapidum Historia*, (ed. Adrianus Toll)

de Bournon, Count (1798) An analytical description of the crystalline forms of corundum, from the East Indies, and from China. *Philosophical Transactions of the Royal Society of London,* **18**, 428–448

de Bournon, Count (1802) Description of the corundum stone, and its varieties, commonly known by the names of oriental ruby, sapphire, etc.; with observations on some other mineral substances. *Philosophical Transactions of the Royal Society of London,* **22**, 233–326

Boyle, R. (1972) *An Essay About the Origine and Virtues of Gems*, reprint of 1672 edn, Hafner Publishing, New York

Chenevix, R. (1802) Analysis of corundum, and of some of the substances which accompany it. *Philosophical Transactions of the Royal Society of London,* **22**, 327–338

Dales, R. C. (1976) *Marious: on the Elements*, Center for Medieval and Renaissance Studies, Los Angeles

Encyclopaedia Britannica, Inc. (1980) *Encyclopaedia Britannica*, vol. 8, p. 306, 'Puranas'

Francis, P. (1983) Ratanpur: the village of gems. *Lapidary Journal*, March, 1980–1987

Greville, C. (1798) On the corundum stone from Asia. *Philosophical Transactions of the Royal Society of London,* **18**, 403–448

Holmyard, E. J. and Mandeville, D. C. (1927) *Kitab Al-Shifa*, Geuthner, Paris

King, C. W. (1865) *The Natural History, Ancient and Modern, of Precious Stones and Gems, and of the Precious Metals.* Bell and Daldy, London

King, C. W. (1866) *Antique Gems: Their Origin, Uses, and Value*, Murray, London

King, C. W. (1885) *Handbook of Engraved Gems*, Bell, London

Kunz, G. F. (1913) *The Curious Lore of Precious Stones*, Lippincott, Philadelphia, PA

Kunz, G. F. (1915) *The Magic of Jewels and Charms*, Lippincott, Philadelphia, PA

Nicols, T. (1652) *A Lapidary: Or The History Of Pretious Stones*, Thomas Buck, Cambridge

Norman, G. (1985) 'Hidden' gems to go on show. *The Times*, 30 March, London

Smith, G. F. H. (1972) *Gemstones*, 14th edn, Chapman and Hall, London

Streeter, E. W. (1892) *Precious Stones and Gems*, Bell, London

Tagore, S. M. (1879) *Mani-Mala, or a Treatise on Gems*, Bose, Calcutta

Warmington, E. H. (1928) *The Commerce Between the Roman Empire and India*, Cambridge

Chapter 2

Chemistry and crystallography

In a series of ground-breaking articles in the *Philosophical Transactions of the Royal Society of London*, Charles Greville (1798), Count de Bournon (1798, 1802) and Richard Chenevix (1802) identified corundum as a distinct mineral species, formally uniting ruby and sapphire under this heading for the first time. Not only were the physical properties measured, but also the chemical and crystallographic features were laid out with surprising accuracy. This finally proved what many had long suspected – ruby and sapphire were actually the same material, corundum.

Chemical composition

Chemically, corundum consists of crystallized alumina, e.g. aluminium oxide. All other elements present within corundum reside as impurities. When it is pure, colourless sapphire results, with the colour of ruby resulting from traces of chromium, and blue sapphire from small amounts of iron and titanium.

Chenevix's (1802) analyses of ruby and sapphire were the first to establish accurately the chemical composition of this new species. They are reproduced below.

Blue perfect corundum, or sapphire		*Red perfect corundum, or ruby*	
Silica	5.25%	Silica	7.00%
Alumina	92.00	Alumina	90.00
Iron	1.00	Iron	1.20
Loss	1.75	Loss	1.80
	100.00%		100.00%

Considering the state of knowledge at the time, this compares very favourably with the modern analyses of corundum, as given in Table 2.1 below, according to Heilmann and Henn (1986), and in Table 2.2.

TABLE 2.1 Electron microprobe analysis of blue sapphires from Elahera, Sri Lanka (in wt%)

Elements	*Sample number*			
	1	2	3	4
SiO_2	0.06	0.03	n.d.	0.06
Al_2O_3	99.43	99.71	99.49	99.48
Na_2O	n.d.	n.d.	n.d.	0.01
K_2O	n.d.	n.d.	n.d.	0.02
MgO	n.d.	0.02	n.d.	0.05
CaO	0.02	n.d.	0.01	n.d.
FeO	0.15	0.17	0.12	0.22
MnO	n.d.	0.03	0.01	n.d.
TiO_2	0.01	0.01	0.01	0.02
Total	99.67	99.97	99.64	99.86

TABLE 2.2 Chemical analyses (wt%) of rubies from various localities

Locality	*Weight percentages*			
	Cr	*V*	*Fe*	*Ti*
Sri Lanka	0.007–0.100 0.20–0.150	0.002–0.010 0.000–0.007	0.002–0.250 0.020–0.830	0.000–0.060 0.000–0.020
Mogok, Burma	0.070–0.100 1.500–1.950 0.890	0.006–0.020 0.050–0.080 –	0.030–0.040 0.040–0.050 0.170	0.006–0.020 0.006–0.020 0
Thailand	0.030–0.600	0.000–0.020	0.030–0.350	0.006–0.020
Jagdalek, Afghanistan	0.550–0.560	0	0.040	0
Mysore, India	0.430–0.560	0.004–0.007	0.030	0
Hunza, Pakistan	0.550 0.100–0.120	0.010 –	0.020 0.010	0.020 0
Umba, Tanzania	0.010–0.100 0.010–0.030 0.020	0.002–0.010 0.000–0.008 <0.050	0.100–0.250 0.004–0.900 0.420	0.006–0.020 0.000–0.020 0.010
Longido, Tanzania	1.400–1.800	0	0.030–0.080	0
Morogoro, Tanzania	0.060–0.090 0.120	0.004 0.030	0.010–0.020 0.020–0.070	0 <0.004
Mangari, Kenya	1.100–1.800 0.940	0.000–0.003 0.230	0.060–0.500 0.190	0 0.030
Fiskenaesset, Greenland	0.190–0.200	0	0.660–1.200	0

The corundum structure

Chemically, corundum is an oxide, meaning a naturally occurring mineral compound in which oxygen is combined with one or more metals. As with haematite, to which it is related, the ratio of metal to oxygen in corundum is X_2O_3.

Haematite group

Haematite	Fe_2O_3
Corundum	Al_2O_3
Ilmenite	$FeTiO_3$

The structures of minerals of the haematite group are based upon hexagonal closest packing of oxygen atoms with the cations in octahedral coordination between them. This is shown in Figure 2.1.

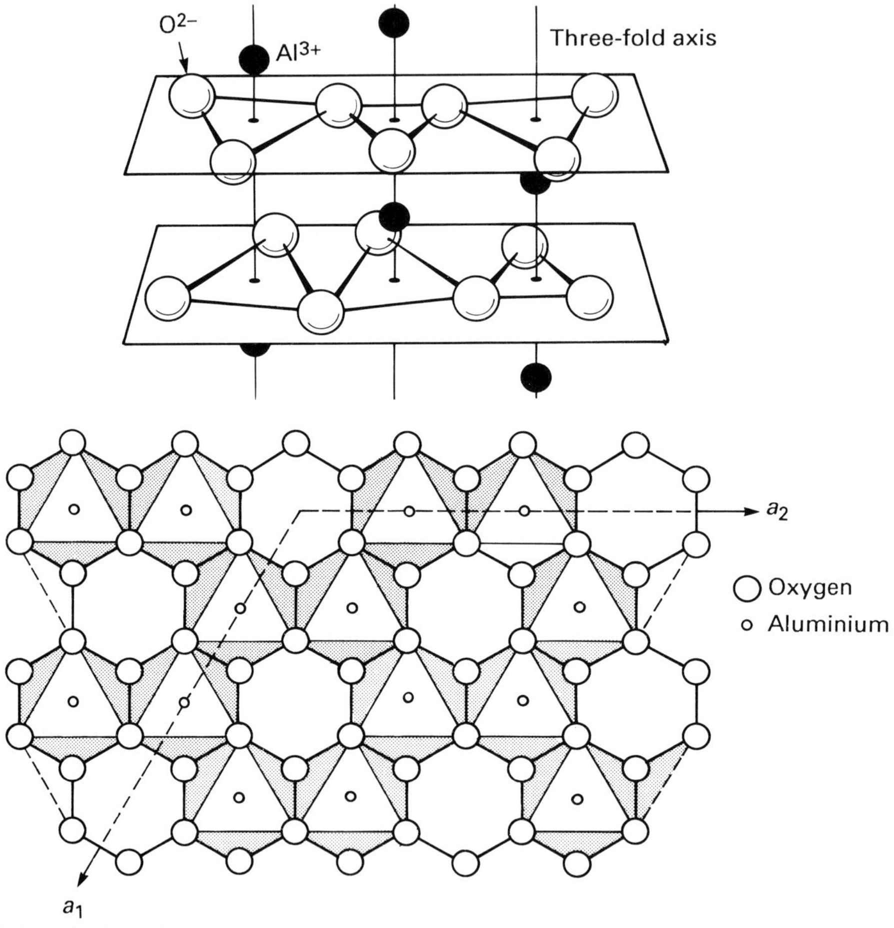

Figure 2.1 A polyhedral model of the structure of corundum

The basal projection of the corundum structure shows that only 2/3 of the octahedral spaces are occupied by Al^{3+} cations. Electrostatic valency (e.v.) or bond strength of the Al^{3+} O^{2-} bonds is related to the presence of 1/3 oxygen octahedra without central Al^{3+} ions. Because the Al^{3+} ions are surrounded by six oxygens, the e.v. of each of the six Al–O bonds equals 1/2. Each oxygen is shared between four octahedra, meaning that four bonds of e.v. = 1/2 can radiate from an oxygen position. Within the basal plane (0001), this allows for only two Al–O bonds from each oxygen, which is indicated by the geometry of two octahedra sharing one oxygen corner. The arrangement of the Al^{3+} ions and of the omitted cations in a vertical section of the corundum structure is shown in Figure 2.1. Each octahedron shares a face between two adjoining layers in the vertical stacking of octahedra. The Al^{3+} cations within the octahedra that share faces will tend to move away from the shared face because of the repulsive forces between them.

Chromium enters the corundum lattice in the form of a trivalent ion, Cr^{3+}, isomorphously replacing some of the aluminium ions. Each Cr^{3+} ion is surrounded by six oxygen ions. The Cr^{3+} ions may form pairs, with the distance between each ion of a pair being 0.273 nm. The Cr^{3+} ions possess a radius (0.065 nm) somewhat greater than that of Al^{3+} ions (0.057 nm). Upon isomorphous replacement of Al^{3+} by Cr^{3+}, certain lattice parameters (*a* and *c*) are increased.

Crystallography of corundum

Corundum crystallizes in the trigonal subdivision of the hexagonal system, belonging to the ditrigonal–scalenohedral class $\bar{3}2/m$ with the following elements of symmetry:

a. One three-fold axis which also represents a three-fold inversion axis.
b. Three two-fold axes perpendicular to the three-fold axis described above.
c. Three planes of symmetry perpendicular to the two-fold axes and intersecting along the axes of the higher order.
d. A centre of symmetry.

For crystals of the trigonal subsystem there are two methods of face indexing: the tetra-axial system of Bravais and the triaxial system of Miller. The indexing system used hereafter is that of Bravais.

Within the $\bar{3}2/m$ symmetry class there are seven simple forms, which are groups of *like* faces. Each face of the same form has the same position in relation to the symmetry elements and crystallographic axes, as well as displaying similar appearances (lustre, striations, etch marks, etc.) and properties, for each is underlain by the same types and arrangement of atoms. These seven simple forms are:

a. Basal pinacoid (0001)
b. Hexagonal prisms $(10\bar{1}0)$ and $(11\bar{2}0)$
c. Dihexagonal prism $(hk\bar{i}O)$
d. Rhombohedrons $(10\bar{1}1)$ and $(01\bar{1}1)$
e. Hexagonal bipyramid $(hh\bar{2}hl)$
f. Ditrigonal scalenohedron $(hk\bar{i}l)$ positive, $(kh\bar{i}l)$ negative

Ruby crystals usually contain the following forms, which are given with the most commonly used letter symbols: *c*(0001), *a*$(10\bar{1}1)$, *n*$(22\bar{4}3)$, *m*$(10\bar{1}0)$, *S*$(02\bar{2}1)$, *R*$(01\bar{1}2)$, *p*$(11\bar{2}3)$, etc.

The unit cell is considered to be the building block of a crystal, i.e. the smallest part which still possesses all the characteristics of the whole crystal. Before the discovery of x-rays, crystallographers determined the dimensions of the unit cell by studying a crystal's external shape, or morphology, and by analysing and measuring faces and angles. Unit cells determined in this way are termed morphological cells, and differ somewhat from structural cell dimensions determined through x-ray studies. Belyaev (1980) gives the dimensions of each as follows:

Morphological unit cell of corundum
Rhombohedral angle = 85°42⅔′
Ratio of axes of hexagonal cell
a:c = 1:1.3652

Structural unit cell of corundum
Rhombohedral cell
a = 0.5128 nm
= 55° 17′

Hexagonal cell
a_h = 0.4758 nm
c_h = 1.2991 nm
c_h/a_h = 2.73

Indexing of crystal planes or faces varies slightly between the morphological and structural unit cells. The choice of which to use depends upon the application. Structural unit cells are used in studies involving the electron microscope or x-ray topography, while the morphological cell is used when studying enlarged manifestations of plastic deformation (twinning or gliding) or cleavage, and determining orientation by the back-reflection Laue technique.

For the gemmologist, of course, it is the external morphology of corundum which is of primary crystallographic interest. The types of faces and their angles of intersection can be of great use in identifying corundum, not only as crystals, but also cut stones, for the internal growth planes and colour zones always lie parallel to crystal faces. A technique for separating natural and synthetic corundum is discussed in Chapter 7.

Belyaev (1980) gives the common faces of corundum and their angles of intersection as shown in Table 2.3.

Twinning in corundum

Under certain conditions, crystals may form symmetrical intergrowths, known as 'twins', in which the crystal lattice of one part bears a definite crystallographic relationship to the lattice of the other section.

Twin crystals may be classified into one of three categories according to the cause:

1. Growth twins: Formed as a result of a mistake during growth, where atoms take up incorrect positions on the growing crystal, resulting in a change in the growth direction, and the formation of a twin crystal.

TABLE 2.3 Angles between face normals of corundum. (After Belyaev, 1980)

C 0001															
90.0	*m* $10\bar{1}0$														
90.0	30.0	*a* $11\bar{2}0$													
32.2	57.8	–	*g* $20\bar{2}5$												
57.6	32.4	43.0	25.4	*r* $10\bar{1}1$											
81.0	9.0	31.2	48.8	23.4	*t* $40\bar{4}1$										
38.2	–	57.6	33.9	47.0	–	*R* $00\bar{1}2$									
–	51.8	–	–	84.2	–	–	*R* $10\bar{1}2$								
72.4	–	34.4	59.3	55.6	58.8	34.2	–	*S* $02\overline{21}$							
–	17.6	–	75.4	50.0	–	–	–	–	*S* $20\bar{2}1$						
42.3	–	47.7	20.5	27.3	–	19.6	–	38.8	–	*P* $11\bar{2}3$					
61.2	–	28.8	–	26.0	–	32.0	–	30.5	–	18.9	*n* $22\bar{4}3$				
–	–	17.3	–	25.7	19.2	50.4	–	37.5	–	–	18.4	*l* $22\bar{3}1$			
–	–	–	–	–	–	–	–	–	–	53.0	–	–	*l* $3\bar{1}21$		
–	–	–	–	–	–	–	–	36.4	–	–	–	–	–	*l* $12\overline{31}$	
69.9	–	–	–	–	–	–	–	49.1	–	21.7	42.1	–	–	–	*w* $11\bar{2}1$

Definitions of faces

C(0001) = basal pinacoid
m($10\bar{1}0$) = first-order hexagonal prism
a($11\bar{2}0$) = second-order hexagonal prism
q($20\bar{2}5$) = positive rhombohedron
r($10\bar{1}1$) = positive rhombohedron
t($40\bar{4}1$) = positive rhombohedron
R($00\bar{1}2$), *R*($10\bar{1}2$) = negative rhombohedron
S($0\overline{22}1$), *S*($20\bar{2}1$) = second-order negative rhombohedron
P($11\bar{2}3$) = hexagonal bipyramid
n($22\bar{4}3$) = second-order hexagonal bipyramid
l($22\bar{3}1$), *l*($3\bar{1}21$), *l*($12\bar{3}1$) = hexagonal scalenohedron
w($11\bar{2}1$) = second-order hexagonal bipyramid

2. Transformation twins: These are 'secondary twins' which occur in a pre-existing crystal. They result when a crystal formed at high temperatures is cooled and subsequently rearranges its structure from the high-temperature form.
3. Gliding or deformation twins: Also a type of secondary twinning, these result from a deformation of the crystal due to mechanical stress. If the stress produces slippage (gliding) of atoms on a small scale, twins may result. If a large slip occurs, gliding may occur without twinning, eventually producing a fracture.

The twinning of corundum is of the first and third types. Growth twins are common in corundum from certain localities, such as Sri Lanka. However, most twinning encountered in rubies and sapphires is secondary twinning due to deformation. When a crystal is repeatedly twinned, with many or all of the twin planes lying parallel to one another, it is known as 'polysynthetic twinning' ('lamellar twinning').

Polysynthetic twinning in corundum commonly occurs on the rhombohedron ($10\bar{1}1$) faces or, more rarely, along the basal pinacoid (0001). According to Belyaev

(1980), the rhombohedral twinning occurs at very low temperatures and very high deformation rates, while the basal twinning occurs within a very narrow range of temperatures and deformation rates. This is why the rhombohedral twinning is much more common.

Twinning in corundum is best observed by immersing the specimen in methylene iodide and examining it between crossed polars. Under these conditions the twin planes will stand out as bright planes fringed with interference colours, while colour zones, which are sometimes confused with twinning, will remain dark.

Boehmite (γ-AlO.OH) needles also indicate the presence of twinning. The long white needles of boehmite are formed through the alteration of corundum at the junctions of the rhombohedron twin planes and, like the twin planes, meet at angles of 87° and 93°. Since they usually only form along twin planes, their presence often indicates the presence of twinning. The boehmite needles and accompanying polysynthetic twinning provide one of the best means of separating natural corundum from the flux-grown synthetic.

Verneuil synthetic corundum also displays a form of polysynthetic twinning due to the rapid cooling of the boule. Termed 'Plato lines', these are found parallel to the hexagonal prism faces (see Chapter 7). Other types of twinning may also be found in the flux synthetic corundums, but without the accompanying boehmite needles. On rare occasions, however, boehmite needles and rhombohedral twinning have been seen in the flame fusion synthetic corundum.

Crystal habit

In normal usage, the term 'form' refers to shape, but in crystallography shape is designated by the term 'habit'; instead, form is used in a restricted sense to designate a specific type of face, or group of *like f*aces. Thus the habit, or outer shape of a crystal, is a term describing the relative development (or lack of it) of a form or forms. At times, habit is indicated by reference to the dominant form. 'Massive' is a habit description for specimens which display no outward evidence of their internal symmetry.

Several different factors may influence the habit, or shape, of a crystal. These include the chemical composition of the mother liquor from which it has grown, as well as the conditions at the time of its growth. Tiny amounts of certain impurity elements present in the growth solution, for example, may produce a change in habit. The temperature and other conditions during crystallization also play an important role.

The habit of ruby often differs from that of sapphire. Rubies tend to grow as tabular hexagonal prisms in which every other corner is truncated by rhombohedron faces. As a result, most rubies, particularly those from Thailand/Cambodia, are faceted with overly shallow pavilions. In contrast, sapphires are generally found as barrel- or spindle-shaped hexagonal pyramids and bipyramids. These are best illustrated by the sometimes enormous specimens unearthed in Sri Lanka. Stones faceted from such crystals differ from most rubies in that the pavilion is often excessively deep.

Prismatic and pyramidal- or bipyramidal-shaped ruby and sapphire crystals frequently display horizontal striations at right angles to the *c* axis. These result from oscillatory growth, where the conditions are just right for the growth of both the pinacoid and pyramid or prism faces. Thus, the growth alternates, with each face growing for just a short time, creating a striated appearance. The horizontal striations create a very characteristic appearance which allows corundum crystals to be easily identified in most cases. Of the corundum lookalikes, only quartz displays similar horizontal striations, but the vast differences in specific gravity between them allows separation.

The most common habits of corundum from major worldwide localities are listed below.

Ruby habits

Burma Tabular hexagonal prisms, often showing a terraced or stepped appearance due to oscillation between the basal pinacoid and pyramid or prism faces. Rarely barrel and spindle shapes. Development of the rhombohedron faces is also common. Crystals are usually somewhat waterworn and silky, and are often confused with spinel, which is found with ruby. Specimens attached to the calcite matrix used to be rare, but now are more common because of the use of dynamite in mining the calcite matrix. Spinel crystals in the calcite matrix are also being recovered in this manner.

Thailand/Cambodia Very thin tabular hexagonal prisms which are always extremely waterworn. Rhombohedron truncations are common at the three opposite corners on each end, creating triangular pinacoid faces. Matrix specimens are not found. Polysynthetic twinning is very common.

Sri Lanka Generally the same as for Sri Lankan sapphires.

Kenya Tabular to blocky hexagonal prisms, often with adhering matrix. Also elongated hexagonal spindles. Polysynthetic twinning is commonplace.

Afghanistan Tabular hexagonal prism and pinacoid combinations with well-developed rhombohedron faces. Also elongated prisms and bipyramids. Crystals are often embedded in a white marble matrix. Polysynthetic twins common.

Tanzania Ruby from Longido occurs as hexagonal plates enclosed within a chromium-green massive zoisite matrix. All crystals show intense cracking due to stress. Crystals are often intergrown with the zoisite at the edges, and these show indistinct boundaries. Polysynthetic twinning is very common.

Sapphire habits

Kashmir Mostly spindle-shaped hexagonal bipyramids of light-blue colour with deep-blue tips and partially corroded surfaces. Crystals are sometimes twisted around one another. Almost all possess a white clay-like crust adhering to the surface. Sometimes crystals are intergrown with dark tourmaline prisms.

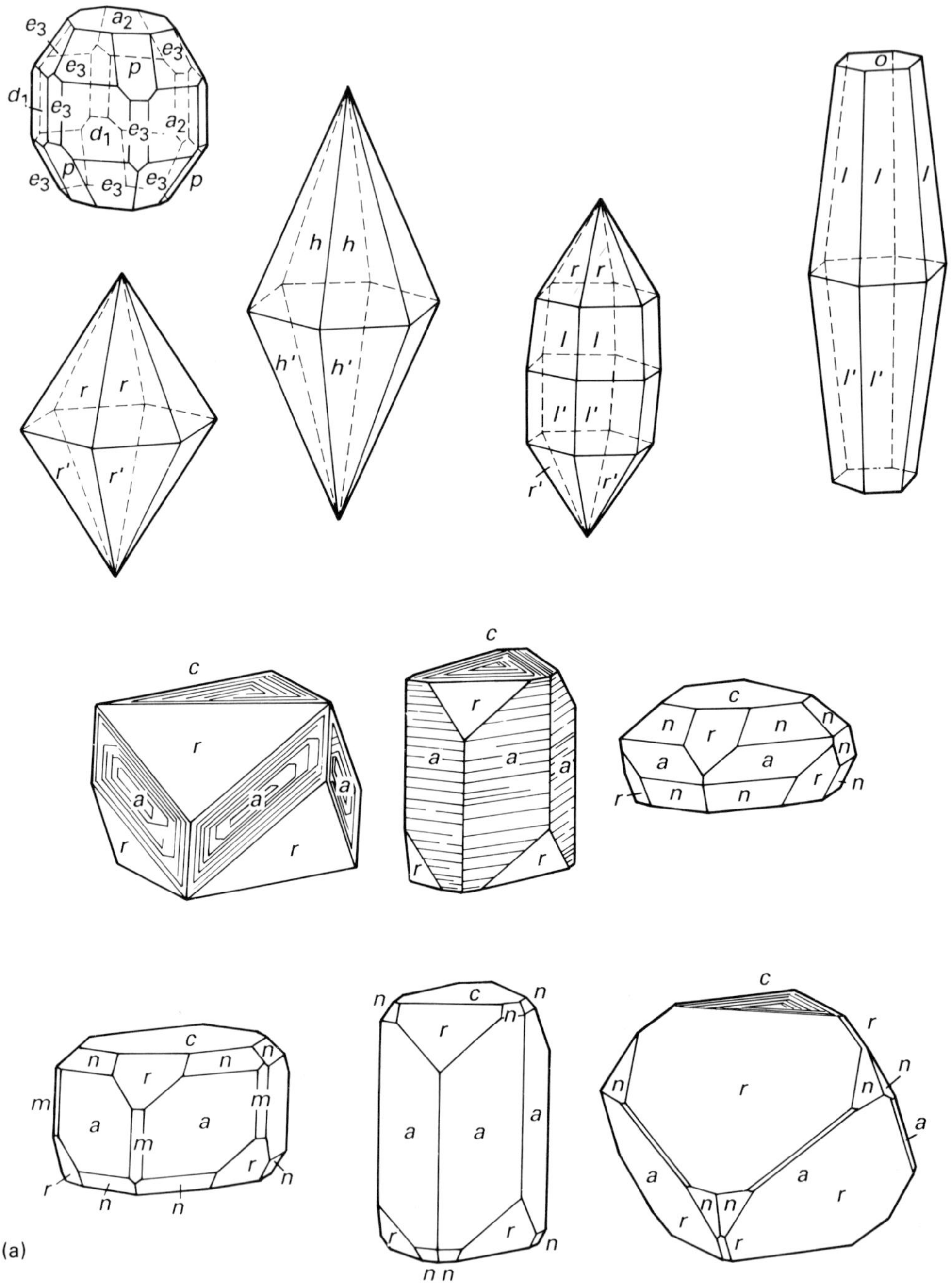

(a)

Figure 2.2 (*a*) and (*b*) Corundum crystals from Burma; (*c*) and (*d*) corundum crystals from Sri Lanka; (*e*) corundum crystals from North Carolina, USA; (*f*) corundum crystals from Yogo Gulch, Montana, USA; (*g*) corundum crystals from Emerald Bar (Missouri River), Montana, USA; (*h*) sapphire crystals from Kashmir, India; (*i*) corundum crystals from other localities. (After Goldschmidt, 1918)

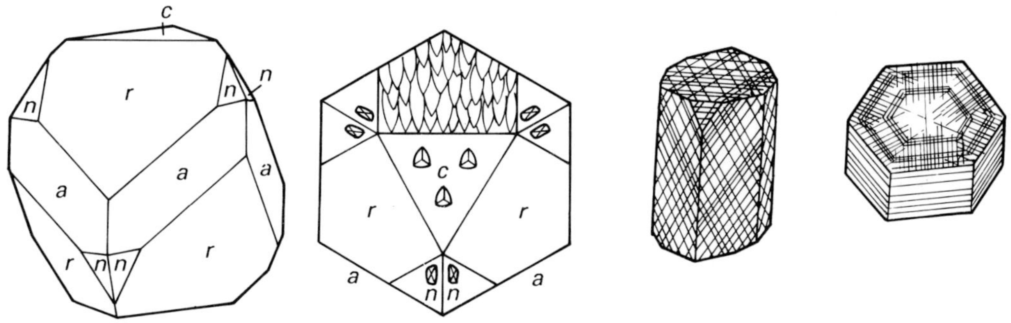

(a) continued

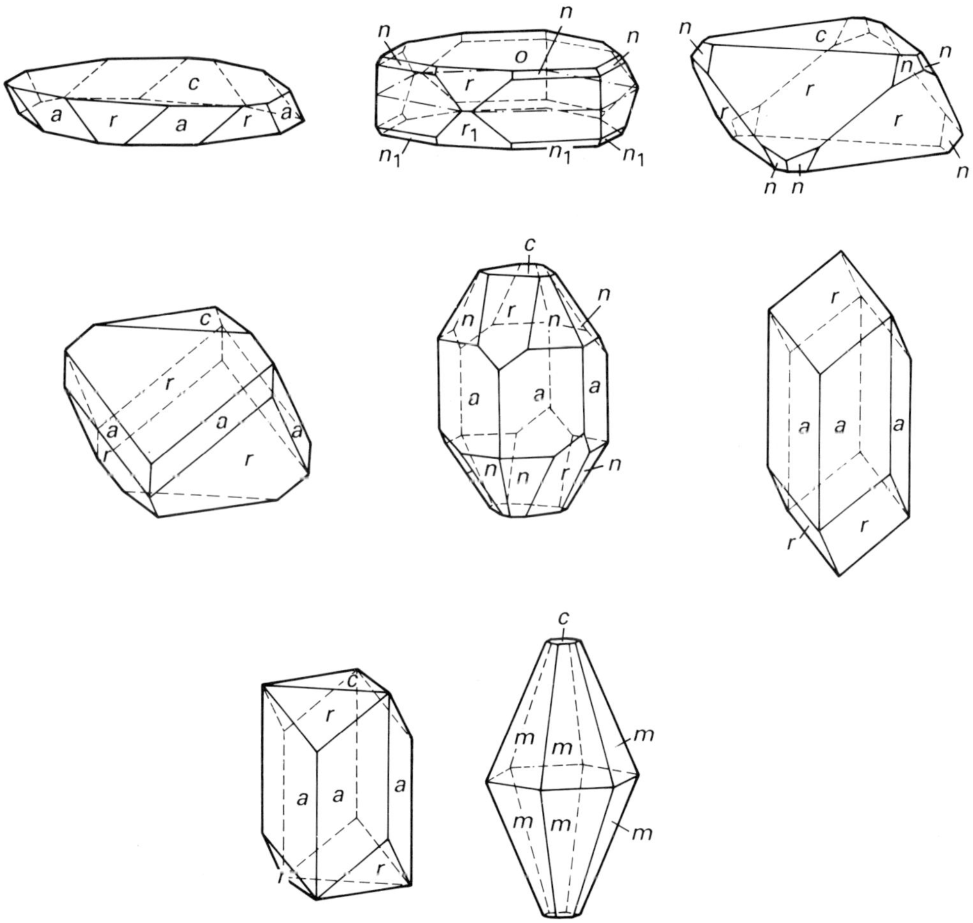

(b)

Figure 2.2 (cont.)

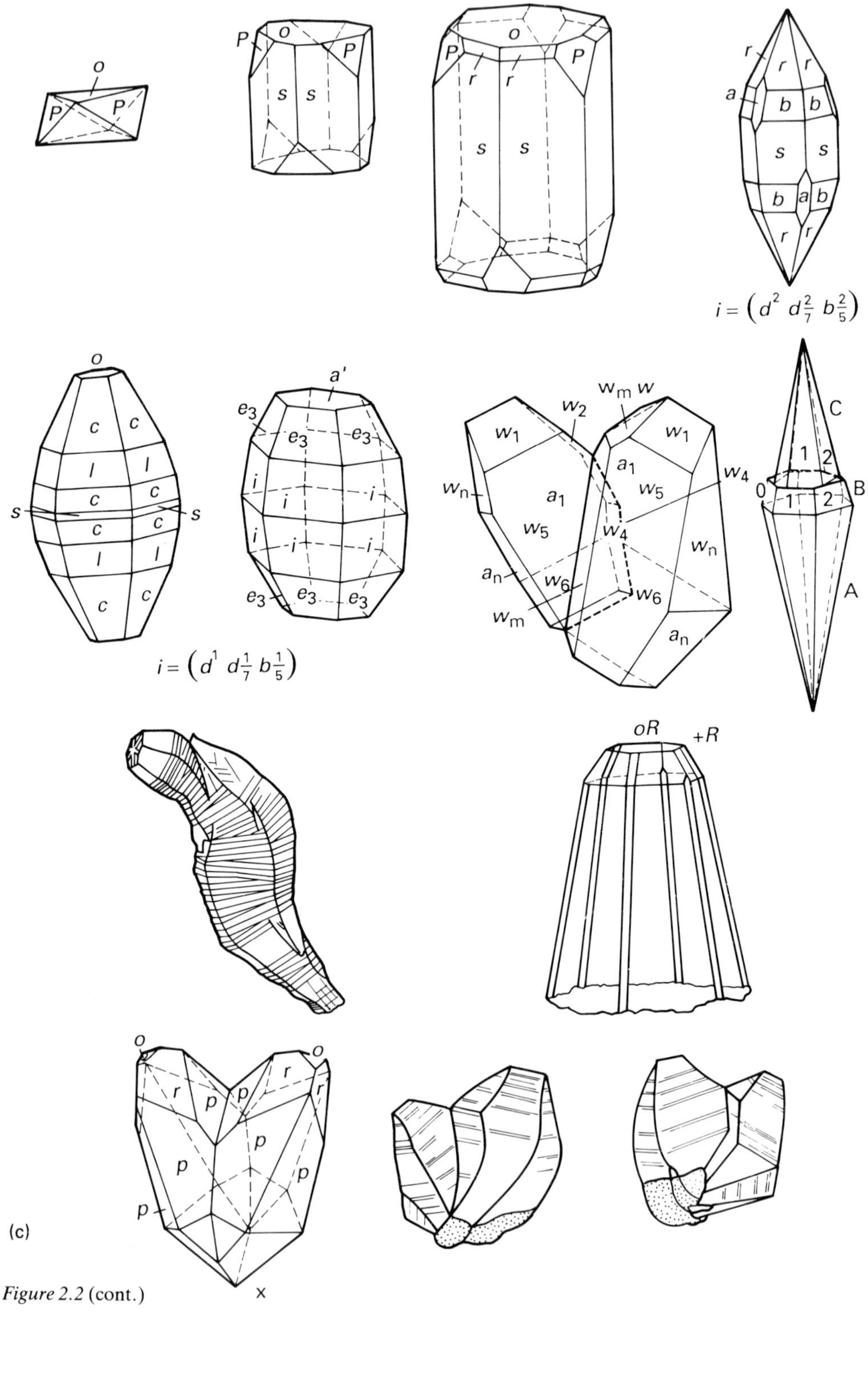

(c)

Figure 2.2 (cont.)

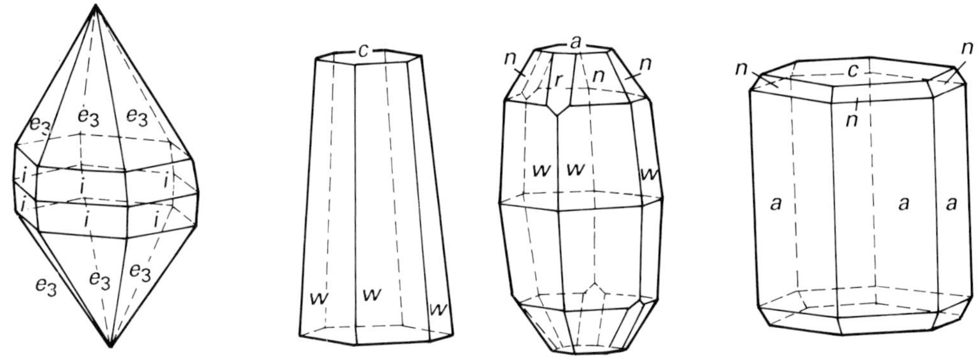

(c) continued

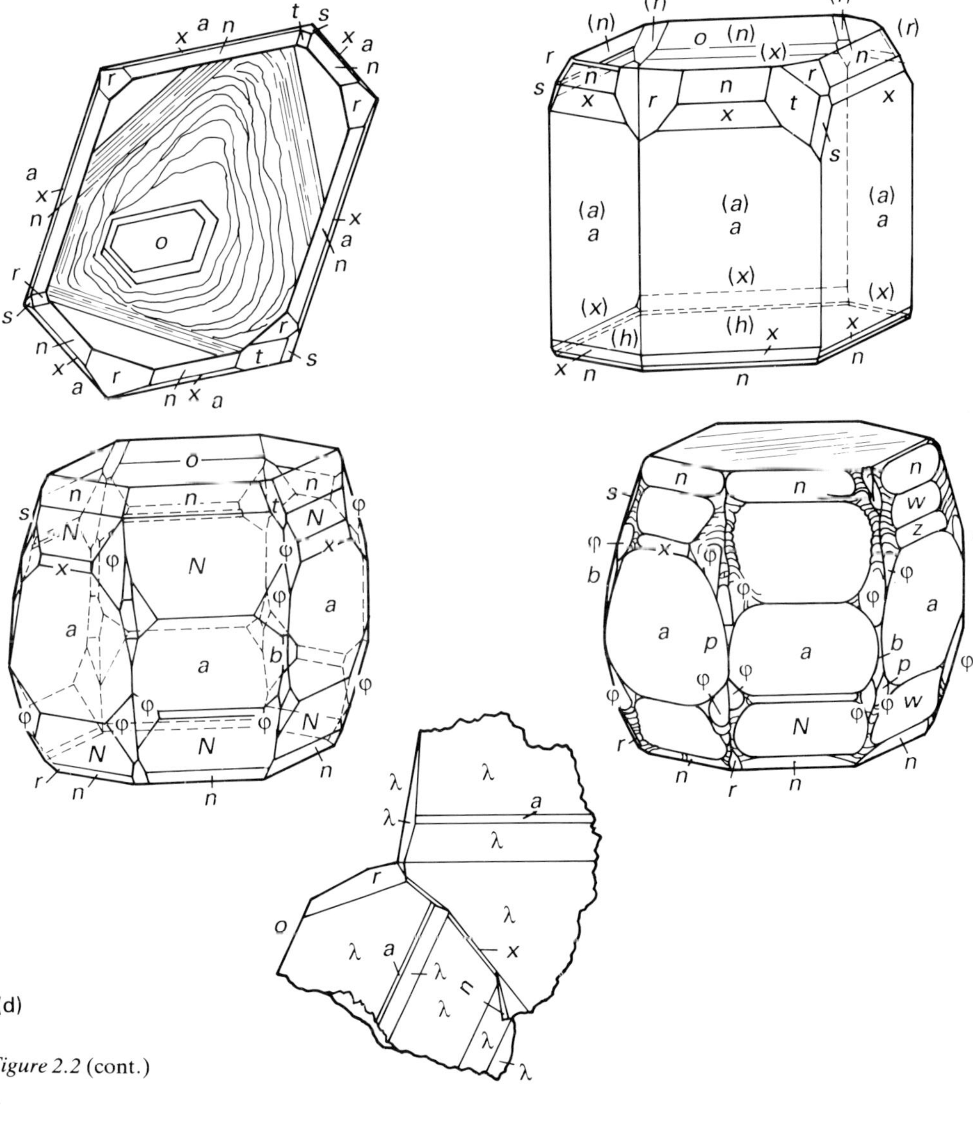

(d)

Figure 2.2 (cont.)

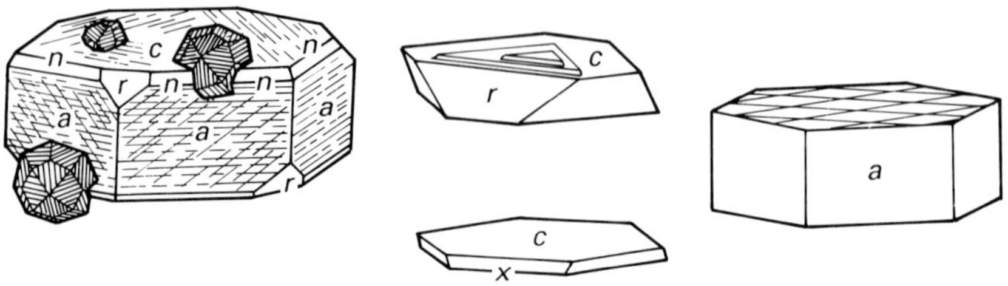

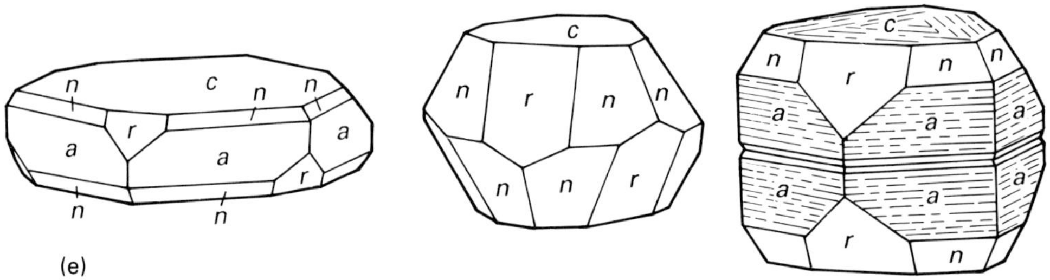

(e)

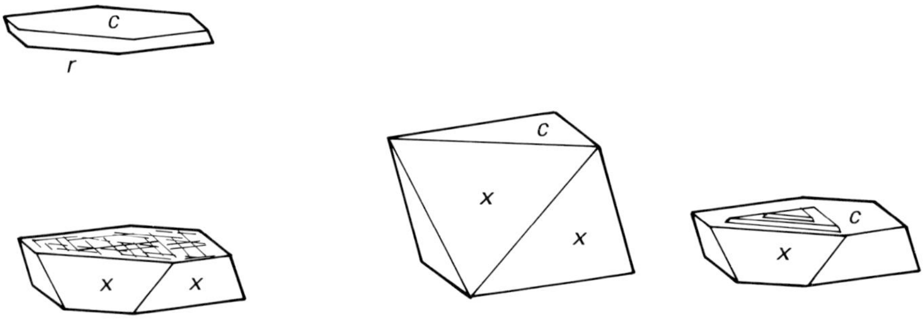

(f)

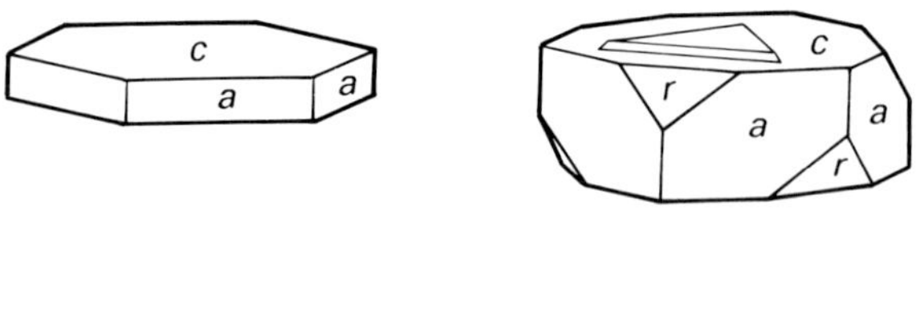

(g)

Figure 2.2 (cont.)

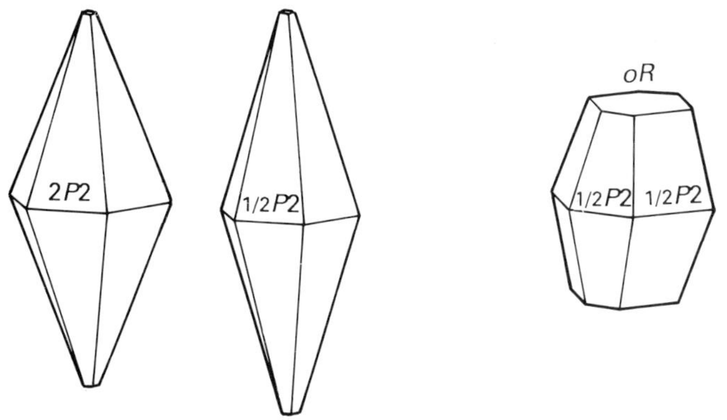

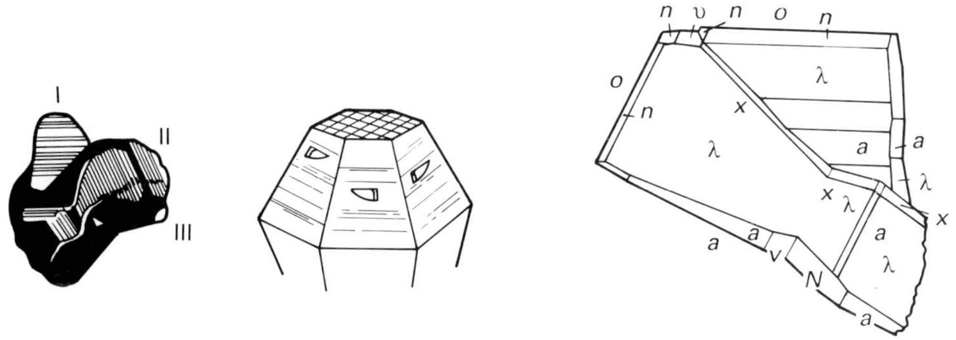

(h)

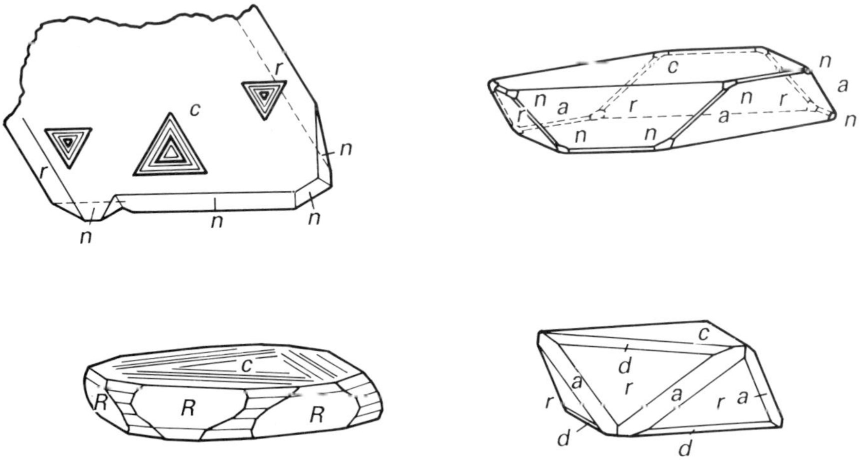

(i)

Figure 2.2 (cont.)

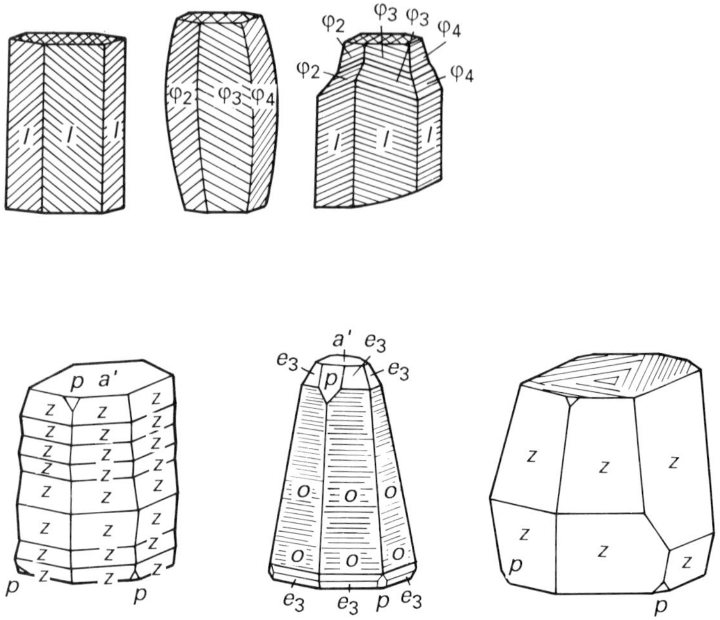

(i) continued

Figure 2.2 (cont.)

Burma Mostly waterworn fragments and blocky to tabular hexagonal prisms and bipyramids of even coloration. Polysynthetic twins are commonplace. Matrix specimens are rare.

Sri Lanka The most prolific source of sapphire crystal specimens. Extremely common are spindle-shaped hexagonal bipyramids, some of which are branching growth twins. Barrel shapes and blocky hexagonal prism/pinacoid combinations also common. Coloration is often irregular and surface oriented. Matrix specimens are practically unknown. Crystal sizes are sometimes enormous.

Thailand Most common are barrel shapes and hexagonal pyramids and bipyramids (spindles). Many crystals are polysynthetic twins. Basal parting often seen, especially in black star sapphires. Material may be euhedral or somewhat waterworn. Matrix specimens are almost unknown.

Australia Similar to Thai material. Barrel shapes and pyramids or bipyramids are most common. Crystals often waterworn. Many display prominent basal parting.

Cambodia Similar to Thailand and Australia. Barrel shapes, pyramids and bipyramids, or waterworn fragments are most common. Basal parting is also common.

Nigeria Similar to the sapphires from Thailand, Cambodia and Australia.

Montana Yogo Gulch material occurs as distinctive tabular plates truncated by rhombohedron faces. Raised triangles often found on basal pinacoid, with reverse orientation on opposite pinacoid faces. Polysynthetic twinning is sometimes seen in Yogo sapphires. Stones from the Missouri River and Rock Creek occurrences are slightly waterworn tabular to blocky hexagonal prisms truncated by rhombohedron faces. Stubby barrel shapes are also common, as well as rounded balls. In contrast to Yogo stones, which are almost all blue in colour, those from other Montana localities are of fancy colours.

Umba Valley (Tanzania) Umba sapphires occur as tabular crystals showing pinacoid and hexagonal prism faces, but without pyramid faces. The basal pinacoids commonly show polysynthetic twin striations that divide them into small triangles. Colour zoning is often as a colourless hexagonal 'tube' parallel to the *c* axis. This tube may be black in rubies.

Bibliography

Belyaev, L. M. (1980) *Ruby and Sapphire*, Amerind, New Delhi

de Bournon, Count (1798) An analytical description of the crystalline forms of corundum from the East Indies, and from China. *Philosophical Transactions of the Royal Society of London*, **19**, 428–448

de Bournon, Count (1802) Description of the corundum stone, and its varieties, commonly known by the names of oriental ruby, sapphire, &c.; with observations on some other mineral substances. *Philosophical Transactions of the Royal Society of London*, **9**, 233–326

Chevenix, R. (1802) Analysis of corundum, and of some of the substances which accompany it; with observations on the affinities which the earths have been supposed to have for each other, in the humid way. *Philosophical Transactions of the Royal Society of London*, **9**, 327–338

Clabaugh, S. E. (1952) Corundum deposits of Montana. *USGS Bulletin*, p. 983

Game, P. M. (1954) Zoisite amphibolite with corundum from Tanganyika. (Periodical unknown), pp. 458–466

Goldschmidt, V. (1918) *Atlas Der Krystallformen*, reprinted by the Rochester Mineralogical Symposium

Greville, C. (1798) On the corundum stone from Asia. *Philosophical Transactions of the Royal Society of London*, **19**, 403–427

Gübelin, E. J. (1982) Gemstones of Pakistan: emerald, ruby, and spinel. *Gems and Gemology*, **18**, 123–139

Heilmann, G. and Henn, U. (1986) On the origin of blue sapphire from Elahera, Sri Lanka. *The Australian Gemmologist*, **16**, 2–4

Klein, C. and Hurlbut, C. S. (1985) *Manual of Mineralogy (After James D. Dana)*, 20th edn, Wiley, Singapore

Sinkankas, J. and Read, P. G. (1986) *Beryl*, Butterworths, London

Solesbury, F. W. (1967) Gem corundum pegmatites in N.E. Tanganyika. *Economic Geology*, **62**, 983–991

Trifonov, D. N. and Trifonov, V. D. (1982) *Chemical Elements: How they were discovered*, Mir, Moscow

Chapter 3

Physical and optical properties

The mineral corundum is among the most durable of all gemstones found on earth. With a hardness surpassed only by diamond and an absence of easy cleavage, it is able to withstand tremendous amounts of punishment through weathering processes. This is borne out by the fact that most gem corundum is recovered from alluvial deposits where, despite the continuous onslaught of the elements, crystals still display much, if not all, of their original shape.

Fracture, cleavage and parting

The properties of fracture, cleavage and parting involve the reaction of a mineral to external pressure or force. Exactly how it reacts is intimately related to its bonding and atomic structure. Cleavage describes the tendency of a mineral to break parallel to flat planes within its atomic structure. As atomic planes form crystal faces, cleavage always takes place parallel to a possible crystal face and is described according to the face it parallels, the smoothness of the cleavage surface and the ease with which the break is effected.

Gemmological texts, such as that by Bauer (1904), have traditionally denied the existence of cleavage in corundum. Instead, the smooth flat breaks have been attributed to parting, a weakness due to structural defects caused by twinning or impurity concentrations. However, recent work performed in the USSR on synthetic colourless sapphire (Belyaev, 1980) has suggested that cleavage is in fact present in corundum. Large plates of sapphire grown via direct crystallization were found, at times, to split along very large mirror planes. Further testing showed that the breaks were not associated with twinning. The cleavage was found to occur along the morphological rhombohedron ($10\bar{1}1$) as well as the hexagonal prism ($11\bar{2}0$). The rhombohedral cleavage was detected more often and the surfaces were also larger, while the cleavage steps of the prismatic cleavage were deeper and visible to the naked eye. At times, both the rhombohedral and prismatic cleavages formed simultaneously, giving macro-step-like relief to the cleaved surfaces. The cleavage in synthetic corundum was found to manifest itself best in crystals grown under conditions ensuring high purity and low defect concentration.

The result of parting is exactly the same as cleavage – a splitting along a plane of weakness. However, parting is due to structural defects rather than the basic structural design, as with cleavage. Thus, the number of possible partings in a mineral is limited to the number of defects present.

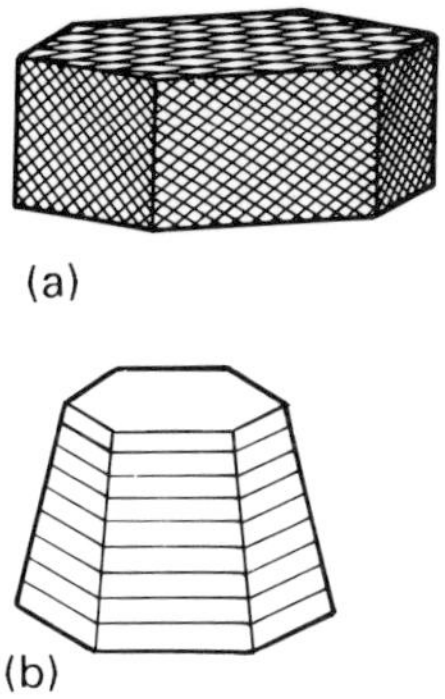

Figure 3.1 Parting in corundum. (*a*) Rhombohedral parting, typical of ruby, arising from the exsolution of boehmite along the rhombohedron faces. (*b*) Basal parting, typical of black star sapphires from Chanthaburi (Thailand), arising from the exsolution of haematite along the basal pinacoid

Not only has the idea of cleavage in corundum been re-examined, but also that of parting. Judd (1895) precisely described the partings of corundum and classified them into three categories: basal (0001), rhombohedral ($10\bar{1}1$) and prismatic ($11\bar{2}0$). This disputed the contemporary view that these partings were actually cleavages; in Judd's view, these partings were attributed to solution planes. The modern view is similar, attributing these partings to twinning. However, research performed by White (1979) suggests that twinning is not the cause of parting in corundum. After examining many types of twins of diverse species over several years, White found that the twin planes were not planes of weakness, as believed. Instead, many twinned crystals tended to break at points other than along twin planes. This led him to question the idea that the easy parting found in many corundum crystals is related to twinning.

White studied specimens of corundum from three different localities in the United States, which displayed prominent rhombohedral and/or basal parting, with both optical and electron microscopy. He found that the parting surfaces were not simply planar fractures, but were actually thin seams of what appeared to be another mineral. This mineral was subsequently identified as boehmite, included in the corundum through exsolution along the rhombohedron ($10\bar{1}1$) planes. Thus, the parting of corundum was found not to be due to twinning, but due instead to the exsolution of thin layers of another mineral. Because the boehmite is normally exsolved along twinning planes, the twinning was incorrectly thought to be responsible.

The black star sapphires of Chanthaburi, Thailand, also display distinct parting, but in this case it occurs along the basal (0001) plane. It seems to this author that the most likely reason for this is the haematite, which is exsolved in the basal plane and which produces the star effect as well as the deep-brown colour. The haematite

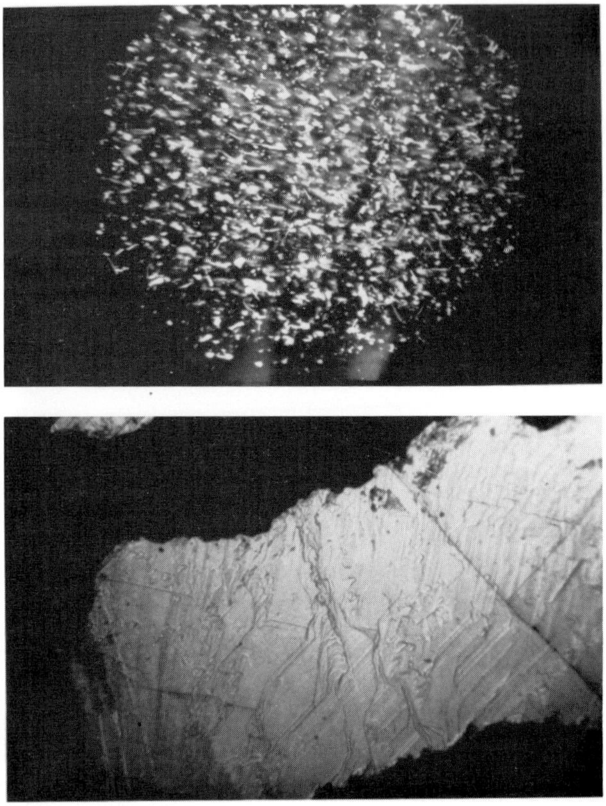

Figure 3.2 Black star sapphires owe their asterism not to rutile needles, but to haematite plates. This haematite, which colours the stones dark brown, results in parting (false cleavage) parallel to the basal pinacoid (parallel to the cabochon's base in properly oriented stones). The upper photograph shows the haematite silk in a yellow sapphire from Chanthaburi (Thailand), while the lower photograph shows the basal parting in a black star sapphire from the same source. (Photographs by the author)

apparently unmixes in thin layers, thus lowering the cohesion in this plane and leading to parting.

In directions other than along the cleavage or parting planes corundum exhibits a conchoidal fracture. In terms of tenacity, rubies and sapphires are relatively brittle, although much less so than other species, such as spinel or zircon.

Hardness

The resistance to scratching of a smooth surface that a mineral possesses is termed its hardness. The classic scale of hardness for minerals was developed by Frederick Mohs in 1824 and is still in use today. Corundum is rated number nine on this scale (see Figure 3.3) and is surpassed only by diamond among the naturally occurring minerals.

Slight differences in hardness have been observed in the different varieties of corundum. Ruby is said to be slightly softer than blue sapphire and the star

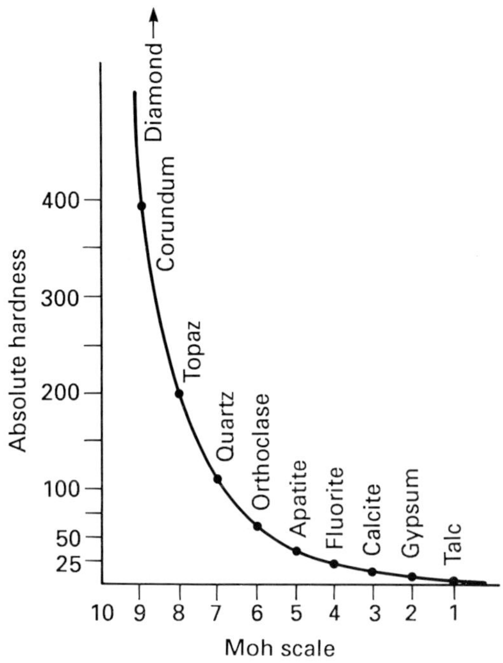

Figure 3.3 The hardness of corundum (9) compared with other minerals on Mohs scale

sapphires are again slightly harder than transparent gems. Corundum crystals also exhibit slight differences in hardness along different directions. Belyaev (1980) reported that in synthetic corundum, the basal plane possessed the least hardness, with the prismatic planes coming next. Maximum hardness was observed in the planes forming an angle of 60° with the optic axis of the crystal. Table 3.1 describes the indention hardness values of synthetic corundum of various types and orientations.

According to Belyaev (1980), the hardness of corundum on Knoop's scale is 1370 kg mm^{-2} with a 1 kg load and random direction of the indentor. Djevahirdjian (no date), however, gives the Knoop hardness as 1800 at 90° to the *c* axis. The modulus of elasticity was stated to be 4.4×10^6 kg cm^{-2} ± 1% Young's *E*. Vickers hardness is given as 2500–3000 kg mm^{-2}.

TABLE 3.1 Microhardness (Knoop scale) of synthetic corundum crystals (kg mm^{-2})

Material	*Orientation of axis of diamond point relative to optic axis of crystal*	*Before thermal treatment*	*After thermal treatment*
Colourless Sapphire (melt grown)	Perpendicular	2720	2720
	Parallel	2350	2150
Colourless Sapphire (Verneuil)	Perpendicular	2280	2370
	Parallel	2350	2450
Ruby (Verneuil)	Perpendicular	2750	2750
	Parallel	2900	2730

Thermal expansion

The thermal expansion (deg^{-1}) of single crystals of corundum is given by Belyaev (1980) as follows:

20–50°C $= 6.66 \times 10^{-6}$ (parallel to *c* axis)
20–1000°C $= 9.03 \times 10^{-6}$ (parallel to *c* axis)
50°C $= 5.0 \times 10^{-6}$ (perpendicular to *c* axis)

Sinkankas and Read (1986) state that this is less than those for diamond and beryl. However, corundum displays a smaller difference between its expansion along the *c* axis and *a* axis than does beryl. Thus, the very small changes in dimensions of both diamond and corundum mean that these gems can be safely left in their mountings while undergoing repairs involving high temperatures. Beryl and quartz, however, expand much more along the *c* axis than along the *a* axis, making the danger of heat-induced fractures far greater. Corundum may also readily be cleaned ultrasonically, unlike emerald where there is a very real danger of breakage, not to mention the loss of clarity caused by the removal of the oil from fractures.

Thermal conductivity

The thermal properties of corundum were given by Hoover (1983) as in Table 3.2 and those of pyrope, almandite and spinel as in Table 3.3.

From these tables it can be seen that the thermal inertia values of corundum are almost twice those of spinel and more than three times those of the garnets. By using a well-designed thermal inertia probe, such as the Presidium/Gem Instruments Diamond Master, corundum can be separated from these ruby

TABLE 3.2

Direction tested	*Thermal conductivity* (cal cm^{-1} °C s^{-1})	*Specific heat* (cal g^{-1} °C^{-1})	*Thermal diffusivity* (cm^2 s^{-1})	*Thermal inertia* (cal cm^2 °C $s^{1/2}$)
c axis	0.0834	0.206	0.101	0.262
a axis	0.0772	0.206	0.0937	0.252
c axis (other)	0.0600	0.206	0.0728	0.222

TABLE 3.3

Mineral	*Thermal conductivity*	*Specific heat*	*Thermal diffusivity*	*Thermal inertia*
Spinel, locality unknown	0.0281	0.216	0.0358	0.148
Spinel, Madagascar	0.0227	0.216	0.0288	0.133
Almandite, New York	0.007 91	0.2	0.0101	0.0789
Pyrope, Arizona	0.007 59	0.2	0.0101	0.754

simulants. Other thermal test instruments, such as the 'Gemmologist', the 'Alpha Test' and the 'Gem-Trak' are designed specifically to identify both diamond/diamond simulants and a limited range of non-diamond gems using this principle (Read, 1983, 1984, 1985). Of the naturally occurring minerals which resemble corundum, only topaz (0.138–0.177) and kyanite (0.171–0.176) have thermal inertia values above those of spinel. Thus this test can prove useful in separating corundum from other minerals.

Thermal inertia probes function by heating the probe tip to a selected temperature and measuring how quickly it cools when the tip comes into contact with the gem material. The more rapid its rate of cooling, the higher the thermal inertia of that material.

Melting and boiling points

The melting point of corundum was given by Belyaev (1980) as 2030°C, but other sources have given the slightly higher figure of 2050°C. According to Belyaev, the boiling point of corundum is 3500°C.

Corundum's solubility in 100 g of water was stated to be 9.8×10^{-5}g at 29°C. It dissolves weakly in boiling nitric acid and in orthophosphoric acid to 300°C, while dissolving well in borax at 800–1000°C and in potassium bisulphate at 400–600°C. Owing to the ability of borax to dissolve corundum, extreme care should be exercised when heating borax in close proximity to corundum, such as when repairing articles of jewellery.

Density and specific gravity

Density and specific gravity are often considered synonymous, but there is a slight difference between the two terms: density describes the weight of a given volume of material (usually grams per cubic centimetre), while specific gravity is a ratio comparing the weight of a substance with the weight of an equal volume of water at 4°C. Thus, specific gravity has no units.

The density of corundum is usually given as 3.98 g cm^{-3}. Webster (1975) gave the specific gravity of colourless sapphire (pure corundum) as 3.989, while ruby and blue sapphire are approximately 3.997, with little variation in specimens from different localities. The exceptions to this are the iron-rich blue and green sapphires from Australia and Thailand, which may be as high as 4.02. Black star sapphires may possess even higher values of specific gravity as a result of the inclusion of haematite silk.

Electrical properties

Single crystals of colourless sapphire are among the best dielectrics (insulators) known. This fact, coupled with their high chemical, mechanical and thermal stability, allows colourless sapphire to be used as an industrial insulating material. Synthetic colourless sapphire is today used as a substrate material for integrated circuits, for the windows of powerful waveguide systems, etc.

The electrical properties of corundum are as follows:

Electrical resistance:	10^{11} $\Omega\,cm^{-1}$ at 500°C
	10^{6} $\Omega\,cm^{-1}$ at 1000°C
	10^{3} $\Omega\,cm^{-1}$ at 2000°C
Dielectric constant:	7.5–10.5
Dielectric loss tangent:	<0.0001
Dielectric strength:	480 000 $V cm^{-1}$ (60 Hz)

Lustre

The lustre of a gemstone refers to the quality and quantity of light reflected by it. Internal brilliance contributing to lustre is a function of the gem's refractive index, proportions and polish. External lustre is surface brilliance and is determined by refractive index and polish, the latter being influenced by the hardness.

Corundum displays a surface lustre which is greater than glass (vitreous) but less than diamond (adamantine), and so is described as being vitreous to subadamantine. Because the quality of the polish may vary from one specimen to another, or even from one facet to another, the lustre should be judged on a smooth crystal face or freshly broken fracture surface. Given a well-polished surface, relative reflectivity meters, such as the 'Jeweller's Eye', may also be used to determine lustre, and because of the relationship between lustre and refractive index, to assist in gem identification.

Optical phenomena

The optical phenomena displayed by corundum include asterism, chatoyancy and change of colour. Of these, chatoyancy is rather rare.

Asterism in corundum is due to the reflection of light from multitudes of exsolved needle inclusions, or silk. The silk in most varieties consists of the mineral rutile, while in black star sapphires from Thailand and Australia it is made up of haematite. It is arranged along three directions in the basal plane which intersect at 60/120° angles. Light is reflected in a band perpendicular to each of these directions, thus forming a six-rayed star in suitably oriented cabochon-cut stones. Twelve-rayed stars are also known; in these the silk forms along both the first- and second-order prism faces (in six directions).

Chatoyancy is possible in corundum where the planes of silk are very closely spaced. To exhibit a cat's eye, the gem must be cut with the base parallel to the *c* axis, unlike star stones where the base is cut perpendicular to the *c axis*. Cat's-eye corundums are rarely encountered, as suitable gems are oriented for asterism instead of chatoyancy.

The change of colour, or alexandrite effect, is found on occasion in sapphires from Thailand, the Umba Valley (Tanzania), Burma and Sri Lanka, as well as in the Verneuil synthetic product. Although the name 'alexandrite-like sapphire' is sometimes used, this term is considered improper because it is so often shortened to alexandrite alone, particularly with the synthetic product. Because of the unusual manner in which these stones absorb light, they display a greyish blue-green colour in daylight (or fluorescent light) and a purplish-red colour in incandescent light. Gübelin and Schmetzer (1982) noted that all colour-changing stones display similar absorption features, namely a balanced transmission of the blue green and red and absorption of the yellow. The greater the difference between the areas of transmission and the absorption of the yellow, the stronger the colour change. Because of the balanced spectral energy distribution of daylight and the high sensitivity of the human eye to green light, the gem appears blue green under this illuminant. However, incandescent light possesses virtually no blue or violet wavelengths, but is very strong in yellow, orange and red energy. Thus, the balance is tipped in this direction and the gem appears purplish red.

Most so-called colour-change sapphires are simply blue-violet stones which become purple in incandescent light. This is not a true change of colour. In the author's opinion, the stone must display a daylight colour on the green side of blue to exhibit this phenomenon. Otherwise, stones such as amethyst and tanzanite would have to be considered as colour-change gems too.

Vanadium is used to colour the Verneuil synthetic gem and may possibly play a role in the natural gem as well, particularly in those from the Umba Valley. However, the colouring agents of most natural colour-change sapphires appear to be a combination of chromium, iron and titanium.

Optical properties

Owing to the fact that corundum crystallizes in the trigonal division of the hexagonal system, its optic character is doubly refractive and uniaxial. Light entering the crystal in all directions except parallel to the optic axis (corresponding to the *c* axis) is split into two rays – the ordinary ray (o-ray) and the extraordinary ray (e-ray). Maximum double refraction occurs when light is incident at 90° to the *c* axis and decreases to zero as the direction of travel approaches this axis.

The refractive index for light travelling parallel to the *c* axis is termed the 'ordinary' or 'omega' (ω) index, while that measured for the other ray when light travels perpendicular to the *c* axis is termed the 'extraordinary' or 'epsilon' (ε) index. Since in corundum the o-ray travels slower (has a higher refractive index) than the e-ray, it is negative in sign. The o-ray always vibrates perpendicular to the *c* axis and so has a constant refractive index, while the e-ray vibrates in a plane containing the *c* axis. When light travels parallel to the *c* axis, the e-ray vibrates at 90° to the *c* axis and so has a refractive index equal to that of the o-ray; when light travels at 90° to the *c* axis, the e-ray vibrates parallel to the *c* axis. At this point, its refractive index reaches its maximum deviation from that of the o-ray.

One method of determining the optic character of corundum is by reference to the interference figure. In most cases, this is fairly straightforward, with the

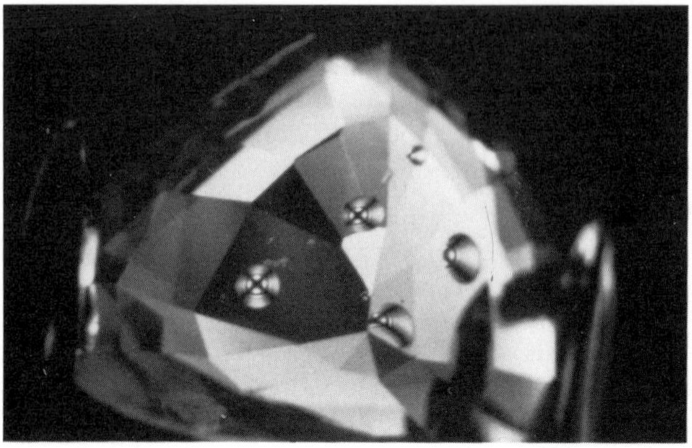

Figure 3.4 Uniaxial interference figures seen on drops of liquid when looking parallel to the *c* axis between crossed polars (Thai ruby). Most (but not all) natural corundums are cut with the *c* axis perpendicular to the table, and so the interference figure will be seen through the table. Shallow-cut stones will show larger figures than those cut deep. (Photograph by the author)

ordinary uniaxial figure being seen. Heavily twinned specimens, however, may display a pseudobiaxial figure. To separate this from a true biaxial figure, we rotate the stone while watching the dark 'brushes'. In uniaxial stones these turn in the same direction as the stone, while in biaxial stones they rotate counter to the stone's rotation.

Optic character may also be determined by plotting the refractive indices of the stone on a graph as the gem is rotated through 180° on the facet being tested. This also allows determination of the optic sign. Depending upon how the facet is oriented with respect to the *c* axis (optic axis), one of the three patterns shown in Figure 3.5 will be seen.

The optic sign is easily determined in examples 2 and 3 in Figure 3.5, as the 1.770 index is constant, but in example 1 another facet must be tested, or the polaroid plate must be used to determine the optic sign. Because the o-ray always vibrates at 90° to the *c* axis, orienting the transmission direction of the eyepiece polaroid at 90° to the *c* axis (parallel to the scale division lines on the refractometer scale) will allow only the o-ray to be seen. In corundum, this will reveal the higher index only, thus showing the stone to be negative in sign.

The refractive index readings of corundum may vary slightly from one locality to another, but the birefringence is extremely consistent at 0.008–0.009. According to Webster (1975), pure corundum gives a reading of 1.7606 for the e-ray (variable index) and 1.7687 for the o-ray (constant index). Generally the e-ray gives readings between 1.757 and 1.772 and the o-ray between 1.765 and 1.780. The iron content of the gem appears to be the major cause of variation. Thus, iron-poor gems from Sri Lanka tend to fall into the lower end of this range, while iron-rich yellow, green and blue sapphires from Thailand and Australia tend to possess higher indices. Despite the slight variation observed in the refractive indices of gems from different localities, refractive index is not a reliable method of determining origin. This holds true for the various synthetic corundums, as well as natural gems.

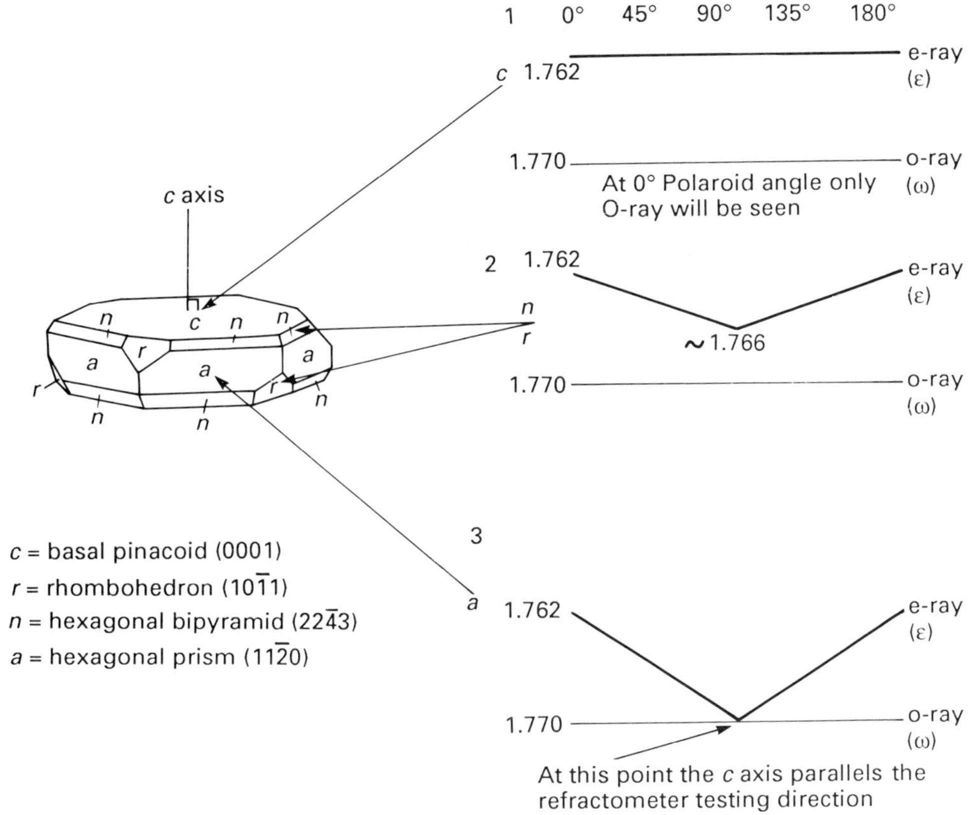

Figure 3.5 Refractive index readings of corundum for facets of different orientations (through 180° rotation). Facets which lie on the basal pinacoid (90° to the *c* axis) will produce two constant readings at 1.762 (e-ray) and 1.770 (o-ray). In order to determine the optic sign of such stones the eyepiece polaroid plate should be oriented with its transmission direction parallel to the scale division lines on the refractometer scale (0° polaroid angle). In this position only the o-ray will be seen (1.770 in corundum, revealing the stone to be negative in sign). When the facet tested lies at an oblique angle to the *c* axis, one constant (o-ray) and one variable (e-ray) refractive index will be seen. The variable curve does *not* meet the constant curve. This pattern is also possible in biaxial crystals, but can be separated by determining the polaroid angle of the constant curve. In biaxial crystals the polaroid angle of the constant curve equals exactly 90°, while in uniaxial gems (such as corundum) the polaroid angle of the constant curve is anything but 90° (0°–89°). If the facet tested is parallel to the *c* axis (parallel to the hexagonal prism faces), one constant (o-ray) and one variable (e-ray) reading will be seen. The variable curve will meet (but not cross) the constant curve at the point where the *c* axis is parallel to the testing direction of the refractometer (parallel to the length and surface of the hemicylinder)

Dispersion is defined as the variation of a gem's refractive index with wavelength and results in the breaking up of white light into its component colours as it passes through the stone. The refractive index of a material is always given for yellow light of a wavelength of 589.3 nm, which corresponds to the mean value of the two Fraunhofer sodium 'D' lines in the solar spectrum. Dispersion is determined by taking the difference in refractive index measured first using light of the Fraunhofer 'B' line (686.7 nm) in the red and then light of the Fraunhofer 'G' line (430.7 nm) in the blue. This gives a numerical value of the stone's ability to spread or disperse light across the visible spectrum. According to Webster (1975), the dispersive power of corundum as measured between the B–G interval is 0.018. This is relatively low when compared with gems of similar refractive index and is visible only in lightly coloured or colourless sapphires. The body colour of most rubies and sapphires completely masks what little fire there is.

Bibliography

Anonymous *Djeva: 1914–1964*, Hrand Djevahirdjian, Monthey, Switzerland

Anonymous *Djeva*, Hrand Djevahirdjian, Monthey, Switzerland

Anonymous (1982) *The Group Pierres Holding and the Corundum*, Pierres Holding, Bienne, Switzerland

Bauer, M. (1904) *Precious Stones*, vol. 2, Dover, New York

Belyaev, L. M. (1980) *Ruby and Sapphire*, Amerind, New Delhi

Gübelin, E. and Schmetzer, K. (1982) Gemstones with the alexandrite effect. *Gems and Gemology,* **18**, 197–203

Hoover, D. B. (1983) The Gem Diamond Master and the thermal properties of gems. *Gems and Gemology,* **19**, 77–86

Judd, J. W. (1895) On the structure-planes of corundum. *Mineralogical Magazine*, no. 11, 49–55

Klein, C. and Hurlblut, C. (1986) *Manual of Mineralogy*, 20th edn, Wiley, Singapore

Read, P. G. (1983) *Gemmological Instruments*, 2nd edn, Butterworths, London

Read, P. G. (1983) The Gemtek Gemmologist – a test report. *Journal of Gemmology,* **18**, (7), 643–650

Read, P. G. (1984) The alpha-test – a test report. *Journal of Gemmology,* **19**, (3), 261–265

Read, P. G. (1986) The Gem-Trak gemstone identifier – a test report. *Journal of Gemmology,* **20**, (4), 242–243

Sinkankas, J. and Read, P. G. (1986) *Beryl*, Butterworths, London

Webster, R. (1975) *Gems. Their Sources, Descriptions and Identification*, third edition, Butterworths, London

White, J. S. (1979) Boehmite exsolution in corundum. *The American Mineralogist,* **64**, 1300–1302

Chapter 4
Colour and luminescence

Colour

When the diverse range of colours to be found among the corundum gems is considered, it is indeed remarkable, and surely must have seemed so to the ancients, that they are all one and the same mineral. In fact, no other gem mineral, save perhaps tourmaline (which is actually a group of related mineral species), displays such a range of colours. From pigeon's-blood red to cornflower blue, almost the only colour not found is a true emerald green, although a less intense olive green is often encountered. But who knows, possibly some day a lucky miner will unearth this 'missing link' too!

The reason for the broad range of colours in corundum concerns its chemistry. Corundum is composed of aluminium oxide (Al_2O_3) which, in its pure state, is completely colourless. However, pure corundum is rarely encountered in nature. Instead, a small number of impurity metals may individually or together replace aluminium atoms in the crystal structure, thus imparting colour. Beryl and tourmaline are also coloured by impurities, a state termed 'allochromaticity'. In contrast, 'idiochromatic' minerals are those that are coloured by essential elements in their chemical make-up, and so occur only in one basic colour. Turquoise is an example of the latter, always being found in the blue-green hue for which it is famous.

In synthetic corundum a number of different dopants have been found to be useful in producing different colours, but with natural corundum the number is rather limited. There is still much to be learned regarding the true causes of colour in corundum, but in many cases a fairly clear picture has emerged. Most varieties of corundum appear to be coloured by a peculiar group of elements termed 'transition elements', which are responsible for the colour of a number of different creations of nature besides minerals. These elements lie at the centre of the periodic table and are peculiar because their electron structure contains inner unpaired 'free' electrons which can be excited, or elevated, to higher energy levels by absorbing *visible* light. Among the first series of transition elements are chromium, iron, titanium, vanadium, manganese and copper. These are the major elements of gemstone coloration and are responsible for the colours of ruby, sapphire, emerald, alexandrite, tanzanite, tsavorite and a plethora of other important gem materials.

Before we discuss the interactions of these elements with corundum, we must first examine the mechanism of colour itself. White light, or sunlight, is made up of a balanced mixture of all the different spectral colours – violet, indigo, blue, green, yellow, orange and red. An object or material does not really possess colour itself; instead, colour is merely a perception based upon a complex interaction between the light source, the object and our eyes. White and black are not spoken of as colours in the true sense of the word. Instead, the sensation of white is created by equal amounts of all the wavelengths of coloured light, from 400 to 700 nm, striking the eye. Black is simply a lack of any visible light reaching the eye. An object appears coloured because of its selective absorption of visible light. This is the manner by which it absorbs some wavelengths, or colours, while freely transmitting or reflecting others. The actual colour we see depends on our eyes' interaction with the wavelengths which strike the eye and is often the complementary colour of the light absorbed. In the simplest terms, grass appears green because, when white light falls upon it, all colours are absorbed except green; or, of the colours not absorbed, a green sensation is produced when they strike the eye together. If the grass was illuminated with a red light, it would appear black because the red light is absorbed by it; thus, no visible light reaches the eye. To someone who is colour blind, the grass would appear red when illuminated by white light because of a lack of green-sensitive optical components, or cones, in the eye. From these examples we can see that the grass appears green not just because of the way it absorbs light, but also because of the physiology of our eyes and, of equal importance, the spectral composition of the light source by which it is illuminated.

With this in mind we will now look at specific situations in corundum.

Ruby

The colour of ruby can be attributed to electron transitions involving the Cr^{3+} ion which substitutes for aluminium in amounts approximating 1%. That chromium is able to absorb visible light concerns its outer electron suborbitals, which lie close enough together that visible light is absorbed by boosting unpaired electrons to a higher excited state.

Electromagnetic radiation is commonly specified by referring to its wavelength, with different types of radiation having different wavelengths, ranging from the very long radio waves to the extremely short x-rays and γ-rays. Another way of specifying the type of radiation is by its energy, usually given in electron volts. The shorter the wavelength, the more powerful its energy.

In order for a chromium atom to absorb visible light, its electron energy levels must lie just the right distance apart. With ruby, two absorption mechanisms occur when white light passes through it. One involves an electron transition from the ground state $4A_2$ to the $4T_2$ level. Radiation of 2.2 eV is thus absorbed, corresponding to yellow-green light. The second involves a transition to the $4T_1$

level, absorbing 3.0 eV radiation, or violet light. These absorption areas are actually bands rather than narrow lines and overlap somewhat. Thus, there is only slight transmission in the blue but strong transmission in the red, giving ruby its rich red colour with slight purplish overtones (Nassau, 1983).

The laws of physics state that energy is never lost, so that which is used in elevating the electron to higher levels must reappear when the electron falls back

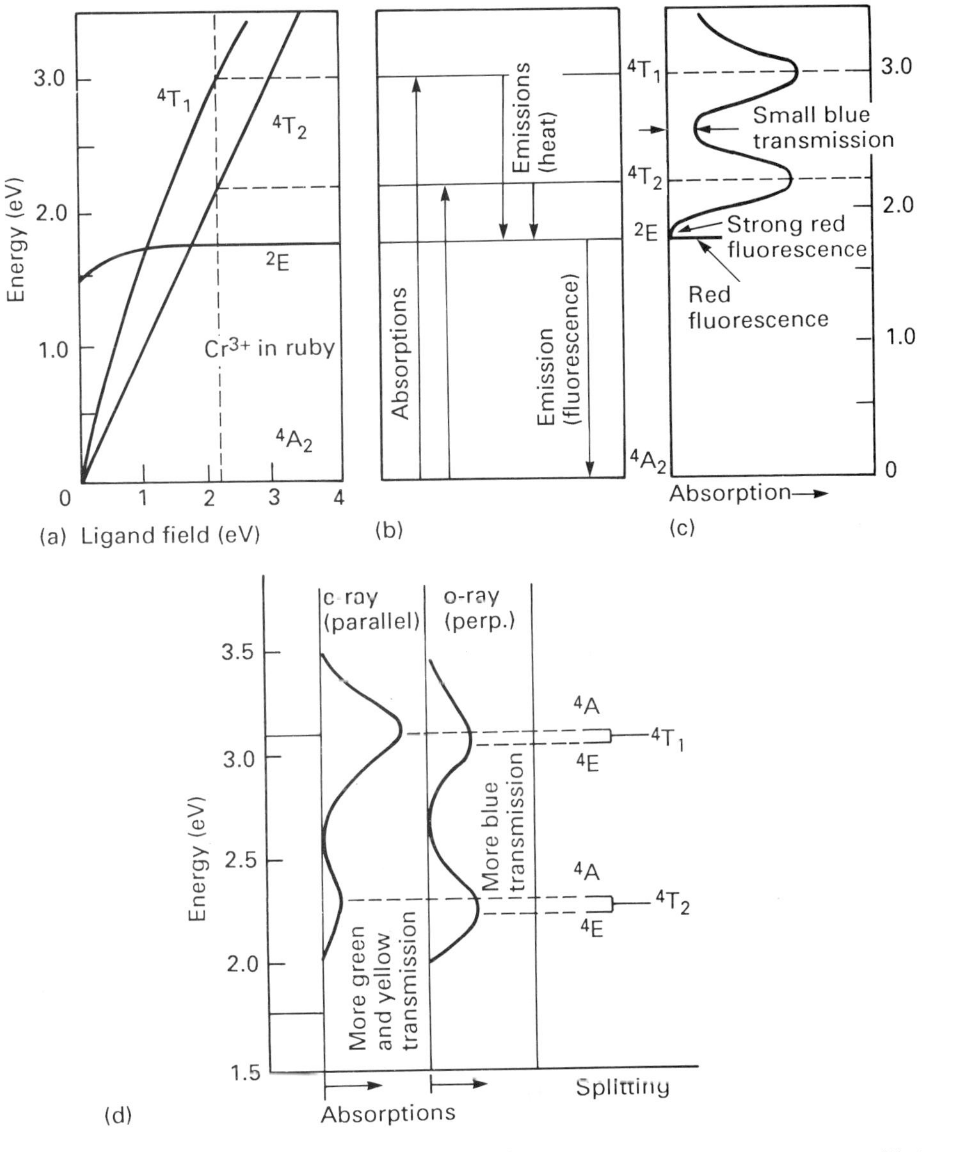

Figure 4.1 Colour in ruby. (*a*) The term diagram of Cr^{3+} in a distorted octahedral ligand field; (*b*) the energy levels and transitions in ruby; (*c*) the resulting absorption spectrum and fluorescence of ruby; (*d*) the absorption spectra for both the o- and e-rays in ruby. (after Nassau, 1983)

down to the ground state. Rather than returning to the ground state in one jump, however, it first falls to the 2E level, giving off an energy of 1.2 eV from the $4T_1$ level and 0.4 eV from the $4T_2$ level. Both of these energies correspond to the infrared portion of the spectrum, so the result is the release of an insignificant amount of heat from the ruby. Now electrons continue their journey from the 2E level to the ground state, involving the release of an energy of 1.79 eV and causing the emission of red light. This produces the rich red fluorescence of ruby which, while not very noticeable under ordinary conditions, can add visibly to the colour.

The best method of observing the strong red fluorescence of ruby is to illuminate the specimen with ultraviolet light in the absence of any visible light. The presence of iron as an impurity quenches this red fluorescence because energy is transferred to the iron and appears as heat. Thai rubies do contain enough iron to reduce this fluorescence, and thus their colour is generally more dull and lifeless compared with rubies from Burma. In ancient times it was believed that a fire burned inside the ruby. Now we can see that this was not that far from the truth. If a ruby is illuminated closely with a strong incandescent light source, such as a fibre optic lightguide, this fluorescence to visible light can be seen, caused by the emission of red light as electrons return to the ground state. Fluorescence will be discussed in greater detail later in this chapter.

To summarize, white light falling on a stone causes electrons to jump to higher states. This produces the absorption of green-yellow and violet light. Thus, blue and red light are transmitted, giving the stone a purplish-red colour. The electrons first fall part of the way down, causing the emission of heat. When they complete their return to the ground state, red light is emitted as fluorescence. It is only a coincidence that ruby is both red in colour and gives off a red fluorescence. However, this coincidence is responsible, in part, for its intense red colour, as the red fluorescence gives ruby an advantage compared with other red stones, such as tourmaline and garnet: it covers up any dark areas (extinction) and actually makes the stone glow with a rich crimson hue.

Blue sapphire

The colour of blue sapphire results from an entirely different mechanism than that of ruby – a process known as *intervalence charge transfer.* Involved in this are traces of both titanium and iron.

The mystery of the magnificent blue colour of sapphire was not discovered until fairly recent times. Auguste Verneuil, father of the synthetic ruby, was the first to discover that titanium *and* iron were responsible for the blue colour, but the actual mechanism involved eluded detection until much later. A few hundredths of one per cent of titanium in corundum produces no colour whatsoever, while the same amount of iron imparts only a pale-yellow colour. However, if both iron and titanium are present together, a beautiful deep-blue colour results.

Iron and titanium both substitute for aluminium in the corundum crystal structure. The iron may be present either in a ferrous (Fe^{2+}) or ferric (Fe^{3+}) state, while the titanium is found as Ti^{4+}. Should both Fe^{2+} and Ti^{4+} be present, they may interact with one another if located in close proximity. A single electron hops from the iron to the titanium ion when stimulated by light rays, absorbing visible light in the process. This is illustrated by the following equation:

$$\begin{array}{c} \text{white light} \\ \downarrow \\ Fe^{2+} + Ti^{4+} \xrightarrow{\text{absorption}} Fe^{3+} + Ti^{3+} \\ \downarrow \\ \text{blue light} \end{array}$$

The intervalence charge transfer mechanism which produces a blue colour in sapphire is a much more efficient colorant than that of the Cr^{3+} ion in ruby, being 100–1000 times more intense. Ruby requires about 1% of chromium for a deep-red colour, while only a few hundredths of one per cent of iron and titanium are needed to achieve a similar depth of colour in sapphire (Nassau, 1983).

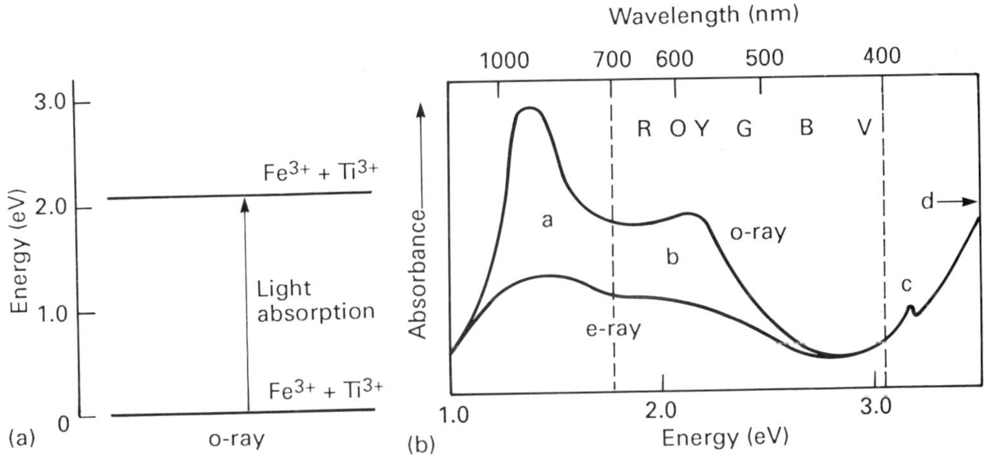

Figure 4.2 (*a*) Energy transitions from the ground state to an excited state in blue sapphire. Through a molecular charge transfer process an electron jumps temporarily from Fe^{2+} to Ti^{4+}, resulting in absorption of light. (*b*) The dichroic o-ray and e-ray absorption spectra of blue sapphire. Band (a) is derived from $Fe^{2+} \rightarrow Fe^{3+}$ charge transfer, band (b) from $Fe^{2+} \rightarrow Ti^{4+}$ charge transfer, band (c) from a ligand field transition in Fe^{3+}, and band (d) from $O^{2-} \rightarrow Fe^{3+}$ charge transfer. (After Nassau, 1983)

Many sapphires possess the necessary impurities for a deep-blue colour, but in the wrong state; thus, they possess little if any colour. Suitable heat treatment can rectify this situation and bring out a beautiful blue where before there was none. This is the process used to treat 'geuda' sapphires from Sri Lanka.

Pale gems containing dense concentrations of rutile silk (TiO) are heated to between 1500 and 1900°C in a reducing atmosphere. The heat melts the rutile, driving the titanium into solid solution, while the reducing atmosphere converts the iron present from the ferric to the ferrous state. Cooling the stone rapidly prevents the titanium from exsolving back into rutile. Now both the Fe^{2+} and Ti^{4+} needed are present. Thus, a cloudy pale-coloured stone becomes clear, with a deep-blue colour. This and other processes for heat treating corundum will be discussed in detail in Chapter 6.

Other sapphire colours

When present by itself, iron can produce a pale-green, yellow or brownish colour, but if titanium is also present other colours may result. These include greens and blue greens and are caused, in part, by the intervalence charge transfer discussed above.

In yellow and orange sapphires the colour can originate in one of several ways. With Thai and Australian yellow sapphires, Fe^{3+} appears to be the cause, although this is not necessarily substitutional iron. Yellow stones from these localities show very prominent iron absorption bands in the spectrum.

In yellow sapphires from Sri Lanka, the cause of colour is apparently due not only to Fe^{3+} but also to colour centres. Colour centres involve a missing electron from an atom, producing a 'hole colour centre', or an extra electron, in a vacancy, producing an 'electron colour centre'. Irradiation causes an electron to be ejected from an atom, simultaneously creating both a hole centre and an electron centre. Either or both of these electrons may then become excited and absorb visible light.

These yellow sapphires from Sri Lanka contain both fading and non-fading colour centres. Apparently the non-fading type was created by millions of years of low-level doses of radiation while in the ground. The non-fading colour centres,

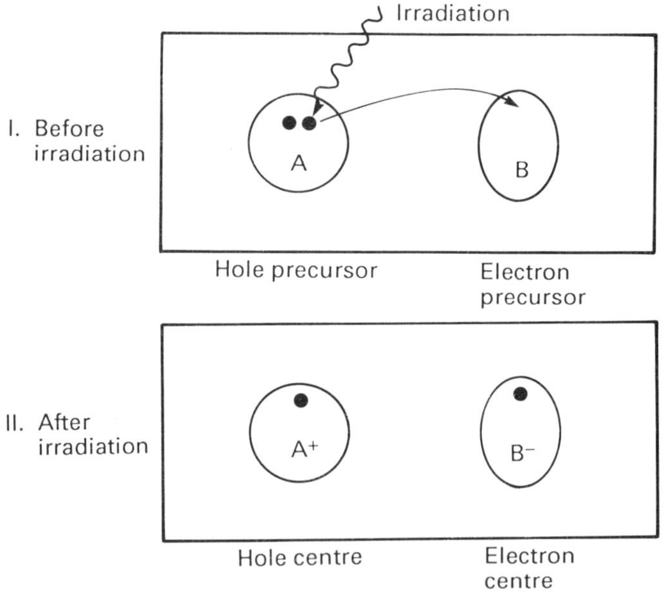

Figure 4.3 The above illustrations show how irradiation simultaneously creates both a hole centre and an electron centre. Irradiation causes an electron to be ejected from an atom, ion, molecule, impurity or other defect (A). This is received by the electron precursor (B) which may be an atom, ion, molecule, impurity or other defect. Now both the hole centre and the electron centre possess unpaired electrons which may absorb visible light when excited. Heating the stone may cause the electron to return to the hole precursor, thus bleaching the colour. Yellow/orange sapphires from Sri Lanka may owe part or all of their colour to these colour centres (fading and/or non-fading types). These colour centres may occur naturally or through man-induced irradiation. That induced by man is of the fading type while both fading and non-fading types (under normal wearing conditions) may occur in nature. (After Nassau, 1984)

coupled with the Fe^{3+}, are believed to be responsible for the colour seen (Nassau, 1984). In the padparadscha, a delicate pinkish-orange sapphire from Sri Lanka, chromium is also present, which adds a pink tint and thus produces the pinkish-orange colour.

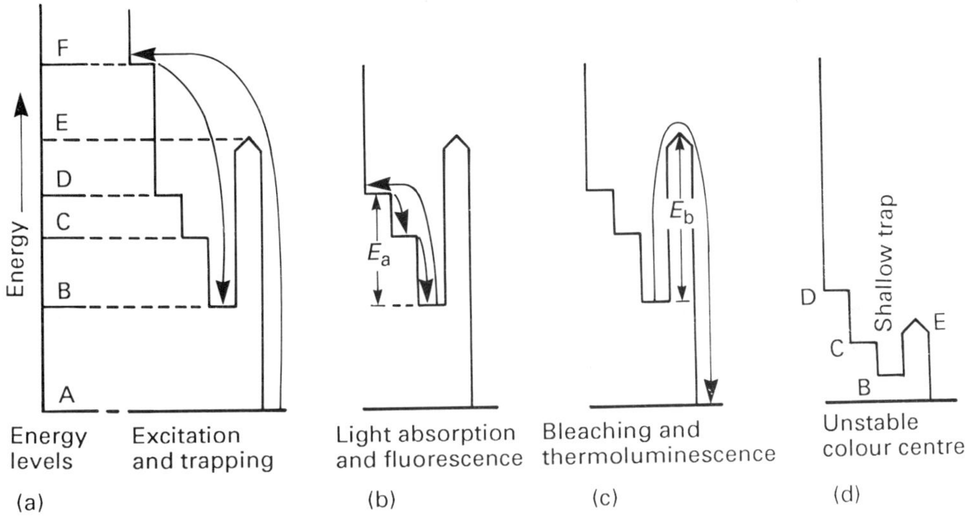

Figure 4.4 The formation of colour centres in gemstones. Colour centres are, in part or in whole, responsible for the colour of non-heat-treated yellow and orange sapphires from Sri Lanka. (After Nassau, 1983)

Heat treatment is currently being used to produce both yellow and orange sapphires from lightly coloured Sri Lankan material. The colours achieved are often so deep as to be totally unlike those produced in nature from this locality. It is believed that their colour is not only due to colour centres, but may also result from the exsolution of iron as the mineral haematite. The reddish haematite then mechanically colours the stones yellow or orange. Evidence of this is seen as these stones always contain multitudes of minute exsolved particles, with the colour concentrated around them.

The colours of the other varieties of sapphire are believed to be largely due to a mixture of the elements that create blue, red and yellow. Violet and purple sapphires are thought to be derived from a mixture of red and blue. Thus the colour would be due to both chromium and the titanium–iron effects. Green, being a mixture of blue and yellow, would be caused by a mixture of the elements that create blue and yellow sapphires. Green sapphires come mainly from Thailand/ Cambodia and Australia. Interestingly, although both blue and yellow sapphires come from Sri Lanka, green sapphires are very rare from this source. Perhaps this is because the conditions for the creation of blue sapphires in Sri Lanka are the opposite for the creation of yellow sapphires. Although sapphires which contain both blue and yellow areas (in a single stone) are found in Thailand and Australia, these are not often found in Sri Lanka.

Colour-change sapphires are sometimes encountered which change from a greenish blue in daylight to a purplish colour under tungsten light. Their colour is believed to be due to a mixture of chromium and titanium–iron, with vanadium possibly being involved as well.

Black star sapphire

The black star sapphire differs from the other varieties of corundum in that its colour is primarily derived not from transition metal impurities or colour centres, but from multitudes of inclusions of exsolved haematite. In a manner similar to the chromium-green mica plates in aventurine quartz, the haematite imparts a deep-brown or black colour to an otherwise blue, green or yellow sapphire. This is termed *mechanical coloration*, or 'colour by inclusions'.

Not only is the haematite responsible for giving some varieties of corundum their colour, but these plate-like inclusions are also the cause of the six-rayed star effect in black star sapphires. Iron enters the corundum lattice at high temperatures in solid solution. As the temperature drops the solubility decreases, literally squeezing the iron atoms out of solution, where they crystallize as tiny plates and needles along three directions meeting at 120° in the basal plane. When this occurs in an otherwise blue or green sapphire, the rays of the star appear white. Rarely, however, the haematite unmixes in a yellow sapphire and the star takes on a rich golden-yellow colour. These 'golden-star' black star sapphires are mined at Ban Ka Cha and Khao Ploi Waen in Chanthaburi Province, Thailand, and are highly prized, as evidenced by the high prices paid for them. Particularly fine pieces of large size may command up to $100 per carat in the local (Thai) market.

Luminescence

Corundum, particularly ruby, displays several forms of luminescence. The most important of these from the gemmological point of view is fluorescence.

Fluorescence is created when a stone absorbs energy from visible light, ultraviolet light or x-rays and then re-emits it at lower energy in the form of visible light of a longer wavelength. A perfect example of the fluorescence of ruby is given by a technique known as the *crossed filter* method, which was expounded by the eminent British gemmologist, B. W. Anderson. The crossed filter technique requires two different filters: a blue filter to absorb all red, orange and yellow light, and a red filter to absorb all green, blue and violet light. When 'crossed', the two filters absorb the entire visible spectrum from violet to red, permitting no light to pass. Plastic or glass filters of the above colours usually work well, as does a strong solution of copper sulphate for a blue filter.

To perform the test, a strong concentrated light, such as a spectroscope or fibre optic light, should be directed through the blue filter onto several rubies. Now the stones are observed by looking through the red filter. What is seen is truly amazing:

the rubies will appear as glowing red coals on a jet-black background! Because of the blue filter, only blue light falls onto the stones. They convert this blue light into a slightly lower form of energy – red light, emitted from the gem at the fluorescent doublet of 692.8 and 694.2 nm. This can be confirmed by examining the stones through the crossed filters with the spectroscope. A bright fluorescent doublet will be seen at the end of the red, silhouetted on a black background. It is a magnificent method of demonstrating exactly the principles of fluorescence, in this case fluorescence to visible light.

Fluorescence can also be stimulated by invisible light such as ultraviolet light or x-rays; however, fluorescence is but one of many different kinds of luminescence. Other types include *triboluminescence*, caused by mechanical disruption such as the red glow when ruby is cut on a diamond saw; *thermoluminescence*, when ruby is excited by one of the above means at low temperature and the energy, or glow, emitted upon heating; and *phosphorescence*, a delayed emission of fluorescence. Other types of luminescence are also known; these are only a few examples.

The most important stimulant for the fluorescence of gemstones is ultraviolet light. Two different wavelengths are commonly used – 366 and 253.7 nm, known as long-wave and short-wave ultraviolet respectively. Viewing cabinets are used in conjunction with these light sources to provide a dark background for the observation of the reactions and to isolate the stone from any stray light. It is best to grasp the stone with tweezers rather than holding it in the hand, again so that the reaction is seen against a perfectly black background. The gem should be grasped table-to-culet and held as close as possible to the light, for the fluorescence diminishes in direct proportion to the distance the stone lies from the light. This allows it to be examined across the girdle, the thickest part of the stone, to see the strongest reaction. A feature which the makers of ultraviolet cabinets would do well to incorporate is a magnifying lens to enlarge the stone during viewing. Magnification allows very small fluorescent features to be picked out, such as zoning, which would otherwise go unnoticed. The removable magnifying eyepiece lens of certain refractometers works well for this task. Care should be taken to check the stone from all sides, as some gems fluoresce in one area only. A note of caution is required with regard to the short-wave ultraviolet light: protective eyewear is needed to prevent damage to the eyes from this more energetic source.

Webster (1939) has reported on the phosphorescence of ruby. Phosphorescence refers to a continuation of the fluorescent glow after the stimulating radiation source has been switched off. Webster found that both natural and synthetic (Verneuil) rubies display a very weak phosphorescence to ultraviolet light. However, as this afterglow lasts but a fraction of a second it is of very limited diagnostic use. Chromium-rich Burmese and synthetic rubies tended to show a slightly longer phosphorescence than rubies from Thailand, where it was believed that the iron had an inhibiting effect on both fluorescence and phosphorescence.

Although fluorescence is a useful gemmological test, some caution must be used in interpreting the reactions. Fluorescence is usually caused by impurities within the stone, with only tiny fractions of one per cent being necessary in many cases. Since the impurity content can vary a great deal from stone to stone, the fluorescent reactions may also vary. Thus, fluorescence should never, never, be solely relied

TABLE 4.1 Fluorescent reactions of corundum

Variety	*Long wave* (366 nm)	*Short wave* (253.7 nm)	*X-rays*
Ruby, Burma, Sri Lanka	Moderate to very strong red or red orange	Moderate to strong red or red orange, slightly weaker than long wave	Moderate to strong red
Ruby, Thailand, Cambodia	Weak to moderate red or red orange	Inert to moderate red or red orange, slightly weaker than long wave	Inert to moderate red
Ruby, Kenya	Strong to very strong red or red orange	Moderate to very strong red or red orange, slightly weaker than long wave	Moderate to strong red
Ruby, other sources	Weak to very strong red or red orange	Inert to very strong red or red orange, slightly weaker than long wave	Inert to very strong red
Blue sapphire, Sri Lanka	Inert to strong red or orange	Inert to strong red or orange, or, in heat-treated stones, weak to moderate chalky blue green, or green (the colourless areas fluoresce)	Dull red or yellowish orange
Blue sapphire, Burma	Inert to moderate red or orange	Inert to moderate red or orange	Inert to moderate red
Blue sapphire, Cambodia, Australia, Thailand	Generally inert	Generally inert, rarely weak chalky blue to green	Inert
Blue sapphire, Kashmir	Inert to moderate red or orange	Inert to moderate red or orange	Inert to dull red or yellowish orange
Blue sapphire, other sources	Inert to strong red or orange	Inert to moderate red or orange, or weak to moderate chalky blue to green	Dull red or yellowish orange
Yellow and orange sapphire, Sri Lanka	Inert to strong orange to orange red	Inert to strong orange to orange red	Weak to strong orange
Yellow and orange sapphire, Thailand, Australia	Inert to moderate red	Very weak to weak red	Inert
Yellow and orange sapphire, other sources	Inert to strong orange to red	Inert to strong orange to red	Inert to strong orange
Purple and violet sapphire, Sri Lanka	Weak to strong red to orange red	Inert to strong red to orange red	Inert to strong red
Purple and violet sapphire, other sources	Inert to strong red to orange red	Inert to strong red to orange red	Inert to strong red
Green sapphire, all sources	Generally inert, rarely weak to moderate red to orange	Generally inert, rarely weak red to orange, or weak chalky blue to green	Inert
Colourless sapphire, Sri Lanka and other sources	Inert to strong orange to orange red	Inert to moderate orange to orange red, rarely weak to moderate chalky blue to green	Inert to moderate red or orange

upon to establish the identity of a gem. Instead, it provides only a rough indication, even under the best of circumstances. Far too many rubies and sapphires have been misidentified by gemmologists who put too much faith in the colourful glows seen under ultraviolet light.

Table 4.1 is an admittedly weak attempt at setting out typical fluorescent reactions for the different varieties of corundum. It should by no means be considered the last word on the subject. Variations in the trace element content can produce entirely different reactions; thus, this must be considered only as a general guide.

Absorption spectra

Among the many contributions to gemmology made by the late British pioneer, B. W. Anderson, probably that which will be remembered the most is his contribution to the field of spectroscopy. Beset with a degree of colour blindness, Anderson was forced to avail himself of other means of identifying gems, rather than relying on characteristic colours. One of his greatest innovations was the introduction of gemmologists to an instrument which has long been used in analytical chemistry – the hand spectroscope. In a series of 40 ground-breaking articles written for *The Gemmologist* in London, he laid down the various absorption spectra for every major gemstone. So comprehensive were these articles that they still remain today the most important reference on absorption spectra in the field (Anderson, 1953–1957).

The spectroscope is one of the most difficult instruments for the novice student to use; however, the information gained through its use can be of great diagnostic significance. Thus, it is vital that anyone aspiring to the title of gemmologist be well versed in the use of this instrument. It is beyond the scope of this book to describe the instrument itself and its techniques, but excellent descriptions can be found in such books as Anderson's *Gem Testing* (1990) and Read's *Gemmological Instruments* (1983).

The purpose of the spectroscope is to analyse the visible spectrum from 400 to 700 nm. It does this by splitting light with a series of prisms or a diffraction grating and directing it so that it can be viewed with the eye. White light is first passed through the stone and then into the instrument. Because of the colouring agents in the stone's structure, certain wavelengths may be absorbed while others are transmitted freely, creating an absorption spectrum consisting of bright colours and dark absorption bands and lines. It is the position of these bands of broad absorption and lines of narrow absorption that have diagnostic value. Their positions can be measured roughly by means of a wavelength scale, or by estimation by reference to the colours or lines of a sample filter or stone. Even with a spectroscope with a built-in wavelength scale, however, this measurement is at best only approximate.

There exist three basic spectra in natural corundum due to the elements chromium, iron and vanadium. Those of chromium and iron are extremely common, while that of vanadium is very rare. Each of these spectra are discussed in detail below under the different varieties of corundum.

Ruby

The absorption spectrum of ruby is derived principally from the element chromium, with iron occasionally playing a small part. Chromium produces a very distinctive spectrum which allows ruby to be easily separated from all other minerals; however, synthetic ruby is also coloured by chromium and shows an identical spectrum.

In ruby, the chief areas of absorption are located in the violet and blue from 400 to 450 nm and a broad band in the yellow green centred at 550 nm. A narrow portion of the blue is transmitted from approximately 450 to 500 nm, and this, coupled with the transmission in the orange and red, gives ruby its purplish-red colour. Three narrow absorption lines are seen in the blue at 468.5, 475 and 476.5 nm, which are diagnostic for chromium-bearing natural and synthetic corundum. The broad absorption band centred at 550 nm and a series of narrow lines in the red complete the ruby spectrum. Two of these lines, at 692.8 and 692.4 nm, will often reverse as fluorescent lines. The crossed filter technique allows them to be seen dramatically on a black background. These fluorescent lines show up best in reflected or scattered light.

Although the chromium spectrum of ruby does not allow separation from synthetic ruby, at times other features, due to iron, will. In some purplish rubies, particularly those from Thailand, iron lines at 451.5, 460 and 470 nm are seen in addition to those lines in the blue due to chromium. Because these 'iron lines' have not been observed in synthetic rubies, their presence is positive proof of natural origin. Of these three lines, that at 451.5 nm is strongest, with the other two often absent.

Blue sapphire

Blue sapphires display an absorption spectrum which is due to iron. A series of three lines is seen, the strongest at 451.5, one slightly weaker at 460 and the weakest at 470 nm. Dark 'inky-blue' sapphires from Australia, Thailand (except Bo Ploi, Kanchanaburi), Cambodia and Nigeria generally show the entire series, with the three lines sometimes merging into one. In contrast, gems from Sri Lanka, Burma, Thailand (Bo Ploi) and Kashmir often display only a single line at 451.5 nm. At times, this too may be absent, particularly in heat-treated gems from Sri Lanka. The iron lines are, by and large, due to the ordinary ray and can be intensified slightly by eliminating the extraordinary ray with a polaroid filter.

As a result of the volatility of iron, most of this tends to burn off in the Verneuil flame. Thus, the distinctive iron lines are usually absent entirely in the flame fusion synthetic blue sapphires. Only on rare occasions has a single faint 451.5 nm line been detected in a Verneuil synthetic sapphire, while the Chatham flux-grown synthetic blue sapphire may also show a faint 451.5 nm line. This means that the presence of the three iron lines together is positive proof of natural origin. In the event that no lines are seen, or just a weak 451.5 nm line alone, the stone could be either natural or synthetic.

In addition to the iron-line complex at 451.5 nm, blue sapphires of either natural or synthetic origin may display a band several tens of nanometres in width centred at about 560 nm. This is due to the ordinary ray and can also be intensified slightly by rotating a polaroid plate over the eyepiece to eliminate the extraordinary ray. Although quite strong at times, the 560 nm band has no diagnostic significance. It is probably due to traces of chromium.

Blue sapphires which contain traces of chromium may also show a weak ruby spectrum superimposed on the sapphire spectrum. The most prominent feature of this is the fluorescent doublet at 692.8 and 692.4 nm. However, this may also be found in synthetic gems with traces of chromium and so is of no diagnostic significance.

Blue sapphires also display absorption lines in the ultraviolet region and these can be useful in separating them from Verneuil synthetic sapphires. As this region is invisible, however, special instruments are required to observe these lines. Lines are present in natural blue sapphires at 379 and 364 nm and are said to be due to ferric iron. These lines can be found even in Sri Lankan sapphires where the 451.5 nm line is not present (Anderson, 1953–1957). The 379 and 364 nm lines are absent in Verneuil synthetic sapphires; however, no information could be found regarding the ultraviolet spectra of the Chatham synthetic product in this colour. Nautiyal and Mukherjee (1958) reported these two lines at 388 and 375 nm for natural sapphires. As they used more sophisticated apparatus to measure these lines than Anderson, their wavelength measurements are probably the more accurate.

Other sapphires

Yellow sapphires from Australia, Thailand, Montana, Tanzania (Umba) and Burma are rich in iron and so show the entire 451.5 nm complex, just like the blue sapphires from these localities. Sri Lankan yellow sapphires, however, contain little iron and thus usually display only a weak 451.5 nm line, or nothing at all. Heat-treated yellow sapphires from Sri Lanka may show a 450–500 nm 'cut-off' (complete absorption from 400 to 450 nm). This is of no diagnostic value, as orangy-yellow Verneuil synthetic sapphires may also show a similar cut-off.

Verneuil-produced synthetic yellow sapphires owe their colour to nickel rather than iron. As a result, we would not normally expect to see the iron lines, but even so, on very rare occasions a weak line at about 455 nm has shown up (Anderson, 1953–1957). Like the natural stones which contain a trace of chromium, the synthetic may display a weak chromium spectrum, particularly the fluorescent doublet in the red. The synthetic may also display a 450–500 nm cut-off, which is of no significance whatsoever. Information on the spectra of the new Chatham flux-grown synthetic yellow sapphires is not yet available. It is of course possible that these stones may display iron lines.

Natural orange sapphires are coloured by a combination of chromium and iron, and so again may show superimposed iron and chromium lines, as with some yellow sapphires. Those from the Umba Valley in Tanzania generally display strong iron lines, while in those from Sri Lanka the iron lines may be absent altogether. Verneuil synthetic orange sapphires are coloured by a combination of nickel and chromium, and so show no iron lines; nor does the flux-grown synthetic orange sapphire produced by Chatham.

Green sapphires originate mainly from Thailand and Australia and are coloured by iron alone, or a mixture of iron and titanium. This is the one sapphire colour which almost always displays the iron lines in their entirety. The Verneuil synthetic is coloured by a combination of cobalt, vanadium and nickel, and thus shows no iron lines. Their absence can then be taken as proof of synthetic origin. Synthetic green sapphires may show lines at 500, 530, 635 and 690 nm.

Natural sapphires are occasionally found which display a change of colour from bluish or greyish green in daylight to purplish red in incandescent light. However, most of the so-called colour-change sapphires merely shift from violetish blue or violet to purplish red. If this qualifies as a colour-change gem then so too should most amethyst quartz. Although in danger of splitting hairs, it is this author's opinion that in order to qualify for the term 'colour-change sapphire', the stone must show a daylight colour on the green side of blue, rather than on the red (violet) side. The vast majority of colour-change sapphires are simply violetish to begin with. As all violetish stones show a more purplish colour under incandescent light, it would hardly seem fitting to accord them a special title.

The colour-change phenomenon in natural corundum appears to result from a mixture of chromium, iron and titanium. This produces the typical ruby spectrum, with superimposed iron lines. In the Verneuil product, however, the colour change results from vanadium alone, creating a single line at 473, a broad band centred in the yellow at 580 and a fluorescent line in the deep red at about 690 nm. Traces of vanadium have been detected in natural corundum through chemical analysis, but apparently not in large enough quantities to show up in the spectrum. In only one case, a sapphire from Burma, has the vanadium line at 473 nm been reported in a natural corundum. Thus, its presence is an almost positive indication of synthetic origin.

The pleochroism of corundum can have a slight effect on the absorption spectrum. By rotating a polaroid plate over the spectroscope eyepiece, the spectra of both the ordinary and extraordinary rays can be viewed independent of one another.

Use of the spectroscope

Space restrictions preclude a detailed description of the spectroscope and its use, but a few brief tips are called for.

Successful use of the spectroscope involves the following factors:

1. position of the light source
2. position of the stone
3. position of the spectroscope itself

It is necessary that each of these three 'links' is correctly adjusted to maximize the chances of seeing the absorption features. The light source itself is of tremendous importance. It must be extremely intense and focusable to a narrow spot. Fibre optic illuminators of 150 W power work very well for spectroscopy. The light may be positioned to allow transmission directly through the stone from below, or reflected from above. In most cases, the reflection method is superior for it allows a longer path through the stone and therefore more absorption. If too much absorption is seen, the path should be shortened by moving the light and/or the stone to lessen the absorption. In Thai rubies, for example, shortening the path sometimes allows the iron lines to be picked out at 451.5, 460 and 470 nm, in addition to the chromium spectrum. Lengthening the path intensifies faint lines, allowing the 451.5 nm complex to be seen in heat-treated Sri Lankan sapphires, where it is normally very weak.

Faceted stones must be carefully positioned so that only light passing through the stone reaches the spectroscope. Owing to the arrangement of facets, light may leave the stone in several different directions; the stone should be positioned so that the maximum amount of light passes out towards the spectroscope. Placing one's hand around the stone in different positions allows the exact determination of where the light is travelling. Oval stones should be positioned so that the broad side faces the spectroscope, if the light is transmitted from below.

The stone's position must also take into account any colour zoning. Absorption lines normally result from colour, so the stone should be positioned so that the light passes through the most deeply coloured areas to maximize the absorption.

The battle to obtain a clearly visible absorption spectrum is nearly over when both the light and the stone are correctly positioned. However, the instrument itself needs to be correctly placed. This amounts to moving the spectroscope into the path of the light leaving the stone. Always make sure that the instrument is firmly anchored with the holder supplied on table models, or with modelling clay or plasticine. Although the instrument is called the hand spectroscope, the fact is that it just cannot be used effectively in the hand alone. Some type of holder to anchor the instrument firmly in place is absolutely essential.

First the instrument is centred on the spectrum by rocking it up and down with the slit only slightly open. Once the spectrum is located, the slit is adjusted just enough to allow the spectrum to appear, *not too much*. Then the drawtube focus is adjusted for each individual area, as different colours come into focus at different points. In locating very faint lines, much skill is needed. As with a pinball machine, much practice is needed to learn how to finesse the instrument. Try not to look directly at a faint line, but instead peer out of the corner of the eye. Open and close the slit rapidly and rock the instrument up and down at the same time. This will help to resolve the very faint lines which are so often encountered.

Pleochroism

The crystal structure of corundum is classified in the trigonal division of the hexagonal system. Lattice points are equally spaced along three directions in the horizontal plane, but unequally spaced in the vertical plane. The manner in which light rays are affected depends upon which of these planes the rays vibrate across.

In uniaxial gems, such as corundum, light entering the crystal in any direction except parallel to the *c* axis is split into two rays, each vibrating roughly perpendicular to the direction of travel and perpendicular to the vibration direction of the other ray. One ray is termed the ordinary ray, denoted by the Greek letter omega (ω). It always vibrates across the equally spaced lattice points and so has a constant refractive index (n_ω = 1.770) no matter what the direction of travel. The other ray is termed the extraordinary ray because, unlike the ordinary ray, it does not obey the ordinary laws of refraction; it is denoted by the Greek letter epsilon (ϵ). Its vibration direction, and subsequently its refractive index, is variable. When light travels parallel to the *c* axis, it too vibrates in the horizontal plane and so has a refractive index equal to that of the ordinary ray (n_ω = 1.770). Its refractive index decreases steadily as the direction of travel approaches 90° to the *c* axis. When light is incident at 90° to the *c* axis, it vibrates in the vertical plane across the unequally spaced atoms. At this point its refractive index reaches its maximum difference from that of the ordinary ray (n_ϵ = 1.762).

Because of the difference in symmetry, and subsequently vibration direction, each ray may be absorbed differently. Thus, one ray takes on one colour while the other takes a different colour. This difference in colour with direction is termed *pleochroism* (multicoloured). Uniaxial materials possess two vibration directions (ω and ϵ) and so potentially two different colours, one corresponding to each vibration direction. As a result, uniaxial gemstones such as corundum are 'dichroic'.

Because pleochroism results from the difference in vibration directions, it will vary in an identical manner to the refractive index as described above. The colour corresponding to the ordinary ray is constant throughout the crystal, while that corresponding to the extraordinary ray is variable. Parallel to the *c* axis (optic axis), it is the same as that of the ordinary ray, so no pleochroism is seen. In intermediate directions, as the direction of travel approaches the *c* axis, its colour gets closer and closer to that of the ordinary ray. The e-ray colour reaches its maximum divergence from that of the o-ray when light travels perpendicular to the *c* axis.

Not all doubly refractive minerals display pleochroism, because in some the difference in absorption between the two rays is not great enough to be detected with the eye. Most varieties of corundum do display strong pleochroism, but its strength may vary with the stone's depth of colour. Generally speaking, the greater the specimen's depth of colour, the greater will be its pleochroism, and vice versa. No pleochroism is observed in colourless stones, or in coloured stones parallel to the optic axis (*c* axis). The strongest pleochroism is seen at right angles to the optic axis.

In Table 4.2 are listed the pleochroic colours of the different varieties of corundum, when judged under incandescent light.

Observing pleochroism

To observe pleochroism, an instrument known as a dichroscope is used. This makes use of either a circle of polaroid cut so that the vibration direction of one half is perpendicular to the other, or a rhomb of calcite. The calcite type is preferable to the polaroid type because it is colourless and so allows finer gradations of colour to be distinguished. Its optics also enable a smaller area of the gem to be sampled.

TABLE 4.2

Variety	*Colour of ordinary ray* (ω)	*Colour of extraordinary ray* (ε)	*Strength*
Ruby	Slightly purplish red	Slightly orangy red	Moderate to strong
Blue sapphire	Violetish blue	Greenish blue to bluish green	Strong to very strong
Green sapphire	Bluish green to greenish blue	Yellowish green to greenish yellow	Strong to very strong
Violet and purple sapphire	Bluish purple or bluish violet	Yellowish purple or greenish blue	Strong to very strong
Yellow sapphire	Slightly orangy yellow or deep yellow	Slightly greenish yellow or pale yellow	Weak to very weak
Orange sapphire	Slightly reddish orange or deep orange	Slightly yellowish orange or light orange	Very weak to moderate
Colourless sapphire	Colourless	Colourless	None

Furthermore, the polaroid types formerly produced by Gem Instruments (USA) tended to deteriorate over time and lose their ability to show pleochroism.

In observing pleochroism in corundum, two colours will be seen in all directions except parallel to the optic axis, where only one is seen. The secret is to examine the stone at right angles to the optic axis, where the extraordinary ray reaches its maximum divergence from that of the ordinary ray.

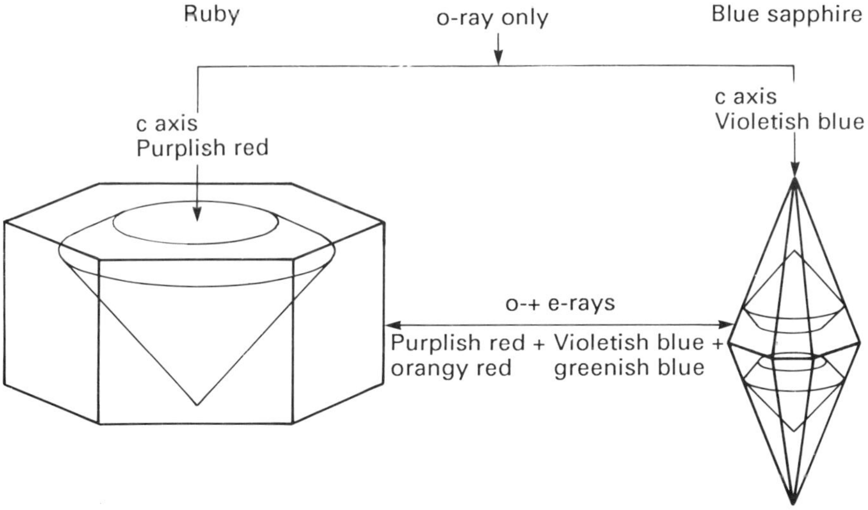

Figure 4.5 Pleochroism of natural corundum. In directions parallel to the *c* axis, only the more intense o-ray is seen. However, at right angles to the *c* axis (along the *a* axis) the o-ray is diluted by the more insipid e-ray. Thus, most natural corundums are cut with the *c* axis at 90° to the table in order to obtain the more intense colour

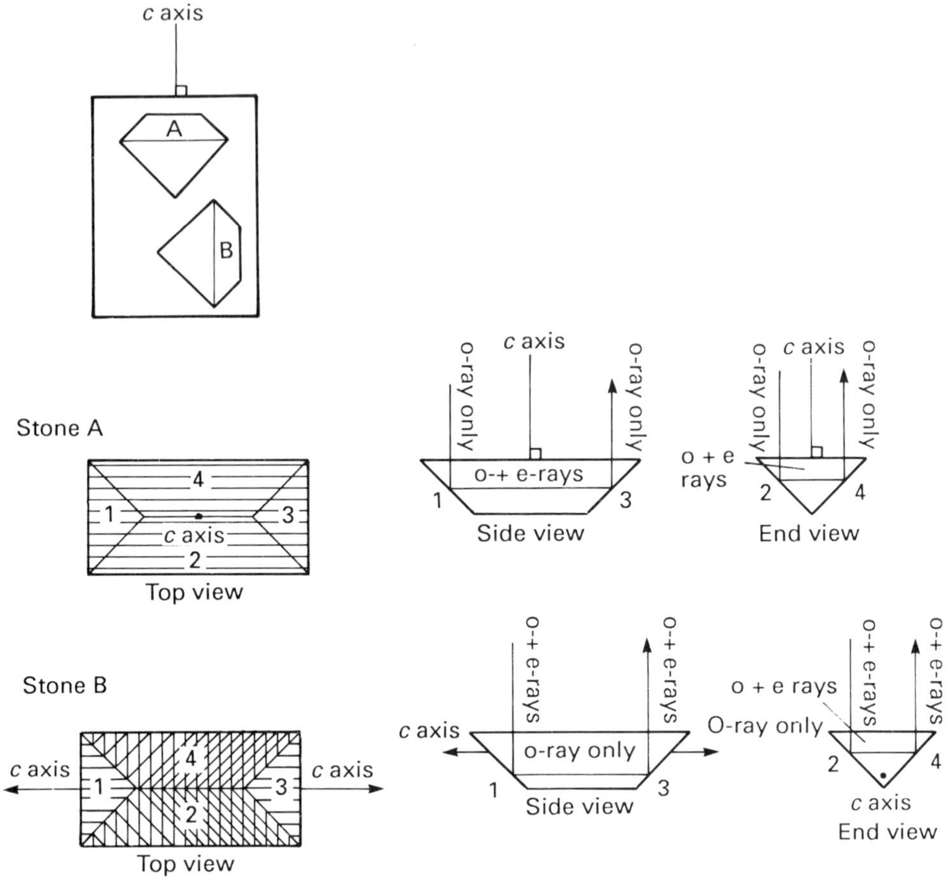

Figure 4.6 The effects of pleochroism on the appearance of corundum. Stone A illustrates a corundum cut with the *c* axis at right angles to the table facet, while in stone B the *c* axis is parallel to the table. Stone A will show a relatively uniform colour on facets 1, 2, 3 and 4 because light travelling along the length (facets 1 to 3) is affected identically to that travelling across the sides (facets 2 to 4). Although it is true that in elongated stones, such as ovals, facets 1 and 3 will display a slightly darker colour than facets 2 and 4 because of the longer light paths involved, this should not be confused with pleochroism. In stone B, however, light travelling along the length (facets 1 to 3) will be richer in the o-ray than light moving across the sides (facets 2 to 4). As the o-ray in corundum is generally a more intense and beautiful colour than the e-ray, this means that facets 1 and 3 in stone B will display a more intense colour while facets 2 and 4 will show a less intense (and less beautiful) hue

In rubies and sapphires, the pleochroism must be taken into account when orienting the gem during cutting. The more beautiful of the two colours is the ordinary ray in almost all cases. For this reason, the cutter must take care to orient the gem so that the *c* axis is perpendicular to the table facet, thus allowing only the ordinary ray to be seen, undiluted by the less intense extraordinary ray. In certain cases, usually due to the shape of the rough or the location of colour zoning or inclusions, the lapidary may adopt a different orientation. By orienting the table parallel to the *c* axis with rubies, the unusual 'cherry-red' colour is obtained.

Bibliography

Anderson, B. W. and Jobbins, E. A. (1990) *Gem Testing*, 10th edition Butterworths, London

Anderson, B. W. (1980a) Distinguishing between natural and synthetic gem corundum. *Retail Jeweller,* **19**, no. 452, 6

Anderson, B. W. (1980b) Special methods for distinguishing natural corundums from synthetics. *Retail Jeweller,* **19**, no. 456, 8

Anderson, B. W. and Payne, C. J. (1953–1957) The spectroscope and its applications to gemmology. *The Gemmologist*, September 1953–January 1957

Anderson, B. W., Webster, R. and Mitchell, R. K. (1958) No new absorption bands in blue sapphire. *The Gemmologist,* **27**, no. 325, 143–144

Hurlblut, C. and Klein, C. (1985) *Manual of Mineralogy (After James D. Dana)*, 20th edn, Wiley, New York

Liddicoat, R. T. (1977) *Handbook of Gem Identification*, Gemological Institute of America, Santa Monica, CA

McClure, D. S. (1962) Optical spectra of transition-metal ions in corundum. *The Journal of Chemical Physics,* **36**, 2757–2779

Nassau, K. (1980) *Gems Made By Man*, Chilton, Radnor, PA

Nassau, K. (1983) *The Physics and Chemistry of Color*, Wiley, New York

Nassau, K. (1984) *Gemstone Enhancement*, Butterworths, London

Nautiyal, S. P. and Mukherjee, B. (1958) Absorption spectrum and color of blue sapphire. *The Gemmologist,* **27**, no. 324, 119–121

Nikolskaya, L. V., Terekhova, V. M. and Samoilovich, M. I. (1978) On the origin of natural sapphire color. *Physics and Chemistry of Minerals,* **2**, 213–224

Read, P. G. (1983) *Gemmological Instruments*, Butterworths, London

Webster, R. (1939) Phosphorescence of gem materials: ruby. *The Gemmologist,* **8**, no. 94, 161–163

Webster, R. and Anderson, B. W. (1983) *Gems, Their Sources, Descriptions and Identification*, fourth edition, Butterworths, London

Wild, G. O. and Biegel, H. (1947) Absorption of sapphire in the ultra-violet region. *The Gemmologist,* **14**, no. 195, 279–280

Chapter 5

Inclusions of corundum

One of the most fascinating and rewarding aspects of gemmology is the study of inclusions. These tiny bits of entrapped foreign debris or structural irregularities reveal much about the gems in which they lie entombed. Formerly regarded as flaws which detract from a stone's value, today they are recognized as valuable clues that help to unravel the secrets of the stone's past. Not only do inclusions speak to us of the exact place of formation, but they also serve as hallmarks of the natural processes which gave rise to precious stones. Flaws? Hardly! Without them the creations of nature would be almost indistinguishable from what Eduard Gübelin (1983) has so aptly termed 'the usurpers from the factory'.

Inclusions in gemstones can be classified according to the scheme proposed by Gübelin (1983), which is based upon their age with respect to that of the host crystal.

1. *Protogenetic (pre-existing) inclusions*
   These are strictly of a solid nature, appearing either as heavily etched or corroded individuals which formed long before the host, or as well-formed crystals which developed just prior to the host.
2. *Syngenetic (contemporary) inclusions*
   Syngenetic inclusions are of widely varying types, including euhedral crystals which developed simultaneously with the host, either in a random or ordered fashion. Also of syngenetic origin are liquid inclusions of two types:
   (a) *Primary liquid inclusions*
   These are generally termed 'negative crystals', and consist of gas- or liquid-filled voids trapped by the rapidly growing host. They may also contain solid phases in addition to liquid and/or gas.
   (b) *Secondary liquid inclusions*
   These consist of cracks which were subsequently filled with growth solution and 'healed' shut. The healing may take place as the host grows or even long after the growth has stopped (which would make them 'epigenetic'

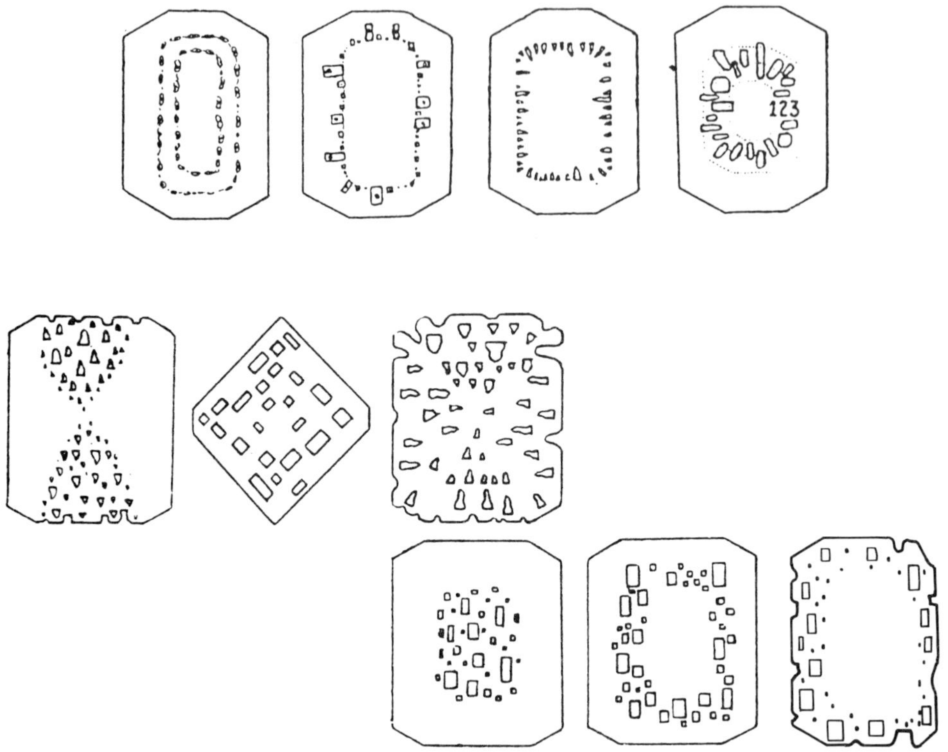

Figure 5.1 Various types of 'zonal' primary liquid inclusions. Such inclusions are often seen in Thai/Cambodian rubies and generally form parallel to the basal pinacoid. They often consist of tiny two-phase negative crystals surrounded by ultrathin liquid inclusions with a hexagonal (distorted or well-formed) or triangular outline. In dark-field or transmitted light only the two-phase negative crystal centre may be seen. However, with overhead illumination (particularly fibre optic lighting) the ultrathin liquid inclusion surrounding the negative crystal can be seen. Owing to the thin 'tabular' shape of most Thai/Cambodian ruby rough, most stones are cut with the table parallel to the basal pinacoid. Thus, in cut stones, these thin two-phase primary liquid films are usually found parallel to the basal pinacoid. Fibre optic lighting often causes these inclusions to show a brilliant iridescence because of their ultrathin nature. Such inclusions (or anything even remotely resembling them) have yet to be encountered in any type of synthetic corundum (including flux grown stones). Thus they provide an absolute positive means of separation. (After Roedder, 1984)

inclusions). Included in this category are the so-called 'fingerprint' and 'feather' inclusions. Solid and/or gaseous phases may also be present.

Other syngenetic inclusions are colour zoning, growth twins and related growth phenomena.

3. *Epigenetic (post-formation) inclusions*
   Epigenetic inclusions are those formed after the host has stopped growing and include secondary twinning, some healing fissures and inclusions produced through 'exsolution' – the unmixing of a solid solution. Examples of exsolved inclusions are the rutile and haematite silk which give rise to asterism in corundum.

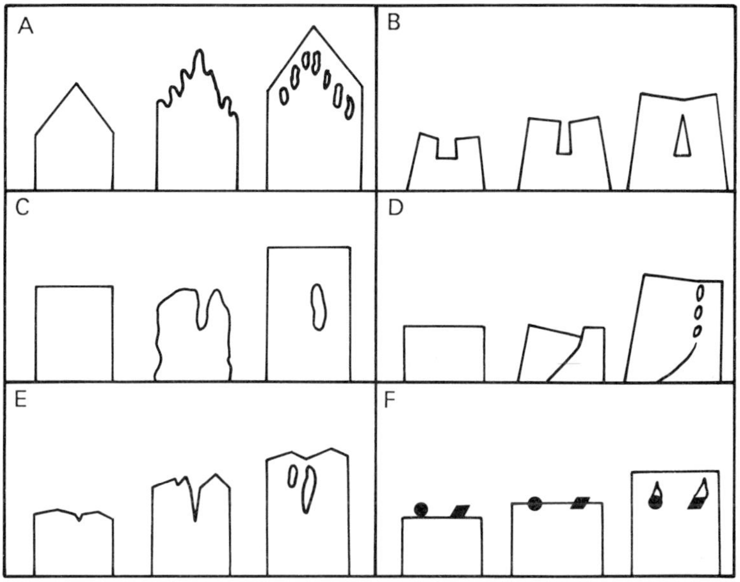

Figure 5.2 Some of the mechanisms by which primary fluid inclusions may be trapped in gemstones. A, rapid feathery growth is covered by later solid growth. B, subparallel growth traps fluid. C, a dislocation in a crystal etched out during partial dissolution is later covered by new growth. D, disturbed growth near a fracture in the surface of a growing crystal results in the trapping of primary fluid inclusions. E, primary fluid inclusions are trapped between or at the centres of growth spirals. F, enclosure of any foreign object on the surface of a growing crystal may include some of the growth fluid as well. (After Roedder, 1984)

Although the above classification system definitely has its merits, it is often impossible for the average jeweller/gemmologist to determine the exact class of a given inclusion. For this reason, the various inclusions found in corundum will be discussed according to the origin of the stone, rather than the inclusion type.

Ruby – Burma

The famed rubies of Burma display an internal picture which is distinctly unique, often allowing separation from rubies of any other source. The major features are both euhedral ('well-formed') and rounded crystal grains, along with dense clouds of rutile silk. Repeated twinning is relatively common, as is hexagonal colour zoning, at times in a swirled pattern termed 'treacle'. Generally absent, however, or in small numbers only are the liquid-filled fingerprints and feathers which are so common in Thai rubies.

Crystalline solids of a number of types are characteristic in Burmese rubies. They typically form clusters of rounded and/or euhedral grains of a light colour (or colourless) concentrated particularly in the centre of the crystal. The most often reported guests are calcite, spinel, corundum, apatite, rutile and zircon. Calcite is present as both rounded and angular rhombs and may be recognized by its cleavage and the polysynthetic twin lamellae which it often contains. Spinel inclusions occur

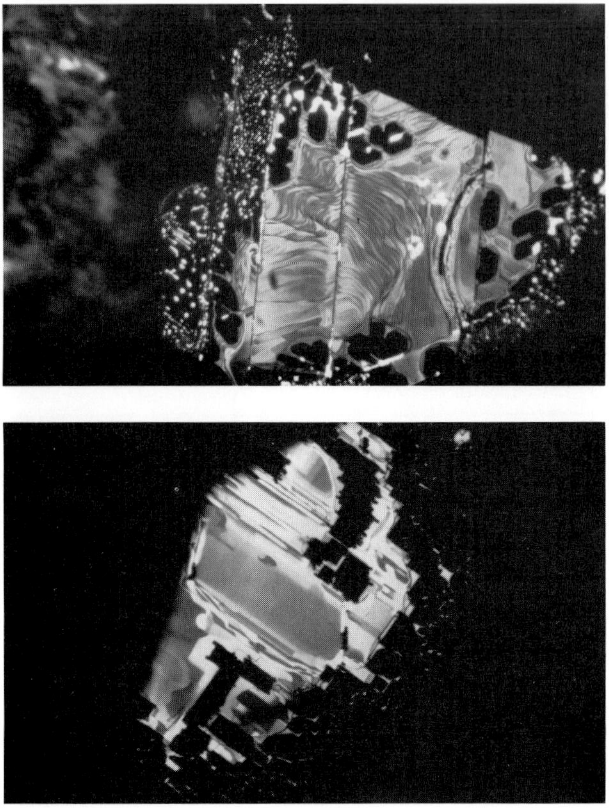

Figure 5.3 Secondary liquid inclusions (healed fractures or 'fingerprints') often display the symmetry of the underlying crystal structure in the healed areas. The upper photograph shows a healed fracture in Thai ruby which formed parallel to the basal pinacoid. As the *c* axis (three-fold symmetry) runs perpendicular to this face, the healed (dark) areas display distorted hexagonal or triangular (60°/120°) outlines. The vertical lines cutting through the fingerprint are repeated twinning striations. The lower photograph shows a fingerprint in a Sri Lankan sapphire. Here the fingerprint has formed parallel to the *c* axis, and so the healed (dark) areas show rectangular (90°) outlines, indicating the two-fold symmetry at right angles to the *c* axis. (Photographs by the author using oblique fibre optic lighting)

as both octahedral crystals and, more often, as rounded irregular forms of low relief. Twinning striations may also be found in included corundum crystals, which occur as tabular or rounded individuals of extremely low relief. These corundums included in corundum typically show a 'terraced' or step-like appearance from multiple development of the basal pinacoid.

In addition to lightly coloured or colourless mineral inclusions, we also find, on occasion, guests with distinctive colours. Rutile crystals of a deep-red colour and metallic lustre stand out in high relief. Their square outline and knee-shaped twin or prismatic habit help to identify them. Bright- to pale-yellow crystals may be either apatite, sphalerite or sphene, while rounded partly resorbed grains of olivine are pale green. Included deep-green prisms of tourmaline have been seen by the author in one spectacular specimen. Dark-brown to opaque slabs or plates suggest

Figure 5.4 A large crystal of what appears to be calcite is seen in a Burmese ruby (non-heat treated) in polarized light. Calcite is suggested because of the intersecting twinning planes visible within the included crystal. Such calcite crystals containing twinning (secondary, repeated) are often seen in Burmese rubies which were formed in a calcite (marble) matrix. (Photograph by the author)

phlogopite mica. In addition, zircon has also been identified, both with and without haloes.

One of the most diagnostic features of Burmese ruby is the dense white clouds consisting of exsolved titanium in the form of rutile. At high temperatures, when atomic spacing is greater, titanium enters into solid solution with the host corundum. As the corundum cools, however, its crystal lattice contracts, literally squeezing the titanium atoms out of solution where they join with oxygen atoms to form minute crystals of rutile. This process is known among mineralogists as 'exsolution' – the unmixing of a solid solution. Because of the constraints placed on their movement by the solid corundum host, the titanium atoms are unable to travel large distances. Therefore, rather than forming large crystals, instead they migrate together to form thousands of tiny slender needles where space permits. In corundum, this space is parallel to the hexagonal prism faces in the basal plane, intersecting in three directions at angles of 60° or 120°. At times, only long slender crystal threads are visible, while in other cases knife- or dart-shapes appear. Closer examination reveals many of these to be twin crystals with tiny 'v'-shaped re-entrant angles visible at the broad end. They are flattened so thin in the basal plane that when illuminated with a fibre optic lightguide from above bursts of iridescent colours are seen, due to the interference of light from these microscopically thin mineral 'lances'.

The needle clouds just described are termed 'silk', in analogy to their thread-like pattern and are responsible for the asterism, or star effect, of some translucent corundums. Not only rutile may form silk in corundum, but also haematite and possibly ilmenite or other minerals may form exsolved needles in rubies and sapphires.

Rubies from Burma usually contain at least some rutile silk. It is found in dense white clouds made up of relatively short individuals, whereas in Sri Lankan corundums the rutile silk tends to be longer and less densely woven.

Along with the rutile silk in Burmese rubies are clouds of minute exsolved particles of an unknown nature. These particle clouds, like the silk, also appear to result from exsolution, since they are arranged in exactly the same pattern (intersecting in three directions at 120° in the basal plane). At times, it has been noticed that heat treatment removes the rutile silk, but not the particles. Thus, it seems possible that, in some cases at least, they are composed of a mineral other than rutile. Because of their arrangement, they also influence the star effect of asterated gems. The author has found that in asterated gems where the 'silk' clouds consist mainly of particles, the star is rather diffuse and lacking in definition. Conversely, in stones where the clouds contain a preponderance of needles, the star possesses better definition. In Burmese rubies and sapphires, both the needle-dominated stars and the particle-dominated stars can be found.

Burmese rubies display still further one additional type of oriented needle-like inclusion – the boehmite (dislocation) needle. These inclusions are found at the junctions of the polysynthetic twinning planes along the rhombohedron faces and are produced by exsolution. Pressure after the corundum has already formed causes certain planes of atoms to slip into a twinned position. This pressure also alters the corundum at the twin boundaries into the mineral boehmite, which runs in three directions (but only two in any one plane) and meets at 87° and 93°.

Within corundum, the polysynthetic twinning usually occurs parallel to the faces of the positive rhombohedron $(10\bar{1}1)$, which intersect each other at angles of 86.1° and 93.9° and meet the *c* axis at 57.5°. Repeated twinning in corundum can be easily

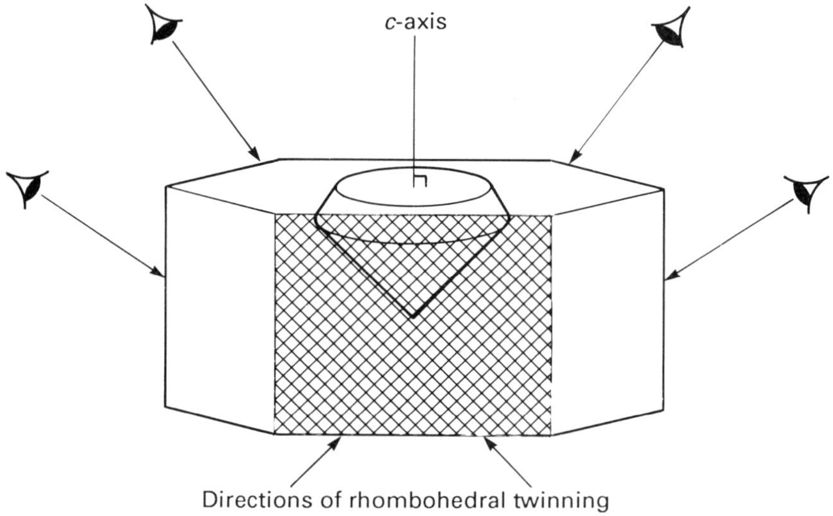

Figure 5.5 Repeated twinning in natural corundum generally occurs parallel to the faces of the positive rhombohedron $(10\bar{1}1)$ which intersect the *c* axis at 32.4°/57.6°. As most natural rubies and sapphires are cut with the *c* axis at 90° to the table facet, this means that the twinning will usually be found at about 30°/60° to the table. Boehmite needles, which form parallel to the edges of the same face (at the junctions of two or three twinning planes), will thus be found in the same directions. They run in three directions (but only two in the same plane) and meet at angles of 86.1° and 93.9°. Repeated twinning is located most easily between crossed polars with the specimen immersed in methylene iodide

observed under the microscope. The planes are only visible when looking exactly along them and generally pass across the entire stone, although not always. Since the twinning almost always takes place on the rhombohedron faces, they are located by looking in directions at about 57° to the *c* axis. Most natural corundums are cut with the *c* axis perpendicular to the table facet, and so they will most likely be found lying about 33° off the table. Immersing the gem in methylene iodide and examining it between crossed polars allows the twinning to be located most quickly, as it appears as bright planes fringed with interference colours against a dark background. It is a stunning effect, and one which the uninitiated must see to believe.

Boehmite needles are long white acicular inclusions which form at the junction of intersecting twinning planes and, as a result, lie parallel to the faces of the positive rhombohedron. Where the planes meet, they may intersect one another at angles of 86.1° and 93.9°. Thus, there should be little likelihood that these boehmite needles be confused with rutile silk, which lies in the basal plane. At times, they may be quite long, running all the way across the stone, and when intersecting in the above manner appear as some sort of 'lattice framework', or 'mother nature's erector set'. Closer examination gives the appearance of very narrow liquid fingerprints, but analysis has indicated that these long white needles actually consist of the mineral boehmite, an alteration product of corundum. Frequently, very narrow stress fractures can be observed, extending outwards from the needles at 45° angles in a spiral fringing appearance. When twin planes run through fingerprints, the boehmite needles often divide the fingerprints or feathers into parallel sections. Together, the polysynthetic twinning and boehmite needles provide the gemmologist with one of the most distinctive and best methods of separation from the synthetic stone, for they are seen in a large percentage of natural rubies and sapphires from all sources, excluding perhaps Sri Lanka. Although repeated twinning and dislocation needles have on rare occasions been found by the author in the flame fusion synthetic corundum, the curved growth lines and gas bubbles allow separation. Nothing resembling this combination occurs in the flux synthetic corundums, making them of extreme diagnostic importance in the battle against the ever more sophisticated products from the factory. Twinning is sometimes found in the flux synthetic corundums, but not with the boehmite needles.

Within Burmese rubies, repeated twinning and the accompanying boehmite needles are often seen, although not as commonly as in Thai rubies. The boehmite needles will, at times, display the stress fringes referred to above.

Along with the solid crystal inclusions, rubies from Mogok also contain negative crystals of primary origin. These generally result where certain areas of the host have grown more rapidly than others, forming and eventually enclosing a void. Flat crystal faces which once faced outwards now enclose a hollow space, which for all intents and purposes looks just like a solid crystal inclusion. These voids usually contain a liquid and/or gas bubble trapped at the time of their enclosure. Scientists study these trapped liquids, for they provide important clues regarding the conditions under which the gem formed, being remnants of the original 'mother liquor', or growth solution.

Negative crystals can often be recognized as such by their high relief, due to the liquid or gas filling. Positive proof is provided by the gas bubbles sometimes trapped within the liquid, creating a two-phase inclusion. It is always a delightful experience to observe a gas bubble which bobs up and down as the crystal is tilted in the microscope. Such 'mobile bubbles' are common in minerals like fluorite, but rather rare in corundum.

The negative crystals of Burmese rubies are frequent in occurrence and distinctive in shape, for they exhibit the same kind of crystal faces and habits as their host. Typical examples show a 'terraced' appearance made up of numerous steps, the result of alternating development of the basal pinacoid and pyramid or rhombohedron faces respectively. Sometimes well-formed examples with flat faces are observed, while others are rounded like waterworn solid crystals. Large cavities

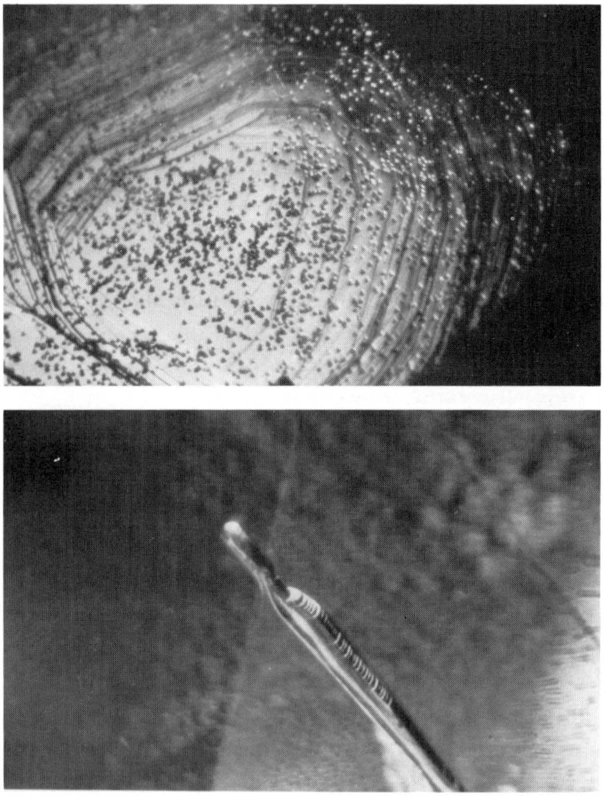

Figure 5.6 Upper photograph: The terraced or stepped basal pinacoid face on a ruby crystal from Burma. Note also the many tiny triangular etch marks (or growth marks?). Such stepped surfaces result from oscillatory growth between the basal pinacoid and prism or pyramid faces, and are often seen on ruby crystals, particularly those from Burma and Sri Lanka. Lower photograph: A negative crystal in a blue sapphire from Sri Lanka. Note the terraced or stepped growth on the basal pinacoid face. Negative crystals often display similar growth features to the host and generally are of high relief. The high relief and stepped growth are two ways by which negative crystals can be separated from solid crystals. Negative crystals are also oriented identically to the host; the *c* axis of the host will lie parallel to the *c* axis of the negative crystal. Identifying the basal pinacoid (with its stepped growth) face of the negative crystal can show its orientation. (Photographs by the author)

of a highly irregular shape have also been observed. All of these can be identified as negative crystals, though, because they show the same orientation as the host. In other words, the basal pinacoid face of each included negative crystal will be strictly aligned parallel to the same face of the host and to any other negative crystals present in the stone. Eppler (1976) has identified the filling of the negative crystals in Burmese rubies as gases containing hydrogen sulphide. This constituent was recognized by its odour when the gems were crushed, opening the cavities. He speculated that gas bubbles within the growth solution perched on a face as the crystal grew. This provided an obstacle to growth at that point on the face, while adjacent areas continued to grow. Eventually the surrounding gem engulfed the bubble completely, trapping it while simultaneously creating the negative crystal. The gas filling is the reason for the high relief of these voids. Illustrated in Figure 5.6 are other possible causes for the formation of negative crystals.

Primary liquid inclusions are those which form simultaneously with the host mineral, liquid-filled negative crystals being one example. Secondary liquid inclusions, however, developed after the host's formation. Included in the latter category are so-called fingerprints and feathers. These represent the remnants of fractures which have been subjected to a process termed 'healing'. They may usually be differentiated from primary liquid inclusions by their curved outline.

Fractures may develop in crystals for a number of reasons, including simply shock or, more likely, from the build-up of strain due to rapid growth. Immediately thereafter, or even millions of years later, the healing process can begin. Should cracks develop as a crystal is growing, the growth solutions penetrate into the open 'wound' via capillary action. If the temperatures of the surrounding environment are high, healing progresses rapidly, while at lower temperatures the pacc is much slower. The inner walls of the curving fracture are dissolved and solutions redeposit this material, as well as any nutrients carried in by the solution itself, forming flat crystal faces. Slowly, inexorably so, the crack is sealed, leaving behind pockets of undigested liquid in fanciful designs which we term fingerprints, feathers, insect wings, etc. Should the conditions be favourable, this ongoing process of solution of the inner walls and redeposition of the nutrients eventually results in highly angular pockets of liquid, actually groups of liquid-filled negative crystals arranged in a fingerprint pattern. Experiments performed by Eppler (1967) and others on different minerals, including Verneuil synthetic ruby, have confirmed the above reconstruction of events leading to the formation of healing fissures and negative crystals in gemstones. The process is illustrated in Figure 5.7.

The healing process just described may take place over millions of years in natural crystals. As a result, the liquid inclusions which form are usually extremely well developed and display the most intricate details of crystal growth. In corundum, the pockets of undigested liquid often surround hexagonal 'islands' of healed material. Tiny growth steps or 'terraces' provide further evidence of the amazing regularity of crystal growth processes. The residual liquid is, in many cases, so thin that brilliant interference colours are seen when it is illuminated from above with a fibre optic lightguide. Although flux-grown synthetic corundum may contain flux-filled inclusions formed by a similar healing process, the tremendous detail of the natural inclusions is missing because of the greater viscosity of the flux

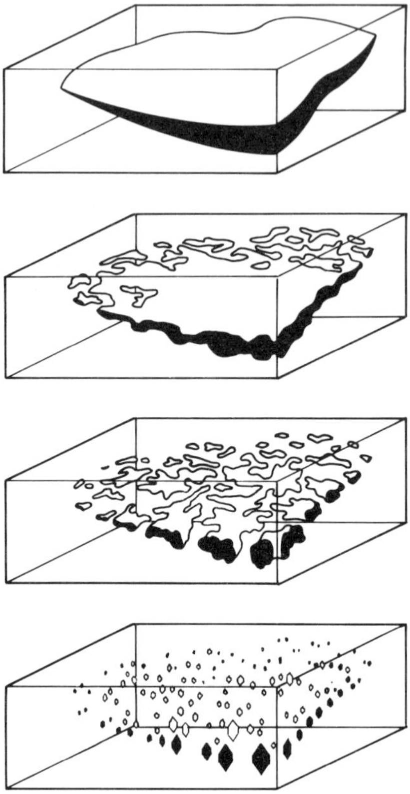

Figure 5.7 The healing or 'necking down' of a crack in a crystal, resulting in secondary 'fingerprint' inclusions. (*a*) A fracture develops during or after the crystal's growth. (*b*) Healing has begun. Growth solutions flow into the fracture and/or the inner walls of the crack are partially dissolved, beginning the healing process. (*c*) The dissolved nutrients are redeposited on the inner walls of the crack as the healing proceeds. (*d*) Eventually the liquid-filled cavities become more angular in shape, turning into liquid- and gas-filled negative crystals arranged in a fingerprint pattern. The liquid that remains behind has, by and large, been leached of its nutrients. These pockets containing undigested growth solutions are smallest along the inner edges and biggest near the outer edges of the original crack. (After Roedder, 1984)

and the far shorter growth times. Thus, the gemmologist is well advised to study carefully under high magnification the healing fissures in natural rubies and sapphires. They can provide important clues in distinguishing between natural and synthetic corundum.

Untreated rubies from Burma contain far fewer secondary liquid inclusions (healing fissures or fingerprints) than rubies from Thailand/Cambodia, Sri Lanka or Kenya. According to Edwin Roedder (1982), crystals formed from metamorphic environments (such as Burmese rubies) are deficient in liquid inclusions. Instead of growing as free crystals, protruding into the liquid from which they grew, crystals in metamorphic rocks have grown in an essentially solid medium by migration of the nutrients via diffusion through other crystals, along grain boundaries, etc. The space for the growth of any given crystal is found by the dissolution or shoving aside of adjoining crystals. As a result, crystals formed in metamorphic rocks often contain many solid inclusions, but few, if any, liquid inclusions. (Heat-treated rubies from Burma may, however, contain many secondary liquid inclusions which formed during the heat-treatment process.)

Thus, the presence and relative abundance of liquid fingerprints and feathers can be quite useful in determining whether or not a particular ruby originated from Burma. In recent years, a large number of suspected Burmese rubies have been brought to the author in Bangkok for examination. A quick look in the microscope, however, often reveals numerous fingerprints and feathers, and the owners' looks of anticipation turn to frowns when they are told that the only thing 'Burmese' about the gem is the nationality of its seller. In the same vein, a story regarding a colleague comes to mind that speaks volumes about the efficiency of modern transportation. This gentleman journeyed all the way from Bangkok to Peshawar, Pakistan, for the purpose of buying Afghan gemstones. He bought several lots of rubies and emeralds from Afghan refugees who had just crossed the border and were eager to raise cash for purchasing weapons to drive the Russian infidels out of their homeland. Shortly thereafter, back in Bangkok I examined his purchases. The emeralds were from Colombia, and the rubies? Well, they came from Thailand.

When secondary liquid inclusions are found in untreated Burmese rubies, they tend to be well healed, with angular negative crystal pockets that sometimes contain gas bubbles. Others may be fractures where very little healing has occurred. Generally lacking are the intermediate-stage fingerprints with narrow liquid tubes. Heat-treated Burmese rubies, however, will contain more fingerprints and secondary liquid inclusions.

In addition to repeated twinning, various types of colour zoning are found in Burmese rubies. The banding is always found parallel to crystal faces, of which that parallel to the hexagonal prism is most common. It may also often be seen along the basal pinacoid and rhombohedron faces, as well as others. When observed parallel to these faces, the bands of colour line up into sharp narrow zones, but in other directions they may appear in irregular swirls termed 'treacle', from their resemblance to the swirls in syrup. Although not found solely in Burmese gems, colour zoning of any sort is very rare in Thai rubies, affording yet another means of separation.

Sapphire – Burma

In terms of inclusions, the sapphires of Burma differ in certain respects from their red relatives. These differences can be accounted for by the different modes of origin for each. Although occurring in close proximity to one another, the sapphires are believed to have originated in pegmatites and nepheline–corundum syenites, while the rubies formed in a metamorphosed crystalline limestone.

Like the rubies, Burmese sapphires contain dense clouds of rutile silk, and a number of fine star sapphires in various shades of blue have been unearthed. Included crystals, however, are much less common in the blue gems than the red, while secondary liquid inclusions are most abundant. Finally, the colour of Burmese sapphires is exceptionally even, and banding is not found in most specimens, even under immersion.

The rutile silk of Burmese sapphires is similar to that of Burmese rubies. Compared with Sri Lankan stones, it tends to be shorter and more densely packed, and can be recognized by its spike or dart shapes. These lie in the basal plane and run parallel to the faces of the second-order hexagonal prism, intersecting at 120° angles. However, the author has also observed in certain Burmese sapphires what appears to be a second type of silk, differing from the rutile silk in several respects.

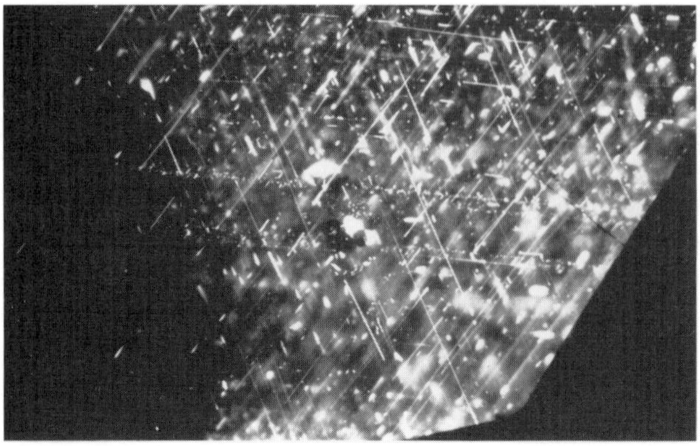

Figure 5.8 Rutile silk and rounded crystals in a blue sapphire from Burma (untreated). Note the dart and knife shapes of many of the rutile needles. (Photograph by the author)

The colour of this new silk tends to be more brownish or yellowish than that of the rutile, and although it is oriented along three directions at 60° or 120° angles in the basal plane, these directions are offset 30° from that of the rutile, running parallel to the faces of the first-order and not the second-order, hexagonal prism. Reference to the hexagonal growth zoning allows separation between these two prisms, as the colour bands and rutile silk follow the faces of the second-order prism. Differences in shape are also apparent, with the new silk occurring as ultrathin elongated plates of a distorted hexagonal outline. Possible identities of this new silk include haematite or ilmenite, both of which have been identified in Thai and Australian sapphires. Rarely, 12-rayed star sapphires and even rubies have been found in Burma. It seems likely that they result from the near equal presence of both rutile and a second type of silk, as described above.

Again as with the rubies, zoned clouds of minute exsolved particles are common in Burmese sapphires, and it seems most possible that in many cases they are merely smaller versions of the rutile silk also present. In other cases, however, differences between the silk and the particles can be seen. At times, the particles produce a pinkish reflection when a fibre optic lightguide is used from above. In others, the reflection is simply white. Thus, it appears that, like the two types of silk, there also exist at least two types of exsolved particles in Burmese sapphires.

With the exception of the exsolved minerals just described, solid inclusions are somewhat rare in the sapphires from Burma. Zircon has been identified as rounded grains, both with and without haloes, as well as magnetite (spinel) octahedra, large single rutile prisms and pyrrhotite (magnetic pyrite) crystals. Moreover, one specimen examined by the author possessed a highly corroded tabular crystal of low relief with a pale-green colour. This might possibly have been olivine. Other crystal inclusions reported are apatite, monazite, fergusonite and phlogopite.

Negative crystals are very common in the sapphires from the Mogok Stone Tract, although most appear to be of secondary, rather than primary, origin. Healing fissures, in all their glory, are usually profusely distributed across the stones. These range from fingerprints with slender worm-shaped liquid hoses, to curving concentrations of angular negative crystals, some two phase in nature. At times, liquid-filled fingerprints are superimposed upon these arrangements of negative crystals, suggesting two separate stages of fracturing and healing. Very characteristic are the fingerprint patterns which appear folded or crumpled like flags in the wind.

The colour distribution of Burmese sapphires is exceptionally even, and this is the most important difference between these and Sri Lankan blue sapphires. In gems from each locality the blue hue is equally fine, but one can never get too much of a good thing and the Sri Lankan stones normally contain substantial areas without colour. Thus, the even coloration of the Burmese sapphire gives it an intensity which is lacking in the Sri Lankan stone. So well dispersed is the colour in the former that, in many cases, even close scrutiny of the gem immersed in methylene iodide fails to yield evidence of the zonal banding. Only with the small blue sapphires from Yogo Gulch, Montana, and those from Burma is the banding sometimes lacking altogether. The blue colour of Burmese sapphires is just slightly more violet than those of Sri Lanka. They are among the few sapphires which today are not heat treated.

Polysynthetic twinning along the rhombohedron faces is relatively common in Burmese sapphires. Accompanying these lamellae are the long white boehmite needles previously described. Again, they differ radically from the orientation of the exsolved rutile silk, lying not in the basal plane but instead along the rhombohedron faces, on the twinning planes. Their angles of intersection are 86.1° and 93.9°, as they follow the edges of the rhombohedron faces.

Ruby and sapphire – Sri Lanka

The corundum gems from the island of Sri Lanka tend to be very similar internally, no matter what the colour. Within unheated stones, rutile silk is commonly present, although high-temperature treatments will generally remove most, if not all, of this rutile. Compared with Burmese gems, the silk tends to be longer and less tightly woven, although not in every case. Sri Lankan rubies and sapphires are noted for the many types of liquid inclusions they contain. Fingerprints, feathers, web and mesh-like patterns – these and so many others await discovery within the Sri Lankan corundums. No other source produces such varied and complex types of liquid inclusions.

Other important features include negative crystals of high relief which are often two phase and mimic the bipyramidal habit of their host. Polysynthetic twinning is somewhat rare compared with other sources, while strong colour zoning and solid inclusions of a number of different types, including zircon, mica and apatite, are characteristic.

Due to the widespread practice of high-temperature heat treatment of corundums from Sri Lanka, the internal nature of these gems can be drastically altered. The effect of such treatments on the inclusions is taken up in detail in Chapter 6, while here are described the features of the untreated gems.

Although rutile silk can be found in Sri Lankan stones of all colours, it is somewhat less common in the yellow sapphires. The prevalence of silk is illustrated by the large number of asterated stones in a variety of hues produced from the island's gem gravels. This rutile silk is composed of long slender needles, rather than the short darts of Burmese stones, and is often loosely spread across the entire stone. At times, the needles are accompanied by elongated or irregular thin plates which also appear to be rutile. In addition to needles and plates, minute exsolved particles of a similar nature are commonplace, forming clouds which follow the growth structure of the gem. Those in most varieties appear to be rutile, but in yellow stones a different type is seen, more coarse in nature. Heat treatment of Sri Lankan yellow sapphires apparently produces further exsolution of these coarse particles, which are closely associated with the coloration of the stone. This is confirmed by the intensification of colour where the particles are in greatest concentration, and the lack of colour in particle-free zones. Nassau (1984) has suggested that these minute exsolved particles in heat-treated Sri Lankan yellow sapphires could be haematite.

In addition to exsolved silk and minute particles, clouds are also found in Sri Lankan sapphires of all colours in which the constituents are so small and evenly distributed that even high-power magnification (100× or more) cannot resolve them. Termed 'texture clouds', they are often a brownish yellow colour in transmitted light and display a roiled milky appearance. Dealers in Sri Lanka refer to this effect as 'diesel', because of its oily yellow look. The 'geuda' sapphires show it best and traders look for the diesel effect when selecting rough for heat treatment. After heating, texture clouds of a somewhat similar nature are commonplace in Sri Lankan blue sapphires. Texture clouds may consist of extremely finely disseminated rutile, but this is not certain.

The rubies and sapphires of Sri Lanka are rich in mineral inclusions. Among the most important of these are rounded grains of zircon surrounded by tiny tension fractures, creating a halo. These haloes are formed by the increase in volume of the zircon as it decays to the amorphous metamict state. Brownish stains may also be seen within the tension cracks.

Other common mineral guests within Sri Lankan rubies and sapphires are dark-red prisms of rutile, hexagonal prisms of apatite, spinel octahedra, and corroded brown slabs of biotite and phlogopite mica. Also identified are crystals of tourmaline, pyrite, pyrrhotite, garnet and calcite. Heat treatment frequently creates round white 'snowball' inclusions surrounded by irregular tension cracks, and, more commonly, glassy circular tension discs around tiny mineral grains, with secondary healing rinds at the outermost edges.

Well-formed negative crystals are a frequent occurrence of Sri Lankan corundum. At times, two-phase fillings are seen, while in one case witnessed by the author a sand-like debris was contained in a negative crystal that stretched from one end of the host crystal to the other. Incredibly, when the spindle-shaped crystal was inverted, the sandy debris ran slowly down to the opposite end, just like in an hourglass! So large was this cavity that the entire performance could be easily observed with the naked eye.

Negative crystals in Sri Lankan stones may be either primary or secondary in origin. Two shapes are common: the hexagonal bipyramid, which is sometimes

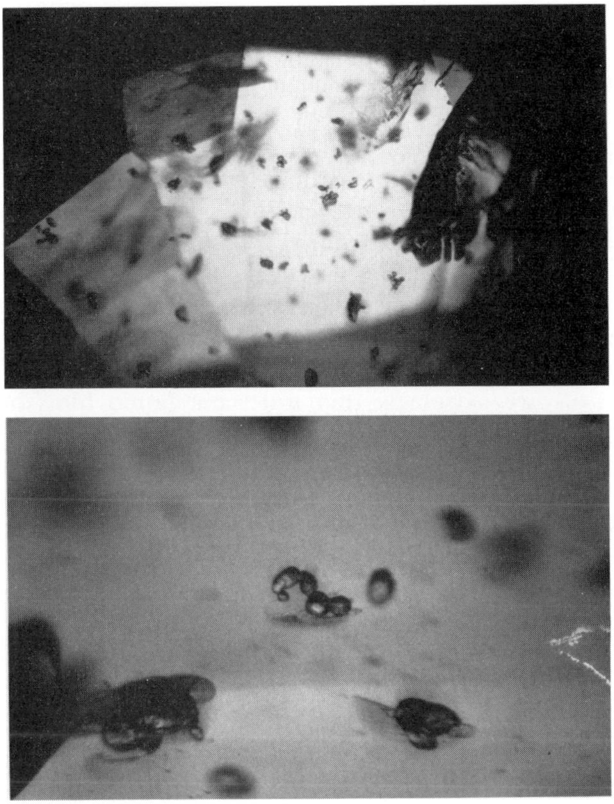

Figure 5.9 Zircon crystals with 'haloes' (tension cracks) in a blue sapphire from Sri Lanka. The upper photograph shows a low-magnification view while the lower one was taken at higher magnification. Tension cracks surrounding some of the crystals are believed to result from the expansion of the zircon due to the metamict (radioactive breakdown) processes (not heat treatment). (Photographs by the author)

extremely elongated, and the tabular hexagonal prism. They can be distinguished from solid crystals by their high relief and identical orientation, with the basal pinacoid faces of each negative crystal running parallel to those of the host corundum (perpendicular to the *c* axis in cut stones).

In terms of inclusions, the type which Sri Lankan rubies and sapphires possess in greatest numbers is the healing fissure. These inclusions occur in every shape, size and pattern possible, turning the stones into tapestries of liquid droplets. On gazing into the interior of the Sri Lankan sapphire, we can see liquid filaments spun into dazzling webs of incomparable beauty. Insect wings, fingerprints, mesh-like healing nets, veils – they are all present, each one a testimony to the natural forces which shaped these gems deep within the earth.

Inside Sri Lankan rubies and sapphires we find the entire spectrum of healing fissures, from open cracks which show no signs of healing, through partially healed cracks or fingerprints, to almost completely healed inclusions consisting of negative crystals in fingerprint patterns. In transmitted light, the liquid filling of these fingerprints may at times appear pale yellow in colour, while reflected light from a fibre optic lightguide reveals details of the healing process on the surfaces of the residual liquid canals. Close examination under high magnification shows each to be a tiny liquid-filled negative crystal in earlier or later stages of development. Those in which the liquid is thin enough show vivid interference colours arranged in geometrical patterns of tiny steps. As more and more material builds onto the inner walls of the fracture, these once irregular voids are gradually reduced in size, eventually taking on crystal outlines of tabular hexagonal shape. These flattened negative crystals in fingerprint groupings frequently display inverted triangular growth steps on the top and bottom basal pinacoid faces, as well as horizontal striations along the prism or pyramid faces. Being liquid filled, however, gives them lower relief than the largely gas-filled negative crystals of primary origin. Also observed in Sri Lankan sapphires are extremely thin iridescent liquid films in parallel planes (usually with their broad side parallel to the basal pinacoid).

The vast majority of all Sri Lankan ruby and sapphire crystals are unevenly coloured, with large areas devoid of colour altogether. In the blue sapphires, the cores of the crystals are often colourless while the blue hue is concentrated just below the crystal faces in thin layers. Such crystals are termed 'ottu', meaning 'to risk', for the colour bands must be very carefully placed during cutting to achieve an even colour, itself a risky operation.

Figure 5.10 Hexagonal colour zoning visible in immersion in an 'ottu'-type Sri Lankan blue sapphire (heat treated). 'Ottu' sapphire crystals possess a colourless core, with the blue colour being concentrated near the edges of the faces. Such stones will be oriented during cutting so that the blue colour lies on both sides of the culet. This often results in the *c* axis lying parallel to the table (as in the stone pictured above). Thus pleochroism will be visible in the face-up position. (Photograph by Wimon Manorotkul, AIGS, Bangkok)

Before heat treatment, these colour bands possess extremely sharp edges. However, after burning the edges become slightly dulled or diffused, with much of the detail of the narrow sharp growth striae disappearing.

Polysynthetic twinning is rather uncommon in gems from the gravels of Sri Lanka. When seen, however, its appearance is identical to that in other corundums, lying parallel to the rhombohedron faces. Long white boehmite needles are also seen associated with the repeated twinning.

Ruby – Thailand and Cambodia

The rubies produced in Thailand and Cambodia originate from identical deposits straddling the border of these two nations, just north of the Gulf of Thailand. Thus, the internal features of rubies from both Thailand and Cambodia are identical. To simplify matters, these stones will hereafter be referred to as Thai rubies, even though they may, in fact, have been mined across the border in Cambodia.

Thai rubies are characterized by a complete lack of rutile silk and other exsolved mineral particles in the basal plane. As a result, star rubies are not produced from these deposits. The long white needles commonly found in Thai rubies form parallel to the edges of the rhombohedron faces at the junctions of intersecting polysynthetic twinning planes. According to Gübelin (1953, 1983), they consist of tubes, filled with a colourless transparent liquid, along dislocation and pressure parting lines. Koivula (1981), however, indicates that in African sapphires stress along the repeated twin junctions alters the corundum to boehmite, AlO(OH). Today it is generally accepted that these long white needles consist of the mineral boehmite.

Most distinctive in Thai stones are the solid or negative crystals surrounded by liquid fingerprints or feathers. These have been dubbed 'saturn' inclusions by gemmologists, because of their unique appearance, and are extremely common.

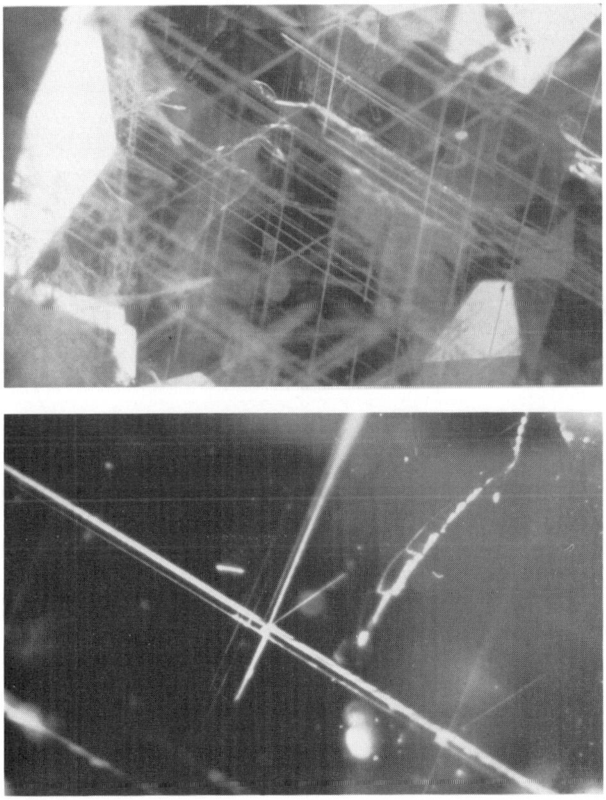

Figure 5.11 Many rubies and sapphires may contain repeated twinning along the rhombohedron faces as a result of pressure after the crystal formed. This pressure often results in the exsolution of thin needles of boehmite at the junctions of intersecting twinning planes. These boehmite needles are shown above. White in colour, they form parallel to the edges of the rhombohedron faces, intersecting at angles of 86.1° and 93.9° (two in any one plane, three directions in total). (Photographs by the author)

Other features include liquid inclusions of both primary and secondary origin, the latter being present in almost every specimen. Colour zoning, however, is very rarely seen in Thai rubies.

Separation of Thai rubies from those of Burma and Sri Lanka can be made by reference to the lack of exsolved silk and/or particles in Thai rubies. The only acicular inclusions seen are the slender boehmite needles found lying at the junctions of the rhombohedral twinning planes. These intersect at 87° and 93° angles and often run through fingerprints, dividing them into banner-like sections. Boehmite needles and the accompanying polysynthetic twinning are extremely common in Thai rubies and among the most useful features for distinguishing them from the flux-grown synthetic rubies. In cases where the twinning planes do not extend across the entire stone, the often irregular edges of these planes are sometimes traced by white boehmite needles as well.

Common solid inclusions in Thai rubies are rounded crystals with flat ends that sit perched upon the rhombohedron or basal planes, surrounded by fingerprints.

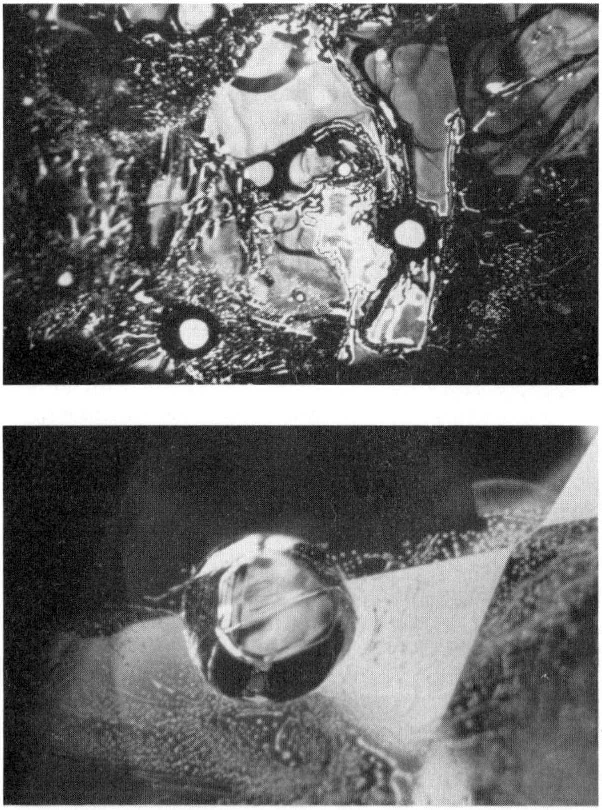

Figure 5.12 Thai/Cambodian rubies often contain opaque crystal inclusions surrounded by fingerprints ('saturn' inclusions). During heat treatment these crystals typically melt, forming glass-filled negative crystals of high relief with flat ends (usually parallel to the basal pinacoid or rhombohedron faces). These glass-filled negative crystals are shown above. In some cases gas bubbles may be found frozen in place (not shown) within the glass filling. (Photographs by the author)

The solid portion of these saturn inclusions may be almandite garnet (brown, opaque), apatite (yellowish), pyrrhotite (black with metallic lustre), plagioclase feldspar (colourless, transparent), olivine, or diopside. Negative crystals also form saturn inclusions. Heat treatment generally melts these solid crystals, leaving behind a high-relief void filled with the glassy remains of the guest mineral, including, at times, a gas bubble frozen in place within the glass-filled void. The brownish stains so often found in the surrounding fingerprint also disappear upon heating. As these saturn inclusions are present in over 80% of all Thai rubies, they make it possible to detect the heat treatment to which virtually all Thai rubies are subjected today. During heating these crystals melt, leaving behind a glass-filled negative crystal.

Gübelin (1971) has suggested that the flat liquid fingerprints in Thai rubies represent the exhausted residual solution left behind after the crystallization of the mineral grain. After forming on a crystal face during a temporary lull in the ruby's growth, the guest mineral sucked out all of the nutrients from the liquid, leaving it to be subsequently trapped when the host again started growing.

Thai rubies possess an abundance of both primary and secondary liquid inclusions. The primary liquid inclusions are found as described above, or as ultrathin films surrounding tiny two-phase negative crystals which lie parallel to the basal pinacoid. They formed when small depressions trapped tiny gas bubbles and liquid on the rapidly growing basal face. The tiny two-phase negative crystals are surrounded by extremely thin liquid films in fingerprint patterns which often possess a triangular or hexagonal outline. Since natural rubies are generally cut with the *c* axis perpendicular to the table facet, these films will usually lie parallel to the table. It is this parallel alignment which allows each separation of these primary liquid inclusions from secondary liquid inclusions produced by healing. Thin primary liquid films may also be found along the rhombohedron planes in Thai rubies, but are much less common than those in the basal plane. Utilizing reflected light from a fibre optic lightguide, the films appear as brightly iridescent fingerprints with a tiny bubble in the centre, stacked one on top of another in the basal plane. Under transmitted or dark-field illumination, only the two-phase negative crystals are seen, not the surrounding liquid films.

Thai rubies may also contain secondary liquid inclusions, or healing fissures, as they are otherwise known. These are commonly found parallel to the basal pinacoid and rhombohedron faces, as well as in random directions. From Bo Rai in Trat Province, Thailand, originates the 'Lai Thai' ruby, a stone whose colour is very fine, being just slightly yellowish red, but whose interior is spoiled by numerous liquid fingerprints and feathers (note: stones such as this, which have fine colour and look very good at first glance, but which upon closer inspection have poor clarity, are termed 'bluff' stones by Thai dealers). The name 'Lai Thai' means literally Thai 'design' or 'motif', because the twisting wispy patterns of the liquid inclusions in Lai Thai rubies resemble the designs seen on Thai silks or other local fabrics. Not only do these secondary liquid inclusions form in wispy patterns, but also they are frequently very thick and thus easily confused with the flux-filled veils and fingerprints of flux-grown synthetic rubies. The author has witnessed many cases where Lai Thai and Kenyan rubies were misidentified by other

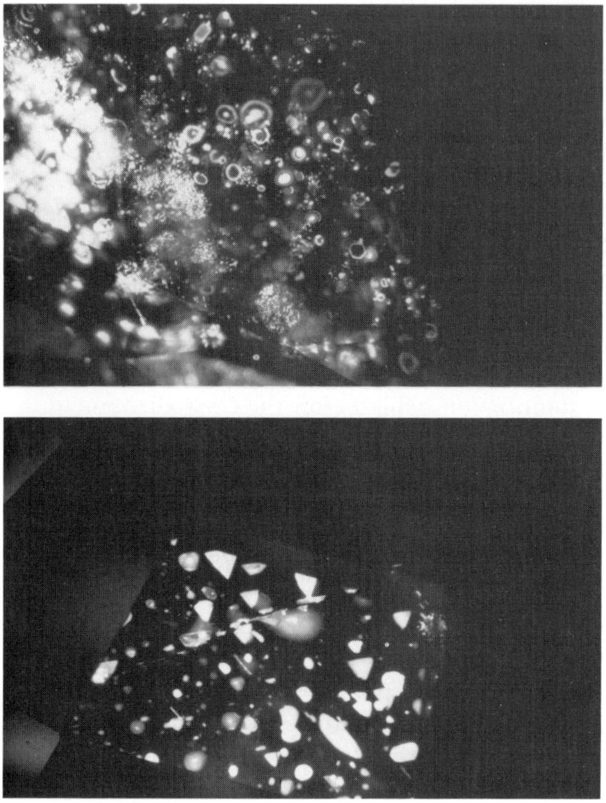

Figure 5.13 Ultrathin primary two-phase liquid films in Thai/Cambodian rubies. Such inclusions typically form parallel to the basal pinacoid; as most natural rubies are cut with the table parallel to the basal pinacoid, they will be seen most often parallel to the table facet. Overhead fibre optic lighting will reveal them as bright iridescent films with triangular or hexagonal outlines. (Photographs by the author)

gemmologists and even major laboratories, probably due to the flux-like appearance of the liquid inclusions. It must be remembered that the patterns of natural fingerprints and feathers may, at times, be identical (wispy, mesh or veil like) to flux inclusions. Thus, separation is achieved not through reference to patterns but by close examination to determine the identity of the filling material (liquid, or a flux glass) and by reference to the other inclusions present. Both the Lai Thai and Kenyan rubies, which are similar in appearance, almost inevitably contain lots of polysynthetic twin lamellae and boehmite needles. Since nothing resembling this combination is found in the flux-grown synthetic rubies, a positive distinction is possible on this basis. Twinning alone is seen in flux-grown synthetic rubies, but not twinning and boehmite needles at the twin junctions.

Lai Thai rubies appear, at first glance, somewhat similar to Burmese rubies. There are several reasons for this. The colour of Lai Thai rubies is slightly more yellowish than that of most Thai stones because the inclusions scatter the light in random directions, which brings in some of the yellowish-red extraordinary ray, even when the *c* axis is perpendicular to the table facet. Rutile silk has the same

Figure 5.14 Repeated parallel twinning ('polysynthetic twinning') in a ruby from Thailand. Such secondary twinning in corundum results from pressure-induced slipping ('gliding') and is found parallel to the faces of the negative rhombohedron $(10\bar{1}1)$. As most natural corundums are cut with the table facet at right angles to the *c* axis, this twinning will usually be found at about 30°–60° to the table. (Photograph by the author)

effect in Burmese rubies. Furthermore, the wispy inclusions in Lai Thai stones effectively cover up the extinction that would otherwise be present. The presence of very obvious extinction in Thai rubies and the comparative lack of it in Burmese gems is the major difference between the two sources. In Burmese gems, the strong fluorescence effectively masks the extinction. The wispy inclusions in Lai Thai gems have a similar effect, but at a price, namely the poor clarity of Lai Thai rubies.

The secondary liquid fingerprints and feathers of Thai rubies may often display the beginning and intermediate stages of healing, but rarely the latter stage in which well-formed negative crystals are seen arranged in a fingerprint pattern, such as within Sri Lankan rubies. Reflected light from a fibre optic lightguide produces bright interference colours that reveal the most intricate surface details of these healing fissures. Narrow growth steps give the appearance of terraced farmlands seen from the air. Between the coarse liquid tubes, extremely minute channels are found, resembling a tiny fingerprint within a larger fingerprint. When the liquid channels are broad and flat, light is easily transmitted through them from below, proving the filling to be liquid and not a flux glass. With Thai rubies in particular, when a fingerprint lies along the basal plane the liquid channels surround 'islands' of healed corundum whose outlines are slightly distorted, or even perfect, hexagons. These hexagonal islands within fingerprints are also common in the primary liquid films.

Other than the ubiquitous repeated twinning, growth features, such as colour zoning, are very rare in Thai rubies. Of the thousands of Thai stones examined by the author, only a handful have ever shown irregularities in colour distribution. None whatsoever have displayed narrow colour banding; instead, all that is found are broad areas of colour concentration with diffuse borders which are, at best, difficult to resolve. Immersion between crossed polars may help to locate banding at times, particularly if the stone is examined *exactly* parallel to the *c* axis. This contrasts greatly with the flux-grown synthetic rubies, most of which display sharp angular growth lines.

Sapphire – Thailand

Although the sapphires from Thailand are usually described as a single entity, they in fact originate from three different major deposits, each located in an entirely different part of the country. For this reason, stones from each deposit possess inclusion suites unique to their particular mine.

The major sapphire-producing localities of Thailand are:

1. Bo Ploi, Kanchanaburi Province (100 km west of Bangkok).
2. Khao Ploi Waen and Bang Kha Cha in Chanthaburi Province and Ban Bo I-Ram, on the border between Chanthaburi and Trat Provinces.
3. Ban Bo Kaeo and Huai Mai Sung, Phrae Province (approximately 500 km north of Bangkok).

Kanchanaburi sapphires

Among the three major sapphire-producing districts of Thailand, the most important today in terms of both production and quality is Bo Ploi. The sapphires of Bo Ploi in Kanchanaburi Province are virtually all of a blue colour and possess distinctive inclusions which allow them to be immediately recognized. To the naked eye, Bo Ploi stones appear much like, and are often sold as, heat-treated Sri Lankan stones, displaying strong and uneven colour zoning. However, the Thai sapphire is slightly milky, due to the concentration of minute exsolved particles along the blue colour zones. These particles are found in every stone, but concentrate only in the blue colour bands, being entirely absent in the colourless areas. Rutile silk is not found in Bo Ploi stones and so no star sapphires are mined. Solid crystal inclusions of a number of types are seen, as well as negative crystals. Additional inclusions are secondary fingerprints and feathers, and polysynthetic twinning with boehmite needles.

The exsolved mineral particles in Bo Ploi sapphires generally allow a quick and easy separation from sapphires of other sources. They give the stones a distinctive turbidity known among local traders as a 'milky and silky' look, although no real silk is found. The particles, which are relatively coarse in size, are arranged in an identical fashion to the silk, intersecting in three directions at 120° in the basal plane. Their distribution is quite distinctive, as they concentrate in dense accumulations which exactly follow the blue bands of colour; colourless regions are generally devoid of particles. Thus, deep-blue stones contain large numbers, while in light-blue stones they are present in fewer numbers. When the stones are observed parallel to the *c* axis along the colour zones with an overhead light source, intense silvery bands can be seen as a result of the light reflecting off the particles. Transmitted light in the same direction produces alternating bands of blue and colourless, with the blue zones showing a characteristic 'powder-blue' milky appearance. The blue colour is pure and deep, equalling that of good quality Burmese or Sri Lankan sapphires, except for the trace of milkiness. Transmitted light in directions at oblique angles to the *c* axis displays the colour in tiny rows of blue dots intersecting in three directions.

Figure 5.15 Minute exsolved particles seen through the table face in a Kanchanaburi (Thailand) blue sapphire. In Kanchanaburi sapphires such particles are always present and lend a certain cloudiness to the stone. Unlike in Sri Lankan sapphires, where concentrations of particles may be found in colourless areas, in Kanchanaburi stones the concentrations of particles follow exactly the blue-coloured areas; where there is no colour there will be no particles, and vice versa. (Photograph by the author)

From the above description it is obvious that these unidentified mineral particles are in some way connected with the coloration of the stone. Probably they contain either iron or titanium, or perhaps both elements. During an extended period at high temperatures, iron and/or titanium atoms may have migrated together, forming numerous local domains of concentration. As the temperature fell, some of these atoms were squeezed out of solution by the contracting host structure, forcing them to exsolve as minute mineral particles along specific directions where the greatest amount of space was available. However, evidently some of the impurity atoms were left behind in the rush to crystallize and remained in solid solution. Thus, they are in a position where they can influence the colour of the stone. This explains why we see tiny colour spots surrounding these particles. Where there are no particles, there is no colour. In the near colourless sapphires from Bo Ploi, the particles are present in only small numbers but, as always, along the blue colour bands. Another possibility is that heat treatment has partially melted the particles, sending the colouring agents into solid solution.

Included crystals of a number of types are found in sapphires from Bo Ploi. Gunawardene and Chawla (1984) identified transparent colourless crystals of feldspar and long prisms of hornblende. Also found were submetallic grains of pyrrhotite. The author has seen on many occasions heavily corroded black crystals of dull lustre, sometimes containing gas bubbles frozen in place. This could be the result of heat treatment, to which virtually all Bo Ploi sapphires are subjected. Heating may melt the solid crystal and leave behind a void containing a bubble within the solidified remains of the former included crystal. Glassy circular tension haloes have also been found around included crystals and are probably another by-product of the heat treatment.

Negative crystals are seen regularly, but appear to be secondary rather than primary in origin, as they are associated with healing fissures. These liquid-filled voids form most often in the basal plane, occurring as flattened individuals with triangular growth markings in some cases. Secondary liquid inclusions in fingerprint and feather patterns are present in most Bo Ploi sapphires. The liquid channels are relatively coarse at times, forming curved and folded patterns of great beauty. Yellowish stains have been seen in some of these healing fissures, while others are iridescent in reflected light where the liquid is thin enough to cause interference.

Growth phenomena in sapphires from Bo Ploi include very uneven colour banding parallel to the faces of the hexagonal prism. This banding appears much like that of heat-treated Sri Lankan sapphires, except for the exsolved particles which always accompany it. In Sri Lankan sapphires, minute exsolved particles may also be found, but the particles in these sapphires are often found in colourless areas, as well as with the colour banding. In Bo Ploi stones, the particles are always found to follow the colour banding exactly. Polysynthetic twinning is also a common feature, along with long white boehmite needles running parallel to the edges of the rhombohedron faces. The vast majority of cloudy sapphires in the Thai market are Bo Ploi stones. This fact is one of the best kept secrets in the gem business.

Chanthaburi/Trat sapphires

The production of sapphires from Chanthaburi and Trat has traditionally been small compared with that of Bo Ploi, but recently increased amounts have been found in the I-Ram area. The mines here yield sapphires of a pure, but very deep-blue, colour, as well as fair amounts of yellow, green and black star sapphires. Bank Kha Cha and Khao Ploi Waen, in particular, are noted for the famous 'Mekong Whisky' yellow sapphires and the 'golden-star' black star sapphire. Very dark-blue sapphires are also found here.

Inclusions in the blue sapphires are very similar to those in sapphires from the famous Pailin mines which lie but 30 km distant across the border in Cambodia. Most distinctive are small red and orange octahedra which, according to Gübelin (1983), consist of the rare mineral, uranium pyrochlore. The extremely intense deep-blue colour may display just the slightest trace of milkiness, due to extremely fine exsolved particles which scatter the light. Other features are secondary healing fissures and polysynthetic twinning with boehmite needles.

The colour of these stones is especially distinctive in the microscope. Growth zoning in a hexagonal pattern is very sharp and well defined, with a lovely deep blue visible parallel to the *c* axis. In transmitted light, slightly turbid yellowish texture clouds are seen with sharp boundaries. Fibre optic illumination reveals the cause of the slight cloudiness to be extremely minute exsolved particles, possibly of haematite. They appear white in reflected light but brownish or yellowish in transmitted light when in thick planes. This yellow colour is diffused throughout the stone, giving an eerie greenish overtone to the intense blue colour in certain positions. So fine are these particles that the sharp zoning takes on a waxy appearance as well.

Zoning in the green sapphires is generally sharp alternating bands of yellowish green and blue, while in the yellow sapphires bands of green and yellow are seen.

The yellow crystals are very tabular and thin, with green banding lying just beneath the upper and lower pinacoid faces. Cut stones are very shallow, with green bands just under the table and culet, as a result. The banding of these yellow sapphires is unusual for corundum in that very sharp parallel zoning is visible parallel to the basal pinacoid, but less often can any colour zoning be found parallel to the *c* axis. The hexagonal zoning parallel to the *c* axis, however, can be determined by the banding of the silk. Australian yellow sapphires are also similar in this respect.

Silk is found only in narrow planes within the green bands, usually at the table and culet. It consists mostly of extremely short 'hairy' haematite needles of a brownish colour, intersecting in three directions at 120° angles. However, the individual haematite needles lie at 90° angles to the hexagonal banding (parallel to the faces of the first-order hexagonal prism). With a fibre optic lightguide, small amounts of rutile silk can also be detected. The rutile needles are longer and more slender than the haematite and lie parallel to the zonal banding (parallel to the faces of the second-order hexagonal prism). This description also applies to the blue, green and black star sapphires. Silk is not so common in the blue sapphires, but when present will also be the same.

The dark-brown to black colour of the black star sapphires results from dense concentrations of the haematite silk, mechanically colouring the stone. When it is included in an otherwise blue or green stone, a white six-rayed star results. However, if the haematite is present in a yellow sapphire, the rare and desirable 'golden-star' black star sapphire is produced, where the rays of the star take on a deep golden-yellow tint. These golden star sapphires are many times more valuable than the ordinary white star sapphires, but are little known outside Thailand.

Close examination of the black star sapphires often reveals a strong six-rayed star, due to the haematite silk, and a weaker six-rayed star offset 30° from the

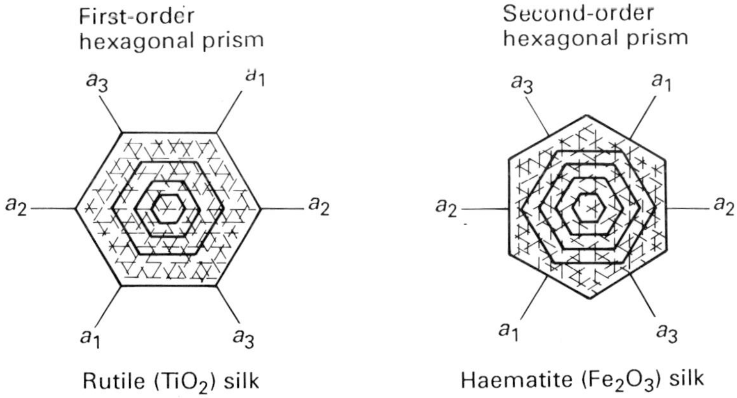

Figure 5.16 Silk in natural corundum is generally exsolved along three directions (parallel to the faces of the hexagonal prism) in the basal plane, meeting at 60°/120° angles. Rutile silk forms parallel to the faces of the first-order prism, while haematite forms parallel to the faces of the second-order prism. As the hexagonal colour zoning is usually parallel to the first-order prism, this means that the rutile silk is usually parallel to the colour banding. In contrast, haematite silk usually runs at right angles to the colour zoning. Rutile silk is commonly found in Burmese and Sri Lankan corundums, while haematite silk is common in Thai and Australian sapphires. Some Thai sapphires (from Khao Ploi Waen and Bang Ka Cha) may contain both haematite and rutile silk, and so can be cut into 12-rayed star stones

other, due to the presence of rutile silk. Many of the black star sapphires from Chanthaburi display this effect, at least to some extent. Should enough rutile be present, then a 12-rayed black star sapphire results. Bang Kha Cha and Khao Ploi Waen produce relatively large numbers of these gems. At times, six of the rays may be yellowish in colour, while the other six appear white.

The heavy concentrations of haematite silk in the black star sapphires weaken them considerably. As a result, they tend to break along flat planes parallel to the basal pinacoid, an effect termed basal parting. Its appearance is identical to that of cleavage, except that the number of possible partings are strictly limited, being possible only where the haematite silk is sufficiently concentrated to affect the structural adhesion. The parting is often visible on the back of the cabochon as extremely flat fractures exhibiting a step-like appearance. Apparently rutile does not have the same effect, for basal parting is not found in the Burmese and Sri Lankan star stones containing concentrations of rutile silk in the basal plane. The pits produced by this basal parting are often disguised by filling with doping shellac. An unusual feature of both the blue and black star sapphires is the extremely tiny discs found scattered through some stones. These discs appear to have resulted from exsolution and are visible only with a fibre optic lightguide. They are flattened in the basal plane, but otherwise appear to be randomly distributed and could easily be mistaken for gas bubbles, which they strongly resemble, except for their flat shape.

Included crystals of various types are found in the blue and green sapphires from Chanthaburi and Trat, but are rarely, if ever, found in the yellow sapphires. Most distinctive are red and orange octahedra of what is probably uranium pyrochlore. These are often surrounded by small fingerprints containing yellow stains, probably due to a partial melting of the crystal during heat treatment. Also seen are slightly rounded colourless crystals surrounded by circular flat tension fractures which are highly iridescent in reflected light. Some stones show small opaque black crystals of triangular outline and metallic lustre which strongly resemble the platinum plates found in certain flux-grown synthetic rubies. Other crystals seen are colourless blocks of what may be feldspar surrounded by flat glassy tension haloes and opaque black blocks of what may be niobite.

Chanthaburi sapphires of all colours display secondary liquid inclusions in the form of fingerprints, feathers, etc., although these are less common in the yellow sapphires. Polysynthetic twinning and the accompanying boehmite needles are also common in all varieties. In addition, the sapphires of this region show basal parting because of concentrations of silk. This is most evident in the black star sapphires, but it is also a prominent feature within the blue and green stones as well. Basal parting is readily observed on the rough material; it is also seen occasionally in cut stones. Some of these parting fractures have been subjected to healing processes, and thus fingerprints and feathers can be frequently observed lying along the basal plane.

Phrae sapphires

The deposits of Phrae Province in northern Thailand are the most recent discovery of the three. Stones from the mines here tend to be dark blue in colour, with some tending towards a deep slightly violetish-blue hue. Other varieties found include some green sapphire, but almost all the production is blue.

Hexagonal colour zoning is very strong and prominent in Phrae stones, similar to those from Australia. Polysynthetic twin lamellae with boehmite needles are also quite common.

Sapphire – Pailin, Cambodia

The famous mines of Pailin in western Cambodia produce only ruby and blue sapphire, with other varieties almost non-existent. Rubies from this district are identical internally to those found across the border in Thailand and so will not be discussed separately. It is the blue sapphires, however, for which Pailin is noted. Of a pure and intense blue colour, they are considered to be among the world's finest.

Internally, the blue sapphires from Pailin bear a strong resemblance to those from Chanthaburi. The most distinctive features are included red crystals of the mineral uranium pyrochlore. When of large size, these octahedral crystals appear black, but smaller examples display a deep-red or orange colour. At times they are found concentrated in rows following the growth structure of the host, but appear mostly as individuals, often in front of comet-like trails of minute liquid droplets. Others may be surrounded by fingerprintts containing a reddish- or orange-stained liquid, which may possibly result from heat treatment.

In addition to uranium pyrochlore, a number of other crystal guests are found. Colourless or white prismatic or tabular crystals of moderate relief are common. These may display striations across their faces and are often surrounded by circular glassy tension haloes, possibly feldspar. Also seen are euhedral hexagonal prisms looking very much like corundum itself, or possibly apatite.

Colour zoning in Pailin sapphires is extremely sharp and follows the hexagonal contours of the crystal. At times, thin concentrations of minute exsolved particles or yellowish texture clouds are also distributed along these zones, similar to stones from Chanthaburi. Rarely encountered are phantom growths formed by the formation of a thin layer of some foreign substance upon earlier crystal faces, producing a perfect outline of the crystal at a previous stage in its development. Rutile silk has not been found in Pailin stones; thus star sapphires are absent from this locality. White boehmite needles, however, are often seen, lying at the junctions of rhombohedral planes.

The liquid-filled fingerprints and feathers of Pailin sapphires are numerous and create delicate patterns of great beauty, somewhat akin to those of Thai and Cambodian rubies. They often exhibit detailed healing, with tiny hexagonal areas surrounded by residual liquid. So thin are these liquid tubes that a vivid iridescence is seen in reflected light, truly a splendid sight.

Coarse negative crystals are also found, but whether they have resulted from actual growth or simply from the dissolution of a solid inclusion via heat treatment is uncertain. As some display gas bubbles frozen in place, it would appear that the latter cause is more likely, at least in some instances.

Ruby – Jagdalek, Afghanistan

Rubies from Jagdalek are only rarely encountered in facet quality, but when clean can be magnificent. In terms of colour, they resemble most the gems of Burma and Sri Lanka, being strongly fluorescent and often of a slightly pinkish or raspberry-red hue similar to the rubellite variety of tourmaline.

The most distinctive features of Jagdalek rubies are the small spots or zones of a fine sapphire-blue colour often found within the stone. In some cases, these blue zones may be hexagonal in outline while in other cases they consist of narrow bands, but all show a sharp division between red and blue, which is diagnostic. The author has not witnessed this effect in rubies from other localities.

Solid inclusions of various types have been found in Afghan rubies. Common are colourless blocks displaying rhombohedral cleavage, most likely of calcite. Inclusions of calcite are not surprising, considering the fact that Jagdalek rubies are found in a marble matrix, just as in Burma.

Transparent plates and books of hexagonal outline are also seen. Due to their anisotropic character between crossed polars and prominent basal cleavage, they are most likely mica. Other plate-like inclusions consist of irregular distorted shingles which are opaque and black or slightly gold in colour. These also display a somewhat micacious appearance.

Several specimens examined by the author contained deep-red-orange prisms of square outline and submetallic lustre. Some were knee-shaped twins with obvious and re-entrant angles, identifying them as rutile. Exsolved rutile silk, however, was not found, but occasionally clouds of minute exsolved particles were. The lack of true silk means that star rubies are not produced. Cabochons may show a silvery sheen, though, when light is reflected off the particles. Additional solid inclusions seen were rounded colourless grains of low relief and, in one specimen, corroded blocks of a yellow colour.

Colour zoning in Jagdalek rubies is very sharp and narrow, and forms in the typical hexagonal pattern parallel to the *c* axis. Rhombohedral polysynthetic twin lamellae are seen in most specimens, inevitably accompanied by long white boehmite needles meeting at 87° and 93°.

Both primary and secondary liquid inclusions are seen, the latter being responsible for the lack of clarity which most of these rubies display. Irregular liquid-filled cavities are seen whose edges appear jagged, much like those in Colombian emeralds. However, the cavities of the Jagdalek rubies are somewhat thicker. The fingerprints and feathers which fill these stones often show a ragged appearance too, with coarse tubes that can be easily be confused with the flux inclusions in flux-grown synthetic rubies.

Sapphire – Kashmir, India

In a period of just 30–40 years, during which the mines were intensively worked, Kashmir sapphires achieved a reputation second to none. Today, fine Kashmir sapphires are virtually unobtainable, with most specimens being found only in museums and private collections, mute testimony of the degree to which they are coveted.

The hallmark of the Kashmir sapphire is its velvety lustre, a slight haziness which, under magnification, is revealed as numerous fine lines intersecting in three directions at 120° in the basal plane. Unfortunately, despite the renown of these inclusions, the virtual absence of samples for study has prevented their exact nature from being determined. Phukan (1966) claimed that the majority consist of exsolved rutile needles, while in some cases decomposition of the rutile had left cavities. He also reported dust-like particles of an unknown nature either scattered amidst the silk or in clouds of dense concentration. This is much like the rutile and particle clouds previously described under Burmese and Sri Lankan rubies and sapphires. In fact, star sapphires are also known from Kashmir, but are said to be somewhat uncommon.

Gübelin (1953), in contrast, considered that the haziness was entirely due to minute canals, or hollow tubes, of a slightly brownish colour which run at roughly right angles to hazy bands intersecting at 120° angles. Although when magnified the canals resemble rutile needles, Gübelin claimed that they do not possess the well-defined outline of the rutile. Additionally, he noted that the haziness in some cases could be traced to fine arrangements of ultramicroscopic fluid droplets, or to a very dense colloidal coloration by titanium and iron intermixtures.

The comments made by Gübelin raise some interesting points. Perhaps the silk found in Kashmir sapphires is not rutile at all, but instead another mineral such as haematite. This might account for the slightly brownish colour which is seen in Thai sapphires containing haematite silk, but rutile silk may show a brownish colour too. Also, the arrangement of haematite silk follows the faces of the first-order hexagonal prism, not the second-order hexagonal prism, as with rutile. The three directions would then lie but 30° from the bands of colour which parallel the first-order hexagonal prism faces. Far fetched? Perhaps, but still a possibility worth investigating. In any event, the inclusions of the famous Kashmir sapphires definitely require further study.

Occurring alongside the sapphires of Kashmir are dark-green and brown crystals of tourmaline. These may be found growing right up against the sapphire and in some cases are actually included within the stone. Most rough material is still partially coated with a white clay-like matrix which fills the pits of heavily corroded surfaces. This clay-like material also appears to be included in many stones with irregular cavities just beneath the crystal surfaces. Phukan (1966) also reported that, rarely, zircon crystals with or without haloes may be seen. Those with the haloes showed the expected brownish staining in these fissures. Opaque black prismatic and rounded mineral guests surrounded by liquid fingerprints were also recorded in two of the nine specimens examined. Additionally, Gübelin (1983) mentioned tourmaline and hornblende as inclusions. Specimens recently examined by the present author have also displayed inclusions of what appears to be mica.

In terms of habit, Kashmir sapphires bear a strong resemblance to those of Sri Lanka, with almost all being spindle-shaped hexagonal bipyramids. Most of the colour is concentrated in a layer at the tips and just beneath the crystal faces, with colourless cores, somewhat akin to the 'ottu'-type blue sapphires from Sri Lanka.

The large fine gems of years gone by were generally cut from the tips of sometimes enormous crystals which measured up to 1 foot in length! Those specimens that possess well-formed crystal faces contain this blue layer intact; however, many pieces are heavily corroded at the surface and thus the blue layer is only partially present, if at all. Owing to the irregular distribution of colour in Kashmir rough, cut stones will also often display very strong colour zoning, similar to Sri Lankan material. However, heat treatment can produce dramatic results with Kashmir sapphires. The author examined one large parcel of stones which were mostly white, with only traces of blue at the tips of some crystals. After burning, nearly every piece in the parcel had been transformed to an even rich-blue colour. The change was by far the most incredible witnessed by this author among sapphires being heat treated. If so, we might think, then why aren't we seeing this material in the market? The answer is simple: no rough. Even the low-grade white material is extremely scarce nowaways, and no mining is being done at present. However, a number of authorities reported 50 years ago that hidden away in the chambers of the Kashmir State Treasury are large sacks and chests containing literally a king's ransom in rough and cut Kashmir sapphires. The material was said to have been culled from over 40 years' production, quite literally the cream of the crop. One piece of rough was said to have been as large as a croquet ball! Although many merchants visited Jammu and Srinagar with the intention of purchasing some or all of these stones, their offers were refused. A caretaker was reported to have tried to steal this treasure in the 1970s, but was caught and the stolen goods recovered. Since then, nothing has been heard of the 'hoard of Kashmir'. All we can do is wait and hope.

Sapphire – Australia

Although Australia is a major producer of sapphires, little study has been done on the inclusions in these stones. There are two major sapphire-producing localities in Australia: Anakie, in Queensland, and the New England district in New South Wales. Generally both produce mostly blue sapphires, with the finer ones coming from New South Wales. Queensland, however, produces better quality yellow and green sapphires. Black star sapphires are also mined in each state.

The occurrence of sapphires in Australia is very similar to that in eastern Thailand and Cambodia. Thus, it is not surprising that the inclusions should be quite similar. Colour banding is a common feature of the blue and green stones, with evenly coloured gems showing very sharp fine banding of alternating blue and colourless or blue and yellow under magnification.

The colour is typically an inky blue in all except the finest stones and follows the usual hexagonal pattern of the crystal when examined along the *c* axis. In some, a triangular zoning is seen at the crystal centre. Fibre optic illumination often reveals concentrations of minute exsolved particles which follow the hexagonal colour banding. These particles appear to be the residue left over after heat treatment to remove the silk that impairs light transmission. Virtually all Australian sapphires are heat treated to remove this silk for, without this treatment, most of the production would be unsaleable.

Figure 5.17 Colour zoning and silk visible parallel to the *c* axis in a blue sapphire from Australia. The outer portions of the crystal are blue in colour, while the core is yellow. Within the yellow core are concentrations of silk in the form of haematite. Such blue/yellow combinations in a single crystal are often found in Australian sapphires. (Photograph by the author, specimen courtesy of Gerald V. Rogers)

Yellow sapphires from Australia possess identical inclusions to those from Chanthaburi, Thailand, so much so that the author has been unable to differentiate between them on this basis. In general, the Thai stones are more of a brownish 'whisky' colour while the Australian stones are slightly greenish yellow, but this is not always the case. Like the Thai yellow sapphires, the Australian stones possess thin layers of haematite (and a little rutile) silk just beneath the top and bottom pinacoid faces. Boehmite needles and the accompanying rhombohedral twinning are also found; these twinning planes and needles often do not penetrate deeply into the stone.

Solid inclusions found in Australian blue sapphires also resemble those in stones from Thailand and Cambodia. Colourless rounded or euhedral crystals of feldspar have been reported by Schubnel (1972). These may be surrounded by glassy tension haloes (presumably due to heat treatment), as well as being accompanied by comet tails. At times, the heat treatment appears to melt these or similar-appearing crystals, for gas bubbles may be found frozen in place within these crystals. Red or orange crystal grains are sometimes found as well, again often accompanied by comet tails. These may be uranium pyrochlore crystals, as identified by Gübelin (1983) in Pailin sapphires. Stubby rods and plates of hornblende and mica, respectively, have been reported in Anakie sapphires by Coldham (1985).

Like the blue sapphires from Chanthaburi, the silk in Australian sapphires does not appear to be solely rutile, but instead is mostly haematite, with only small amounts of rutile present. The haematite occurs as thin plates or very stubby needles of a gold colour parallel to the faces of the first-order hexagonal prism, while the rutile is found as longer needles aligned parallel to the second-order hexagonal prism.

Rhombohedral twinning is found in Australian sapphires along with the boehmite needles formed at the edges or junctions of these planes through the exsolution of excess hydrous alumina or alteration of the corundum itself. Secondary liquid inclusions such as fingerprints and feathers are also common in Australian sapphires.

Sapphires of primarily a blue colour are also found at Lava Plains in northern Queensland. These possess a deep superior blue when compared with most other Australian stones, but it remains to be seen whether enough facetable material will be forthcoming to make a success of mining at this locality.

Sapphire – Montana, USA

The sapphires of Montana in the United States come from several different areas scattered across the state. However, only those from Yogo Gulch are of major commercial importance.

Yogo sapphires are notable for their amazing uniformity. Over 90% of the mine's production is of a consistent medium-blue colour, the remainder tending towards violet and lilac shades. Virtually no other colours are found here. The only factor stopping Yogo Gulch from becoming a major force in the market place is the disappointingly small sizes of the sapphires produced. By far, most weigh in at less than one carat in the rough, and the flat tabular shape of the crystals means relatively low yields for cut stones. Thus, faceted Yogo sapphires of greater than one carat are rarities. This is most unfortunate as their colour is truly beautiful, especially in the scarce large sizes, and their colour consistency makes them ideal for putting together matched sets for jewellery.

So even is the colour of Yogo sapphires that the vast majority of stones display no trace of the colour zoning which disfigures many a stone from other localities. Even when stones are immersed in methylene iodide and examined under magnification, zoning is rarely discovered.

Inclusions of other types are also scarce in Yogo stones, many being entirely clean. Most common are tiny crystal grains of various types. Dark-red crystals of metallic lustre which strongly resemble rutile have been seen. These occur in rod shapes and knee-shaped twins, often surrounded by fingerprints. Colourless crystal grains have also been found surrounded by very thin fingerprints of hexagonal outline in the basal plane. These resemble very much the primary liquid inclusions of Thai rubies. Tabular colourless crystals of what may perhaps be corundum or negative crystals are surrounded by partially healed fractures containing reddish stains.

At times, irregular cavities filled with a whitish or yellowish liquid are seen. Additionally, polysynthetic twin lamellae along the rhombohedron planes are also found on occasion in Yogo sapphires.

The author has not examined many Montana sapphires other than those from Yogo Gulch. However, several blue to green sapphires seen from Rock Creek showed features unlike any others. Within these stones were both large and small round globes of a blue colour floating in a colourless body. Each blue sphere surrounded a small crystal and the colour appeared to diffuse outward from these crystal grains. Unusual, to say the least, and most certainly a sight to remember! These features apparently result from heat treatment which melts the crystal and sends the colouring agents into solid solution.

Ruby – Kenya

The rubies of Kenya are mined primarily in the Tsavo National Park region. Very little study has been done on these stones, which is unfortunate, for fine pieces are sometimes found. Most of the material from Kenya is suitable only for cabochons, and virtually all today is heat treated to improve the appearance.

Kenyan rubies are noted for their fine red colour and intense fluorescence, similar to many Burmese rubies, with which they are often confused. However, they can be easily separated from unheated Burmese stones because of the large numbers of liquid fingerprints and feathers which they contain. In contrast, Burmese rubies are normally quite deficient in liquid inclusions, except those that have been heated. Heat treatment introduces many fingerprints and feathers into Burmese rubies which are not found before treatment.

Rutile silk is not found in Kenyan rubies, so star stones are not produced. Kenyan rubies do, however, contain concentrations of minute exsolved particles arranged in patterns identical to rutile silk. Actual rutile needles, though, are absent. In some cases, these particles are relatively coarse, while in other cases they may be so fine that individual particles cannot be discerned. Instead, some show only diffuse white 'texture' clouds somewhat similar to those found in certain heat-treated blue sapphires from Sri Lanka. These texture clouds follow the hexagonal zoning of the crystal structure.

Colour zoning is common and appears both as large diffuse zones and as sharp and narrow colour bands. Rhombohedral twinning is also very common and is accompanied by long white boehmite needles along the edges and junctions of these planes.

Large numbers of liquid fingerprints and feathers are the hallmark of Kenyan rubies. Coupled with the clouds of exsolved particles, these features give a slightly turbid appearance to most stones, making them suitable only for cabochons. Even faceted gems generally possess this sleepy overall look, somewhat like 'Lai Thai' rubies, but with extremely strong fluorescence. Many fingerprints appear thick and wispy, making confusion with the flux-grown synthetic products a very real possibility. However, appearing with the liquid inclusions are boehmite needles and lamellar twinning, allowing a separation to be made. In heavily included gems such as the Kenyan rubies, in order to examine a stone's interior it may be necessary to open the microscope's light source to light field and close the iris diaphragm until it is just smaller than the stone.

Close examination of the liquid pockets of these fingerprints reveals a two-phase appearance in some cases. Others are seen to be tiny negative crystals which have resulted from the healing of cracks. Primary negative crystals are also found, both as well-formed individuals and as irregular cavities. These again may be confused with flux inclusions.

Solid inclusions are less common in Kenyan rubies. White or colourless crystals are seen, either euhedral or heavily corroded. Their identity is unknown at present.

Overall, the internal appearance of rubies from Kenya is characterized by large numbers of liquid inclusions bounded and criss-crossed by rhombohedral twin lamellae and boehmite needles. This is similar to the case in many Thai rubies, but without the saturn-type crystal formations and with turbid areas of exsolved matter and strong fluorescence. Immersion reveals colour zoning; this and the strong red fluorescence allow separation from Thai rubies. At times, chalky-white patches among the red areas are also seen under short-wave light. The colour of Kenyan rubies may resemble that of Burmese stones, but the large numbers of liquid inclusions common to the former are lacking in the latter. Despite the major internal differences, the author continues to see Kenyan rubies being sold as Burmese ones. In the world of coloured gemstones, there is so much in a name!

Sapphire – Umba Valley, Tanzania

The fancy sapphires of the Umba Valley in Tanzania are famous for their wide variety of hues, of which the colours of some almost defy description. Practically every colour in the rainbow is found, with the possible exception of a true ruby red.

Umba sapphires possess a very distinctive internal appearance. Most are criss-crossed by rhombohedral twinning planes and the accompanying boehmite needles. Together with these long boehmite needles are multitudes of tiny thin plates or films which have also formed along the rhombohedral planes. These features appear to be microscopically thin plates and needles of another mineral formed through exsolution and give a dazzling iridescence when illuminated by a fibre optic lightguide. Haematite has been suggested as the substance of these needles and plates.

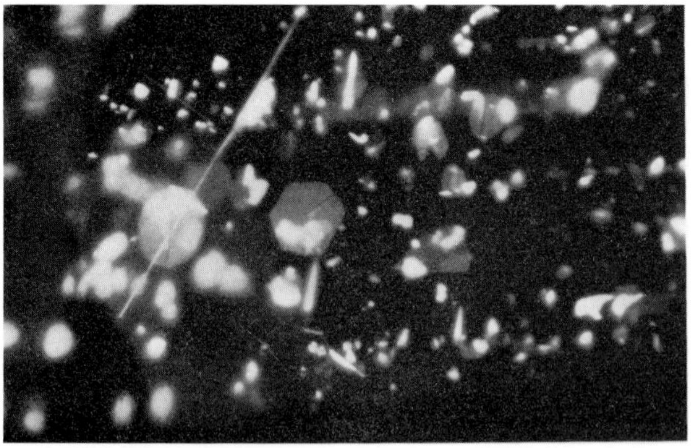

Figure 5.18 Ultrathin haematite plates and needles seen in a sapphire from the Umba Valley, Tanzania. (Courtesy of the Gemological Institute of America)

Crystals of apatite are a common sight in Umba stones and may range in size from large euhedral or rounded individual crystals to tiny plates or fragments. Some stones may feature large numbers of these hexagonal apatite prisms.

Bibliography

Anonymous (1979) *Gemstone Characteristics Magnified*, Gemological Institute of America, Los Angeles

Atkinson, D. and Kothavala, R. Z. (1985) Kaschmir-saphire. *Lapis,* **10**, no. 10, 11–22

Bowersox, G. W. (1985) A status report on gemstones from Afghanistan. *Gems and Gemology,* **21**, 192–204

Brown, J. C. (1956) Sapphires of India and Kashmir. *The Gemmologist,* **25**, no. 298, 77–80

Chikayama, A. *Gem Identification by the Inclusion*, Gemmological Association of All Japan, Tokyo

Coldham, T. (1985) Sapphires from Australia. *Gems and Gemology,* **21**, 130–146

Eppler, W. F. (1959) The origin of healing fissures in gemstones. *Journal of Gemmology,* **7**, 40–66

Eppler, W. F. (1966) The origin of negative crystals in gem stones. *Journal of Gemmology,* **10**, 49–56

Eppler, W. F. (1972) Needles in corundum other than rutile. *Journal of Gemmology,* **13**, 41–44

Eppler, W. F. (1976) Negative crystals in ruby from Burma. *Journal of Gemmology,* **15**, 1–5

Gübelin, E. J. (1953) *Inclusions as a Means of Gemstone Identification*, Gemological Institute of America, Los Angeles

Gübelin, E. J. (1957) A contribution to the genealogy of inclusions. *Journal of Gemmology,* **6**, 1–47

Gübelin, E. J. (1969) On the nature of mineral inclusions in gemstones. *Gems and Gemology,* **13**, 42–56, 74–88

Gübelin, E. J. (1971) New analytical results of the inclusions in Siam rubies. *Journal of Gemmology,* **12**, 242–252

Gübelin, E. J. (1972) Inclusions in gemstones. *The Australian Gemmologist,* **11**, no. 8, 3–14

Gübelin, E. J. (1983) *Internal World of Gemstones*, ABC Edition, Zurich

Gunawardene, M. (1984a) Reddish-brown sapphires from Umba Valley, Tanzania. *Journal of Gemmology,* **19**, 139–144

Gunawardene, M. (1984b) Contributions towards the identification of treated corundums: heat and diffusion treated rubies. *Journal of Gemmology,* **19**, 298–310

Gunawardene, M. and Chawla, S. S. (1984) Sapphires from Kanchanaburi Province, Thailand. *Journal of Gemmology,* **19**, 228–239

Halford-Watkins, J. F. (1934) New facts about Siam rubies. *The Gemmologist,* **4**, no. 41, 147–149

Halford-Watkins, J. F. (1935) Kashmir sapphires. *The Gemmologist,* **4**, no. 42, 167–172

Koivula, J. I. (1981) Photographing inclusions. *Gems and Gemology,* **17**, 132–142

Koivula, J. (1985) More on needles. *Lapidary Journal*, December, 25

Nassau, K. (1984) *Gemstone Enhancement*, Butterworths, London

Phukan, S. (1966) Studies on inclusions in some Indian gemstones. *Journal of Gemmology,* **10**, 1–7

Roedder, E. (1982) Fluid inclusions in gemstones: valuable defects. In *International Gemological Symposium Proceedings 1982*, Gemological Institute of America, Los Angeles, pp. 482–501

Roedder, E. (1984) Fluid inclusions. *Reviews in Mineralogy,* **12**

Schubnel, H. (1967) Determination of solid inclusions in gemstones. *Journal of Gemmology,* **10**, 189–193

Tait, A. S. (1955) Asterism in corundum. *Journal of Gemmology,* **5**, 65–72

Webster, R. (1961) Corundum in Tanganyika. *Gems and Gemology, Fall*, 202–205

White, J. S. (1979) Boehmite exsolution in corundum. *The American Mineralogist,* **64**, 1300–1302

Zwaan, P. C. (1967) Solid inclusions in corundum and almandine garnet from Ceylon, identified by x-ray powder photographs. *Journal of Gemmology,* **10**, 224–234

Zwaan, P. C. (1974) Garnet, corundum and other gem minerals from Umba, Tanzania. *Scripta Geologica,* **20**, 19–30

Zwaan, P. C. (1982) Sri Lanka: the gem island. *Gems and Gemology,* **18**, 62–71

Chapter 6

Treatments

The mining of gemstones differs greatly from that of most metals and industrial minerals. Unlike rough gem materials, which are generally useless unless of high clarity and cuttable sizes, industrial minerals can be processed to extract the desired product. By means of crushing and smelting it is possible to separate usable material from the waste, or 'slag', as it is known. Gems are far too fragile to withstand such abuse and thus, out of a mine's entire yield, only a tiny fraction is of sufficient size and quality to warrant cutting. Defects in clarity and colour and insufficient size preclude use of the remainder.

Because of the scarcity of gem-quality materials, various attempts have been made over the years to improve the appearance and/or stability of inferior samples. Heating or irradiation to improve the colour and oiling to hide fractures are but the tip of the gem treatment 'iceberg', a virtual pot-pouri of different recipes aimed at increasing the beauty and saleability of the gems and ornamental stones. By developing treatments, traders have, in part, solved the dilemma with regard to the inability to process lower grades, as is done for industrial minerals. Thus, treatments allow a larger percentage of the total production to be utilized as gemstones, in effect finishing the job that nature started.

The legitimacy of certain treatments has in recent years been called into question because, in many cases, the treated gems cannot be distinguished from those which are completely genuine. This has created an unfortunate situation, with colour- and/or clarity-enhanced stones competing directly against the natural product. Although many treatments have been practised for centuries, modern technology today produces far more dramatic changes than those obtained in the past. The question then arises of where to draw the line between acceptable treatment and simply producing a totally synthetic gem. This issue of whether or not treated gems should be sold as natural gems is currently an item of hot debate in trade circles. Clearly this is something that the gem and jewellery trade must come to grips with if government regulation is to be avoided. The pros and cons of this matter will be taken up at the end of this chapter.

Heat treatment

> The blue tinge which detracts from the value of the pure ruby, whose colour should resemble "pigeon's blood", is removed by the Singhalese, by enveloping the stone in the lime of a calcined shell and exposing it to a high heat.
> (Sir James Emerson Tennent (1859) *Ceylon: Physical, Historical and Topographical*)

The above account is a description of one process used in former times to improve the colour of corundum by heat. Far from being a recent discovery, as we might believe, the heat treatment of rubies and sapphires has been practised in one form or another for many centuries, quite possibly dating back to Roman times. However, within the last 15 years new discoveries, particularly regarding the potential for heating Sri Lankan gems, have resulted in the practice of heat treatment on a far greater scale than ever before. Today it is estimated that 90% or more of all rubies and sapphires used in the gem trade have, at some point, been exposed to the heat of an oven. Furthermore, the changes in colour and clarity achieved nowadays are much more dramatic than in former times. Indeed, certain pieces of rough sapphire that 15 years ago would have been considered absolutely worthless can now, with luck and the proper heat treatment, be transformed into gems worth tens of thousands of dollars.

Heat-treatment techniques vary from the modern, using expensive electric ovens, to the primitive, with little more than a fire, a crucible and a blowpipe. Certainly among the more unusual methods, however, is that described to the author by a Sri Lankan gem dealer. He claimed that most sapphires and rubies are best treated in a fire or oven, but for a small minority more gentle heating is needed. Eager to learn more about this highly secret art, I pressed him for details. After glancing over his shoulder to make sure that no one else was listening, he leaned forward and in a deadly serious voice whispered, 'These we feed to our chickens.'

Now this is not quite so ridiculous as it first sounds. For over 2000 years man has been treating pearls by feeding them to chickens, the acidic conditions and grinding action of the crop producing improvements in both whiteness and lustre. However, as far as corundum is concerned, although I don't wish to be a 'doubting Thomas', frankly it appears to be of questionable use. Suffice it to say that confirmation from other quarters is still awaited.

The modern era of corundum heat treatment began in the mid to late 1970s with the entrance into the Bangkok market of large quantities of burned 'geuda' blue sapphires from Sri Lanka. Geuda, a Singhalese word, is used to describe a certain grade of low-quality sapphire which is found mainly in Sri Lanka. Typically pale blue, yellow or pink in colour, geuda stones are unsuitable for gems 'as is', due to an abundance of exsolved titanium in the form of rutile silk clouds. Before the secret of the geuda was unlocked, their usefulness was limited to employment as abrasives, ornamental stones for rock gardens and as good-luck charms to be buried under the foundation posts of village homes. However, today's ugly duckling is tomorrow's graceful swan, or, more properly, yesterday's geuda is today's cornflower-blue sapphire. At some point in time, someone, somewhere, discovered that the wretched little geuda sapphire, pariah of precious stones, could, with the proper application of heat, be changed into a true beauty. Once fit only for fishtanks, it now festoons the fingers of the wealthy around the world. Why it is even possible that the Sri Lankan sapphire presented to Lady Diana by Prince Charles on the occasion of their engagement first entered this world as the humble geuda. Yes indeed, the geuda sapphire is truly the Cinderella of sapphires!

No one knows for certain who had the good fortune to discover the potential of the geuda first. Among the many stories going round Bangkok, that repeated most often credits an enterprising Swiss couple with first heating geudas. They are said to have profited quietly for a number of years by acquiring the rough for next to nothing in Sri Lanka, heating it in Bangkok, and then selling the cut gems as untreated blue sapphires. However, in this 'Adam and Eve' tale, enter temptation and exit common sense. Deciding to step up production, they hired local help to run the ovens – and the secret was out. Their once tranquil 'garden' rapidly took on the look of a squatter colony as nearly every dealer in town arrived to try their hand at 'burning geudas'. From this point on, the facts are known. In the late 1970s, large numbers of heat-treated Sri Lankan sapphires began streaming out from the ovens of Bangkok's new gem chefs. In the early days of the 'Great Geuda Rush', before the miners in Sri Lanka caught on, the rough could be had for next to nothing and fortunes were amassed overnight by those Thai burners fortunate enough to have gotten in on the ground floor. However, with planeloads of Thais arriving in Colombo to buy geudas, it was only a matter of time before the miners in Sri Lanka learned what was happening. They first attempted to heat the stones themselves, but with little success, and so they opted for the next best thing: prices of geuda rough were hiked to levels only slightly less than that of the already-blue stones! With detection methods lagging far behind the technology of the treatment, the gem market then committed what was, in the mind of this author, a crucial (but understandable) error: the treated stones were accepted and sold as completely natural, with no distinction made between treated and naturally coloured gems. Why this occurred is extremely complicated. In the diamond market there has never been any such question. Naturally coloured diamonds, by virtue of their rarity, are worth more than those coloured artificially. However, today the coloured gem market faces exactly the opposite situation – the treated stones, by virtue of their generally finer colour and clarity (on the average), fetch more than the humble natural colours with their obvious defects. The ridiculous is nowadays reality in the ruby and sapphire market.

It should be mentioned that the heat treatment of geuda sapphires is by no means an easy task. Included among the many calamities that may befall a 'burn' are the melting of stones and the explosion or cracking of certain pieces. Some lose whatever colour they had to begin with while others may develop unsightly white clouds. If the rough is cheap this causes little concern, but if large sums are paid just to obtain the rough, such events can terminate the business. It is no laughing matter to have a $20 000 investment turn into a puddle of fused alumina!

Today, geuda rough brings very high prices in Sri Lanka. Certain pieces which Thai buyers are confident will burn well can sell for as much as $50 000 or more. Some burners in Sri Lanka have had success, but the business still remains largely in the hands of Thais, who have proven more adept at coaxing the colour from the geuda.

All geuda rough is not created equal, with some types considered more valuable than others. The key feature displayed by geudas, as opposed to ordinary rough, is an oily golden reflection inside the stone when examined in transmitted light. It is termed 'diesel' because of its resemblance to diesel oil and is an indication that the stone will burn to a blue colour.'Young milky' is the term used to describe the well-formed pale-blue crystals with a cloudy appearance and a peculiar waxy lustre. In transmitted light they display a strong diesel effect and are highly sought after. Another category of rough is 'ottu', a Tamil term meaning 'to risk' or 'take a chance'. This is an allusion to the hazardous nature of buying ottu stones, for they possess blue colour only at the tips or the faces of the crystal, the core being largely devoid of colour. The cutting of ottu rough is quite difficult, as the lapidary must attempt to position the coloured portions such that they lie at, and extend across, both sides of the culet. When correctly oriented the entire stone will face up blue, but if the colour lies on one side alone, a large 'hole', or window, will result. Some ottu stones are oriented with a single narrow band of colour extending across the table or crown. Both types, with the colour at the crown or culet, may be mistaken for doublets, as the crown appears coloured while the pavilion is colourless when seen from the side. Other terms used to describe geuda rough include blue geuda, silky geuda and waxy geuda.

Beginning in 1981, Thai burners made another major discovery with regard to the heat treatment of Sri Lankan sapphires. They found that many of the sapphires which did not take well to the blue burning could, by varying the treatment conditions, be burned to a yellow or orange colour. This started another boom and soon deep-yellow to orange sapphires began pouring from the ovens of Bangkok and Chanthaburi. The sapphires display an intensity of colour which is very rare or even completely unknown in untreated stones from Sri Lanka, and with suitable rough being relatively abundant the yellow sapphire has been promoted from the realm of a mere collector's stone to that of an important market force.

Following their great success achieved in heating Sri Lankan sapphires, Thai burners quickly began pursuing rubies and sapphires from other sources as well. Many of these, such as the Australian blue sapphire and the Thai ruby, have had a long history of treatment, predating that of the stones from Sri Lanka, but for others, like the Burmese ruby, little had been done. Nowadays, gems from every major locality are heat treated to improve the colour or clarity, not only by Thais, but also by people around the world. With the possible exception of the sapphires from Kashmir, none show the dramatic changes of the Sri Lankan material, but this is definitely not for lack of effort on the part of the burners. Today, the heat-treated ruby or sapphire is most certainly the rule rather than the exception.

The heat treatment of rubies and sapphires involves a number of complex changes to the stones. It is influenced by the following:

Time:	The length of time that the stones are heated at the maximum temperature and the rate of temperature ascent and descent
Temperature:	The maximum temperature reached

Atmosphere:	The chemical nature (oxidizing or reducing) of the atmosphere in which the heating is performed, and its pressure
Chemical nature of the stones:	The chemical elements (major elements and impurities), and their states, present in the stones themselves
Surrounding material:	The nature of the material in contact with, or in close proximity to, the stones

There exist five major heat-treatment processes that can be performed on rubies and sapphires. They are as follows:

1. *Removal of silk*
   Heat to between 1000 and 1900°C, cool rapidly.
   (a) Can be used in combination with any process except number 2. Cloudy material becomes clear.
2. *Development of silk*
   Heat to approximately 1300–1900°C for 1–14 days or more.
   (a) Can be used with any process except number 1. Allows the development of a star in titanium-rich material.
3. *Development of blue colour*
   Heat to between 1600 and 1900°C in a reducing atmosphere.
   (a) Geuda sapphire (Sri Lanka), Kashmir sapphire – used in combination with process 1. Pale silky stones become clear and deep blue.
   Note: A small percentage of Sri Lankan stones turn blue when heated in an oxidizing atmosphere.
4. *Removal of blue colour*
   Heat to between 800 and 1900°C (may be as low as 450°C) in an oxidizing atmosphere.
   (a) Lightens the colour of dark 'inky'-blue Australian, Thai, Cambodian and Nigerian sapphires. Higher temperatures may be used in combination with process number 1.
   Note: A reducing atmosphere has also been reported for this type of heating.
   (b) Removal of the blue component from purplish Thai, Cambodian, Burmese, Sri Lankan, Kenyan and Tanzanian rubies. Higher temperatures may be used for Burmese, Sri Lankan and East African rubies in combination with process number 1.
   (c) Removal of the blue component from green or greenish-yellow Australian and Montana (non-Yogo Gulch) sapphires, thus intensifying the yellow colour.
5. *Development of yellow colour*
   Heat to between 1600 and 1900°C in an oxidizing atmosphere.
   (a) White and pale-yellow Sri Lankan sapphires become an intense yellow or gold colour. Pink stones change to an orange colour.

At this point a note of caution should be sounded with regard to the heat-treatment processes described. Although the author has attempted to secure accurate information, it is all secondhand and strictly of a speculatory nature. Burners each have their own methods, and great pains are taken to keep them secret. No guarantee of any kind can be given regarding the reliability of these descriptions and anyone attempting to follow them in treating gems does so at their own risk.

Just how dangerous burning can sometimes be is illustrated by the following experience of a Bangkok dealer. Around 1981, a fine untreated Sri Lankan blue sapphire of large size was purchased. This was at a time when sapphire prices were at record levels and the amount paid for the stone was in excess of $50 000. It was an outstanding piece, to be sure, but with one slight defect: just beneath the table facet was a small cloud of silk, not really distracting, but noticeable none the less. The solution? A quick trip to the ovens of his number one burner. No problem, right? Wrong. Big problem. Colour no more. Gone, vanished without trace. We can readily imagine that the ensuing conversation between dealer and burner must have been a goldmine for the student of Thai idiom.

But the story is not yet complete. Our dealer decided to give the dice another throw and burn the stone again, this time with his number two burner. After all, what else could possibly go wrong? Right? Wrong again. Now, no colour plus one large crack. At the last report, our dealer was said to be considering forsaking the material world and becoming a Buddhist monk.

In the heat treatment of corundum, two elements are of primary concern – iron and titanium. A third element, chromium, also plays a major role in the colouring of rubies and sapphires, but is apparently unaffected by heat. Treatment of geuda sapphires from Sri Lanka requires changes to both iron and titanium, the colouring agents of blue sapphire. The blue colour of sapphire is derived from a charge transfer process involving the jumping of an electron from an iron ion to one of titanium, as follows (Nassau, 1984):

$$Fe^{2+} + Ti^{4+} \xrightarrow[\text{absorption}]{\text{light}} Fe^{3+} + Ti^{3+}$$

Geuda sapphires are typically cloudy, with the clouds consisting of rutile silk, a form of crystallized titanium dioxide (TiO_2). This is soluble in corundum at temperatures above 1600°C. Thus, the heating of any rutile-containing ruby or sapphire to, say, 1800°C, essentially melts the silk, sending the titanium into solid solution. In order to prevent it from reforming, the stone must be cooled rather quickly (in a few hours). This not only removes the cloudiness from the stone, but also provides the titanium necessary for the coloration of blue sapphires.

The second part of the geuda equation involves reducing the iron to the ferrous (Fe^{2+}) state. If the heating above is carried out in a reducing environment, any ferric iron (Fe^{3+}) will be converted to ferrous. Now the gem possesses both the titanium and ferrous iron required for the charge transfer mechanism. Hence, a single heating to between 1600 and 1900°C in a reducing atmosphere not only clarifies the stone, but also produces a blue colour in suitable material. This is the theory behind the heating of geuda sapphires.

On paper, it is also possible to perform the reverse of process number 1 – the recrystallization of the silk in titanium-rich stones where the titanium is held in solid solution. Prolonged (1–14 days or more) heating at 1400–1900°C and slow cooling will allow the titanium atoms to unmix from solid solution and form rutile needles.

Thus, star stones, in which the star is weak because of the lack of silk, might be heated to strengthen the star. In essence, this is what is done to produce synthetic star rubies and sapphires, but so far burners have not been successful in duplicating this process in natural material.

Many mined rubies contain a bluish or purplish tint that lowers their value, particularly those from the Thai–Cambodian border. Heat treatment can reduce or eliminate this bluish secondary colour in what is basically the reverse of process number 3. This entails heating the stone in an oxidizing atmosphere, which produces the following reaction:

$$Fe^{2+} \xrightarrow[\text{atmosphere}]{\text{heat in an oxidizing}} Fe^{3+}$$

With the conversion of the iron to the ferric state, a pale-yellow colour replaces the blue, giving the gem a much purer red colour. Normally, only relatively low temperatures (800–1300°C) are needed to bring about this change, unless it is necessary to remove the silk simultaneously, as is sometimes the case with Burmese or Sri Lankan rubies. Then, higher temperatures (up to 1900°C) would be indicated. In recent years, practically the entire ruby production from Thailand has been burned in the above manner, with the result that the Thai rubies being traded today are of much higher quality than those in previous times. A very definite improvement is effected by such a treatment, which is not limited to Thai stones alone. Large numbers of Burmese, Sri Lankan and Kenyan rubies are similarly treated.

Australian, Nigerian and some Thai blue sapphires generally suffer from being too dark and 'inky' in colour. This is largely due to the presence of silk, which in stones from these countries may be either haematite (Fe_2O_3), rutile (TiO_2) or another iron–titanium compound. The silk affects the colour by reflecting and scattering light to such an extent that the pure blue and blue-green dichroic colours are mixed together. Heating to between 1000 and 1700°C removes this silk, effectively improving the colour in the process. Since both oxidizing and reducing conditions have been reported for such heating, it would seem that the change is primarily due to the removal of the silk, rather than to any lessening of colour from the partial oxidation of the iron present.

Green and yellow sapphires from Australia and Thailand are also heated to remove the silk, which in the yellow sapphires is frequently present in thin green planes just beneath the upper and lower basal pinacoid faces. In addition, some improvement may also occur as a result of the oxidation of the iron, producing a richer gold or yellow colour.

One burner, Mr Gerald V. Rogers, has found that a small percentage of Sri Lankan material will turn blue when heated in an oxidizing atmosphere. This is in direct contrast to the geuda sapphires, for which a reducing environment is used. Furthermore, the blue colour obtained in oxidizing conditions is said to be superior to that produced by the reducing atmosphere. So far, no explanation regarding how or why this change takes place has been forthcoming.

Besides that used for the geuda sapphires and the purplish rubies, the other major type of heating performed on Sri Lankan corundum is the development of yellow and orange colours. Success with this process was first achieved in Bangkok during the year 1981. Prior to this, deep-yellow and orange sapphires were extremely rare in Sri Lanka, but with the discovery of a method for deepening the colour of pale material intensely coloured gems in these colours soon flooded the market. The treated stones could be produced in depths of colour totally unknown up to this time, so much so that prices descended to one-fourth or less of the pre-1981 levels. This is rather a shame, for in contrast with the other heated corundums a simple test, that to the best of the author's knowledge is 100% accurate, does exist for separating the treated and untreated gems. It was discovered by the author at the Asian Institute of Gemological Sciences in Bangkok, in 1981.

Before the test is begun, a stone should be selected which approximates as closely as possible the body colour of the stone in question. This stone will serve as a means of comparison to detect any change in colour during the test, a control stone so to speak. The gem being tested is placed on a fireproof surface, such as a metal box, and an intense 150 W spotlight brought within a centimetre of its surface. After approximately 10–15 minutes the lamp is turned off and the colour of the control stone is immediately compared with that of the test stone. Should the test stone be a heat-treated Sri Lankan yellow or orange sapphire, its colour will be found to have darkened temporarily, due to the heat of the lamp. As it cools, its colour returns to pre-test levels, so it is important that the comparison is made immediately after the lamp is turned off.

Since 1981, the author has tested thousands of heat-treated yellow and orange sapphires from Sri Lanka in the above manner and has yet to encounter a single heat-treated piece which did not behave in this way. The change is most dramatic in deeply coloured stones and less so in light pieces, but all show a noticeable darkening, or 'browning', of their colour, after a few minutes' exposure. The colour of untreated yellow or orange Sri Lankan sapphires generally remains unchanged, while in irradiated yellow or orange Sri Lankan stones the colour will fade. (Some naturally coloured Sri Lankan stones may also fade slightly, but this is not a big problem as they would fade under normal wearing conditions anyway. The fading of some naturally coloured yellow/orange sapphires has been reported by miners in Sri Lanka, but this fading is said to occur soon after the stone is taken from the ground.) With yellow or orange sapphires from other localities, such as Thailand or Australia, the colour remains unchanged by the fade test. Sri Lankan gems can be separated from the Thai and Australian yellow sapphires by their lack of strong iron lines (451.5, 460, 470 nm) in the spectroscope, which are always present in the latter.

The heat treatment of Sri Lankan yellow and orange sapphires is normally performed under oxidizing conditions at temperatures ranging from 1600 to 1900°C. This apparently produces a change in the hydration or aggregation of iron impurities contained within, forcing them to exsolve, possibly as the mineral haematite (Nassau, 1984). When present in large enough quantities, these minute haematite particles impart a pale- to deep-yellow, orange or even reddish-orange colour to the stone, often rendering it slightly cloudy or silky in the process. Under microscopic examination the true nature of the gems' coloration can be seen, for these heated sapphires are strongly zoned, with the most intensely coloured regions displaying the greatest concentrations of mineral particles. Of the thousands of burned Sri Lankan yellow sapphires examined by the author, all have shown these same inclusions, to a greater or lesser degree, arranged in an identical fashion to the rutile silk of star rubies and sapphires. Similar to aventurine quartz, which is coloured green by inclusions of fuchsite mica, these microscopic guests 'mechanically' add colour to their corundum host.

Colour stability of heat-treated corundum

The stability of the various colours produced by heating corundum has, in the past, been called into question by members of the trade. Certainly the subject of colour permanence is an important one, for gems are generally considered as heirlooms to be passed from one generation to the next.

Every so often a report surfaces regarding heat-treated sapphires which have allegedly faded years after being purchased. In most instances the sources are well-meaning laypeople or dealers who have simply erred in their recollection of the original colour of the stone. This is somewhat akin to an experience every jeweller has doubtless had – that of a customer believing their stone had been switched when in fact the piece had simply been cleaned. However, when coloured stone dealers spin this yarn about the fading of heat-treated sapphires it makes one wonder at the validity of their claim. The following is a typical tale. Famous gem dealer James Millionaire (a fictitious name) announces to the trade press that heated sapphires should be avoided because their colour is unstable. As 'proof', he cites a parcel of heated stones which were purchased five years before and which had lain unnoticed in his safe until recently. Originally all were of a similar fine yellow colour, but when examined last week several were found to have faded to near colourless. (These stones were obviously irradiated Sri Lankan yellow sapphires. Sometimes it is claimed that the stones darkened in colour with time. In one instance, they changed from fine blue 'Cambodian' sapphires to almost black Australian-type stones (Voynick, 1985).)

In terms of scientific process, the above cases must be taken with a liberal dose of salt, since they rely solely on the memory of the dealer. When we consider that temperatures ranging from 800 to 1900°C are necessary to modify the colour of a ruby or a sapphire in the first place, it seems highly improbable that sunlight alone could produce a reversal, let alone a change occurring at room temperature with the stones kept in a stone paper. Although the same cannot be said about irradiated gems, where fading in sunlight is a possibility, with respect to heated rubies and sapphires their colour would indeed appear to be permanent.

Methods of heat treatment

The heat treatment of corundum is, by its very nature, a risky operation and those people involved often display a penchant for taking risks, just like the alchemists of days gone by. Although slowly evolving into a science, heat treatment still involves a considerable slice of experimental chemistry, with practices at times more akin to alchemy than modern industry.

As with the early alchemists, secrecy is the rule. Each burner possesses his own special recipe which was discovered through past experiments and is jealously guarded. With some it is a magic liquid; for others a paste or powder; but for many it seems as vital as the flame itself.

Material may be prepared for burning in one of several ways. Generally the first step is to remove any foreign (non-corundum) minerals from the lot by gemmological tests (refractive index, specific gravity, etc.), hand selection, or both. This is extremely important, as certain types of stones, particularly topaz, beryl, tourmaline and chrysoberyl, may spoil an entire load, causing the corundum to break or, even worse, to melt. Any matrix adhering to the corundum must also be removed for the same reason. This is done with acid or by grinding.

Contrary to what is often reported (Kane, 1984), burning is usually done on rough stones. The only grinding performed is to trim away matrix, or dangerous inclusions such as large crystals and negative crystals that might crack the gem when heated. It would be most foolish to facet the gems first, before burning, as it is never known which areas will achieve the finest coloration. Additionally, it would certainly be a pity to cut a stone, burn it and repolish it, only to find its weight to be 1.99 carats. This is not to imply that cut stones are never burned, for they often are, but it is more desirable to perform heat treatment on rough, rather than cut, specimens.

Once the stones have been cleaned and trimmed they are ready for the oven. In order to avoid cracking during the treatment, the gems are placed into a set of nested alumina crucibles which can withstand temperatures up to 1900°C. This set is often sealed in the presence of the customer, with the seal being broken only by the customer after the burn is completed, to prevent the theft or switching of material.

One problem with high-temperature treatments involves stones adhering to one another; this often results in fractures when they are later split apart. It can be avoided by embedding the stones in an inert substance, such as alumina powder or ruby sand. Not only does this allow the gems to be freed without damage, but also it provides insulation against rapid temperature changes that may produce cracking. The crucibles too may be insulated for the latter reason. One ruby burner in Chanthaburi wraps the crucibles in wet clay obtained directly from the mines themselves, possibly in the belief that it will help the gems to 'ripen'.

In addition to preventing adhesion and providing insulation, the substances placed in or around the crucibles may be used to create the desired atmospheric conditions. Charcoal or graphite can be used to create reducing conditions, while various gas combinations can produce either oxidizing or reducing conditions. Other substances used in the crucibles include borax, caustic soda, sodium or potassium hydroxide, glycerin, citric acid, potassium nitrate, buffalo dung, urine, snake legs . . . and anything else that goes. The effect of fluxes, such as borax, is said to glaze the surface areas of the stones, producing brighter appearing rough.

Furnaces used in Thailand for the heat treatment of corundum are mainly of

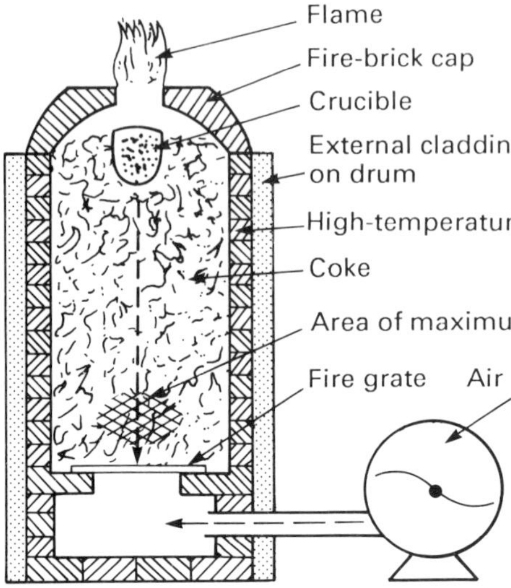

Figure 6.1 The construction of a typical furnace used for heat treating Australian (and other) sapphires in Thailand. (After Coldham, 1985)

three types: electric, gas or petroleum-fired, and charcoal or coke-fired. Electric ovens are the most costly and thus used for expensive rough, while the others are used for all types. Most simply consist of converted oil drums insulated with fire bricks.

A typical 'burn' may run from 1 to 24 hours or more, with the more efficient high-temperature ovens requiring less time. After the oven is shut down, the crucible is allowed to cool slowly to prevent cracking, a process which can take from 1 to 8 hours. When it is completely cool, it is then removed and the stones examined. Preliminary burns are often performed on Sri Lankan sapphires to determine how the material will respond. Depending on the results of this 'trial run', the gems will then be sorted into groups for developing either blue or yellow colours. Should the results be unsatisfactory on any particular stone, it may be heated several times, possibly by more than one burner, for the cardinal rule of heat treatment is 'if at first you don't succeed, try, try again'.

Detection of heat treatment

The increasing use of high-temperature heat treatments to improve the colour and clarity of the corundum gems makes the development of detection methods an area of vital concern. Fortunately, techniques are in existence that will allow the separation of many heated rubies and sapphires, and thus prevent unfair competition between natural and treated gems.

To date, in only one instance does there exist a simple and completely reliable test for the identification of heated corundums. This is for the heat-treated yellow and orange sapphires from Sri Lanka, which can be separated by a fade test, as described earlier. When placed within 1 cm of a 150 W spotlight for 10–15 min, the colour of the treated gems shows a temporary darkening, being more pronounced the deeper the original colour. As the gem cools, the original colour returns. The beauty of this test lies in its simplicity and reliability. It can be performed by almost anyone, and during tests carried out at the Asian Institute of Gemological Sciences in Bangkok, it has proven to be 100% effective. Most natural untreated yellow and orange Sri Lankan sapphires show no change, while some natural (very few) and all irradiated Sri Lankan gems will fade under the lamp. This test is only effective for treated Sri Lankan sapphires of the yellow and orange colours, not for those from Australia or Thailand.

Other features which serve to confirm the identification of the treated Sri Lankan yellow and orange sapphires are fluorescent reactions, body colour and inclusions. Untreated yellow/orange sapphires from Sri Lanka typically display a strong orange ('apricot') fluorescence under both long- and short-wave ultraviolet light. Heat treatment lessens or completely eliminates this reaction, particularly in deeply coloured stones.

The body colour of the treated gems often gives clues to their identification, being unnaturally deep with brownish overtones. It may grade into 'burnt'-orange shades completely unknown in nature from Sri Lanka.

Treated yellow sapphires from Sri Lanka always contain multitudes of minute iron-rich particles arranged in an identical fashion to the rutile silk. It is believed that these particles are responsible for the colour of the stones, being exsolved during the heating process. The colour banding will be found associated with these inclusions; deeply coloured stripes occur where concentrations of particles are located. Although particles of a similar nature may also be seen in untreated Sri Lankan stones, the colour is not necessarily associated with them.

Other inclusions in the treated Sri Lankan yellow sapphires are typical of all sapphires subjected to high-temperature colour alteration. These features include circular glassy fractures surrounding negative or solid crystals and colour zones

Figure 6.2 Clouds of minute exsolved haematite particles seen in a heat-treated yellow sapphire from Sri Lanka (parallel to the *c* axis). When the stone is viewed along the *c* axis, such clouds will line up into a hexagonal pattern, as above. Heat treatment caused exsolution of the iron as haematite particles, arranged in a similar pattern to the silk (three directions at 60°/120° in the basal plane). These particles 'mechanically' colour the stone. (Photograph by Wimon Manorotkul, AIGS, Bangkok)

showing slightly diffuse 'fuzzy' edges. In contrast, the zones in naturally coloured Sri Lankan stones are extremely sharp.

Due to the high temperatures commonly involved in heating corundum, and the fluxing agents added to the crucible, surfaces become pockmarked and require repolishing after burning. Occasionally lapidaries will miss small areas during the repolishing operation, leaving tiny patches displaying this dimpled or melted appearance. Most common at or near the girdle, they are an absolutely positive indication of heat treatment.

Still on the subject of surface features, a few words should be said about the presence of 'double girdles'. In gemmological publications, double or multiplane girdles are frequently cited as evidence of heat treatment, the logic being that they indicate repolishing or recutting, a necessity after burning. Unfortunately, almost all of the gems faceted in Bangkok that this author has examined have been found to display the double girdle, even those cut directly from the rough. Thus, the double girdle cannot be considered as indicative of recutting or repolishing, and consequently is of no use in identifying heated stones.

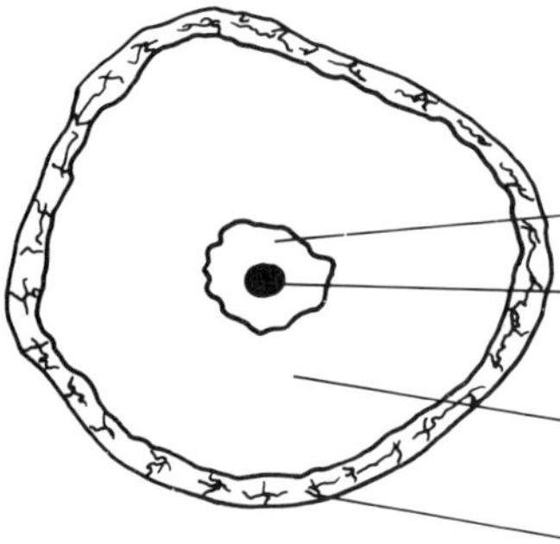

Figure 6.3 A discoid fracture resulting from heat treatment. When a gem containing crystal inclusions is heat treated, the crystals often expand, creating a distinctive circular tension crack with a glassy (shiny) lustre. If the melting point of the crystal was reached during treatment, a spillage zone will also be seen surrounding the melted crystal. At the outermost edge of the tension fracture healing may have already begun, creating a secondary healing rind. Inclusions such as this are indicative of heat treatment. (After Koivula, 19??)

Indications of heat treatment in blue sapphires from Sri Lanka are, in terms of inclusions, somewhat similar to those of their yellow and orange cousins. Untreated blue sapphires possess extremely sharp growth lines, and these are noticeably more diffuse at the edges in heated gems. Certain types of crystals often turn into round white 'snowballs' after treatment. These snowballs may or may not show glassy circular fractures around them. Also present in treated gems are white 'texture clouds' which follow the hexagonal structure of the crystal. At times, these clouds may show a substructure of minute particles arranged in an identical manner to the rutile silk of untreated gems, while in other instances the clouds are completely smooth throughout. Somewhat similar texture clouds may also be seen in unburned stones on occasion.

Probably the most important feature for identifying untreated Sri Lankan blue sapphires is the presence of rutile silk – not just exsolved mineral particles, but

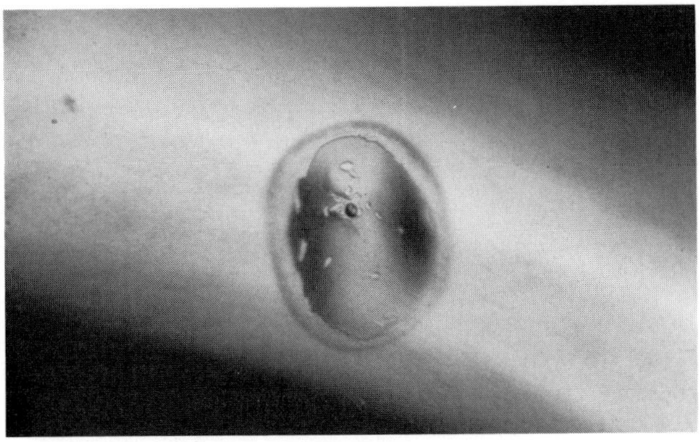

Figure 6.4 A circular tension disc resulting from heat treatment in a heat-treated yellow sapphire from Sri Lanka. During heat treatment the included crystal at the centre expands, producing a circular tension fracture of often glassy lustre. In some cases (as above) the crystal melts, creating an irregular spillage zone immediately surrounding it. As a result of the high temperatures involved, the disc begins to heal, creating a secondary healing rind at its outermost edges. Such tension discs surrounding crystals are typical of heat-treated stones from all localities. (Photograph by the author)

actual needles intersecting in three directions at 120° in the basal plane. Before burning, both particles and needles may be present, but after heating the needles will be gone. Therefore, the presence of actual rutile needles indicates an untreated stone. With the exception of yellow sapphires, most unburned corundums from Sri Lanka display at least some rutile silk. During heating, rutile silk may not dissolve completely, but instead often coalesces into small droplets arranged in the pattern of the silk. This 'partially dissolved' silk is another indication of heat treatment.

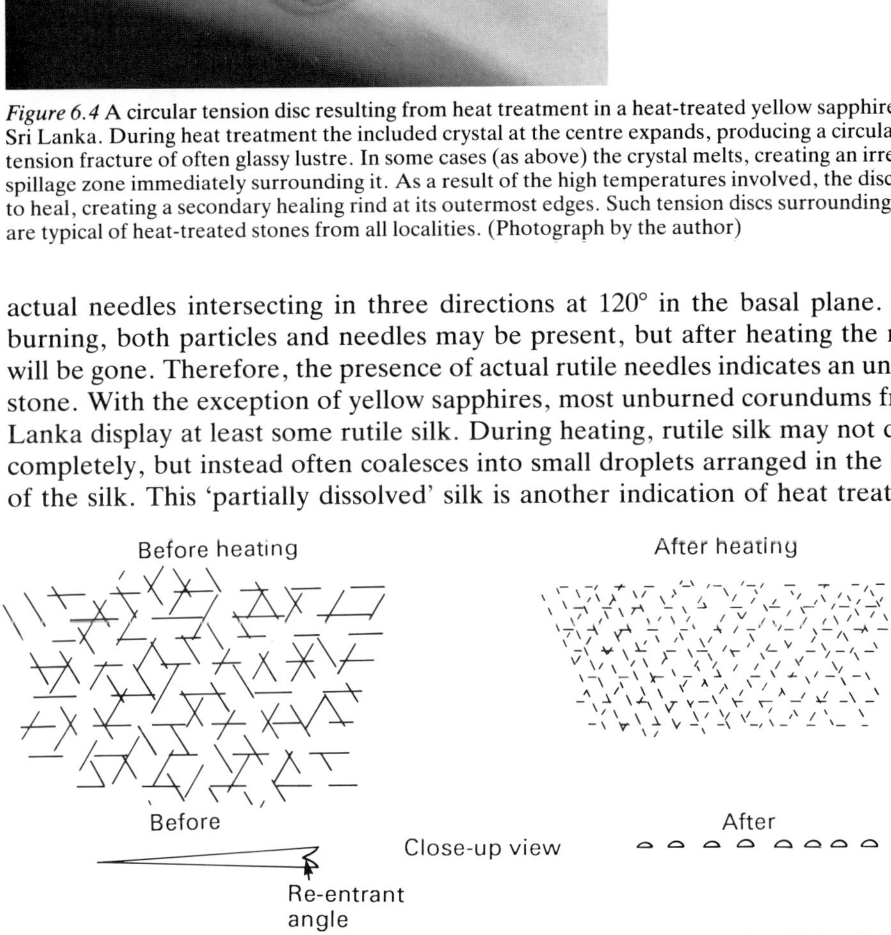

Figure 6.5 The effect of high-temperature heat treatment on rutile silk in corundum. Before heating the rutile silk consists of needles which, when highly magnified, are often tiny 'arrow'- or 'dart'-shaped twins with small re-entrant angles at the wide end. The rutile needles are extremely thin in cross-section and may be flattened in the basal plane. Heat treatment causes the rutile to melt. Some of it may be absorbed by the corundum (thus clarifying the gem), but traces usually remain behind, visible under fibre optic lighting. Rather than coalescing into one globule, each needle melts into a series of tiny droplets arranged in the same pattern as before heating. This 'partially dissolved silk' may be indicative of high-temperature heat treatment. However, minute exsolved particles which occur naturally may be confused with the partially dissolved silk. The presence of long needles and dart- and arrow-shaped rutile is a strong indication that the gem has *not* undergone high-temperature heat treatment

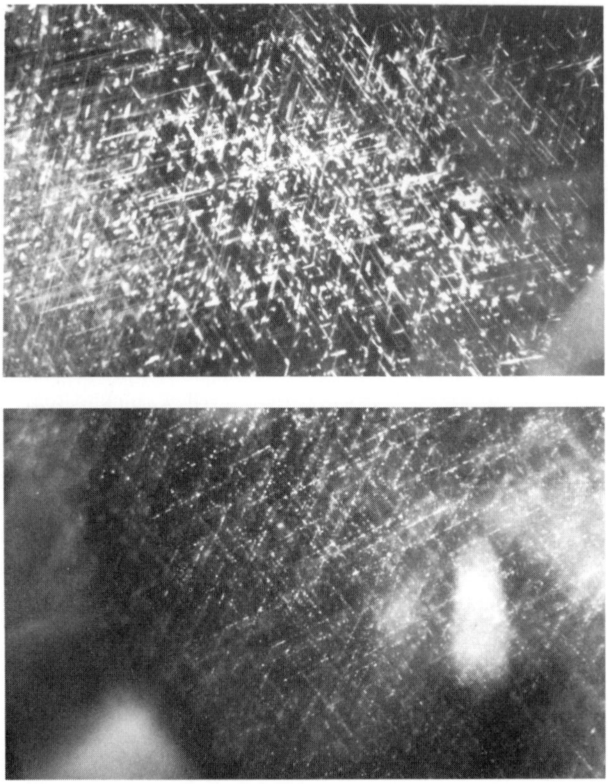

Figure 6.6 Upper photograph: Rutile silk in an untreated Burmese ruby. Lower photograph: Rutile silk in a Burmese ruby after high-temperature heat treatment. After heat treatment the individual needles have been partially absorbed by the gem, leaving behind little droplets arranged in a similar pattern. (Photographs by the author)

Fluorescent reactions can also be of use in identifying heat-treated Sri Lankan blue sapphires. Untreated stones are generally inert, with some chromium-bearing stones showing a weak to moderate orange-red colour, a reaction slightly stronger in long-wave than short-wave light. In contrast, many heated gems display a fluorescence much akin to the Verneuil-produced synthetic blue sapphires. This is a weak to strong chalky-blue-green glow under short-wave light only. Magnification using ultraviolet light reveals this reaction to be associated with the colour zoning of the gem, again like the synthetic. It is the colourless portions of the gem which fluoresce, while the blue areas are inert, and so the fluorescence is strongest in gems with large colourless areas. Should chromium also be present, orange-red fluorescent areas may be seen as well. The chalky-blue-green fluorescence has not been observed in untreated Sri Lankan blue sapphires and so is diagnostic for heated gems.

Although the temperatures used for the treatment of Thai/Cambodian rubies are, at times, less than those for Sri Lankan sapphires, they still produce changes that can identify the gems as treated. Only one major factor of use in identification has been found to date, but fortunately it is present in most stones. This telltale sign is the alteration of the solid crystal grains which form the centre of the so-called 'saturn' inclusion.

A quick glance through the section on Thai rubies in Eduard Gübelin's classic book, *The Internal World of Gemstones*, or the *Photoatlas of Inclusions in Gemstones* by Gübelin and Koivula, will reveal the types of crystals so typical of the untreated stones of this locality. Opaque brown grains of garnet, yellowish apatites, metallic black pyrrhotites and, more rarely, colourless plagioclase feldspars are all to be found. However, these distinctive features are rarely present in the Thai rubies circulating today. Heating melts the crystals, leaving behind high-relief negative crystals filled with a colourless or, more rarely, a brownish-stained glass. These glass-filled voids often contain a gas bubble, frozen in place by the solidified remains of the melted crystal. So common has the heating of Thai rubies become that the present author cannot recall having seen more than a handful of unburned rubies with the solid crystals still intact over the last five years. Approximately seven out of ten stones will display these glass-filled negative crystals, proving that they were heat treated. Untreated Thai rubies also typically show rusty-red stains in their cracks. Heat treatment removes these stains, which result from the red soil in which the gems are mined.

Rubies from other localities are also heated in large amounts, but their separation from untreated gems is more difficult. The lack of rutile silk in Burmese and Sri Lankan stones is generally a strong, but not absolutely positive, indication of heat enhancement. Unusual fluorescence in gems from Burma, Sri Lanka and Kenya is another. Under short-wave light, chalky-whitish zones may be seen in the colourless areas, together with the red fluorescent areas. Heated corundums from all localities may display internal and surface characteristics previously described for treated blue and yellow sapphires from Sri Lanka.

Previously, heat-treated Burmese rubies were an uncommon occurrence. However, starting in about 1984, quantities of heated Burmese rubies came onto the market and many of the Burmese rubies available today have been heated to remove silk. This burned material contains inclusions that are very different from those in untreated material. Fingerprints are normally an uncommon occurrence in Burmese rubies. After heating, however, the stones may be filled with fingerprints and feathers, and overall bear a strong resemblance to the treated rubies from Kenya. The rutile silk, such a distinctive feature of unburned Burmese rubies, is gone. In place of the knife-shaped needles are little droplets of partially dissolved silk, similar to those seen in heated Sri Lankan corundums. Thai burners have not had the same success with Burmese blue sapphires, which are not presently treated on a large scale. However, given a little time . . ..

One final word on the detection of heat-treated geuda blue sapphires from Sri Lanka: Olle Fjordgren (1986) of Sweden has reported that infrared photography can be used to detect heat treatment in certain cases. By photographing a suspect stone under infrared light with an infrared film, differences can be seen in the developed photographs between untreated and heat-treated geuda blue sapphires. The treated gems are said to appear black or blackish, while untreated stones tend to look rather reddish to violetish on the infrared film. This report has yet to be confirmed, however, by other gemmologists.

Surface diffusion

To the layperson, gem materials such as ruby and sapphire appear to be absolutely solid impregnable fortresses. However, this is but an illusion, for if an object is of a small enough size, as with certain atoms, it can pass through a gem as easily as sand through one's fingers. The process whereby atoms may move through a material is termed diffusion and is responsible for a number of the different changes in corundum subjected to heat treatment.

When a crystal is heated, interatomic spacing increases by an amount proportional to the temperature reached, making it possible for atoms to move, or diffuse, through the structure. Atoms of a small size, such as hydrogen or oxygen, diffuse rapidly through corundum. Thus, colour changes involving these elements can be effected throughout the entire gem in as little as a few hours by heating under the proper conditions. However, the diffusion of transition metal colouring agents, such as iron, titanium and chromium, is far slower, so even heating the gem at 1800°C for several days will produce a movement of only half a millimetre or less.

Diffusion of atoms by means of heat treatment has been applied to rubies and sapphires on a commercial basis for the past several years with the aim of imparting a thin layer of colour at, and just below, the surface. The process used is based upon that patented by the Union Carbide Corporation in 1975. Colourless to lightly coloured corundum gems are readied for treatment by preforming and cutting, but not polishing, as this inhibits the diffusion somewhat. They are then embedded in a crucible containing a powder that consists of alumina, titania and metallic oxide colorants. Alumina itself has no effect, while the titania aids surface penetration and lessens the formation of colour bands (Yaverbaum, 1980). The colouring agents used are generally the same as those employed for doping Verneuil synthetic corundum. Now the crucible is heated to between 1600 and 1800°C for up to 200 hours. This allows the atoms within the powder to diffuse into the surface areas, creating a thinly coloured skin which, after a light repolishing, is rarely deeper than a few tenths of a millimetre. Thus, natural rubies and sapphires in which the colour is too light, or uneven in distribution, can be treated by the above method to correct these deficiencies.

Due to the relatively large quantities of clean lightly coloured sapphires available from Sri Lanka, this material is most often used for surface diffusion treatments. Stones from other localities may also be used, but less frequently. In terms of the colours produced, the majority are blues, with some rubies (reds), padparadschas (oranges) and yellows also being treated. When the process is done properly, the colours resulting from surface diffusion can be nearly equal to nature's finest efforts. However, because the colouring agents have been artificially induced into the gems, and because the colour is only at the surface, diffusion-treated stones have not been accepted by the trade in the same way as those involving heat treatments alone. Thus, it is imperative that gemmologists are able to identify these stones.

A visual inspection of diffusion-treated gems yields comparatively little of use with regard to their separation from untreated stones. Deeply coloured pieces may at times appear slightly darker around the girdle, because of the concentrations of colour at the surface. The effect is most pronounced when the stone is immersed in methylene iodide and examined under magnification with diffused light-field illumination. Use of the immersion cell clearly reveals the differences between the treated and untreated gems because it allows light to traverse the gem unrefracted and it eliminates blind spots caused by surface reflection. For an immersed stone, the depth of colour is simply a function of the amount of colour light rays pass through on their way across the gem. Evenly coloured natural gems will appear lightest at the girdle, for this is the thinnest part of the stone. In contrast, diffusion-treated gems appear darkest at the girdle because, with most or all of the colour at the surface, the shape of the stone causes light rays to pass through more colour at the girdle than anywhere else.

Immersion can reveal other important features in the treated stones. Repolishing is necessary after the treatment, but inevitably more material is removed from some facets than others, and so the depth of colour varies from facet to facet. Facet junctions appear highlighted for the same reason, as do surface pits and cavities that are beyond the reach of the polishing wheel. In addition, facet junctions appear darker because greater penetration of the colouring agents is achieved at the corners than in the centre of a facet. Fractures and feathers which break the surface may appear darker as well, due to a 'bleeding' or colour into these openings.

In most cases, immersion will readily identify diffusion-treated stones. However, on occasion, untreated stones are encountered in which the colour is also distributed primarily at the surface. Examples of this are the 'ottu'-type blue sapphires from Sri Lanka, which when uncut possess a blue skin around a colourless core. Stones of this type should be carefully examined while immersed for colour zoning in the surface areas. Although zoning may be found in the treated stones too, it will not be sufficient in quantity to account for the depth of the stone's colour, nor will it be located in the deeply coloured surface regions. In addition, ottu stones do not display colour on every cut surface, like most treated stones. If all of the colour of the treated gem is removed from some facets during repolishing, it can still be identified by means of the highlighted facet junctions and the patchy nature of the remaining coloured areas.

Surface repair

Among the connoisseurs of gemstones, few envy the task faced by the lapidary. The job involves the creation of the most beautiful gem possible and, at the same time, recovering the maximum amount of weight from the rough – an enormously difficult proposition when bearing in mind that each of these ideals can only be achieved at the expense of the other. Thus, success for the lapidary is measured in terms of an ability to strike a delicate balance between beauty and size. When working with costly materials such as ruby and sapphire, particular attention must be given to retention of weight. Because of pressures to squeeze every last point from the rough, cutters may be forced to retain small surface defects, such as pits and cavities. It is simply a case of the lesser of two evils, for although grinding the defects away completely will make the gem more attractive, this improvement is often not enough to offset the loss in weight that results. Thus, the presence of surface pits should not indicate a lack of skill on the part of the cutter, but instead serve as evidence of the difficulties in turning a profit within this highly competitive business.

Competition is of course vital to the well-being of any business, for it is what fuels the search for the better mousetrap. However, not everyone plays by the rules. During the first six months of 1985, gemmologists in both London and Bangkok discovered rubies in which the surface blemishes had been filled in with a transparent colourless glass (Scarratt, 1984; Hughes, 1984a,b). Dubbed 'surface repair', it represented a new method of dealing with an old problem. However, the treatment not only reduces or eliminates the visibility of surface defects, but also increases the weight of the stone at the same time. As a result, it cannot be considered an acceptable treatment, and it is important that the trade is both aware of, and able to detect, this latest charade.

To date, the vast majority of surface-repaired stones encountered have been rubies, with a handful of blue sapphires also turning up. However, there appears to be no reason why gems other than corundum could not be treated in a similar manner. Details regarding the actual method of treatment were recently supplied to the author by someone who had talked to a man performing surface repair on rubies. This gentleman from Chanthaburi was primarily involved in the heat treatment of yellow sapphires from Sri Lanka and claimed to be the first person in Thailand to treat rubies by surface repair. According to him, surface repair involves applying a silica-based gel to the surface of cut stones with a paint brush in the areas to be repaired. The stones are then heat treated, turning the gel to glass and fusing it into the pits and cavities in the process. He claimed that his surface repair jobs were superior to all others as the glass adhered tightly to the surface and did not fall out like those of other burners.

It should be mentioned that the idea of filling in surface cavities with a foreign substance is not totally new to the trade. In the past, a number of different materials including epoxy and doping shellac have been used to hide unsightly defects, particularly on cabochon-cut stones. Patents have also been taken out on processes very similar to surface repair, such as those for the glossing of corundum and spinel rods mentioned by L. H. Yaverbaum (1980) in his compilation of gemstone patents.

The detection of surface-repaired rubies and sapphires can be readily accomplished by noting the following factors:

1. The difference in lustre between the glass filling and the surrounding corundum under an overhead light.
2. The difference in relief between the glass (refractive index $\approx$ 1.52) and the corundum (refractive index = 1.762–1.770) when immersed in methylene iodide.
3. Inclusions of spherical gas bubbles within the glass filling.
4. The colourless nature of the glass filling.

Initially, the surface of the suspect stone should be examined under magnification using an overhead light source. Repaired areas will be visible because of their different lustre when the stone and light source are positioned such that light falling on the surface is totally reflected to the eyes. This discrepancy in lustre is due to differences in refractive index and polish between the glass filling and the surrounding corundum.

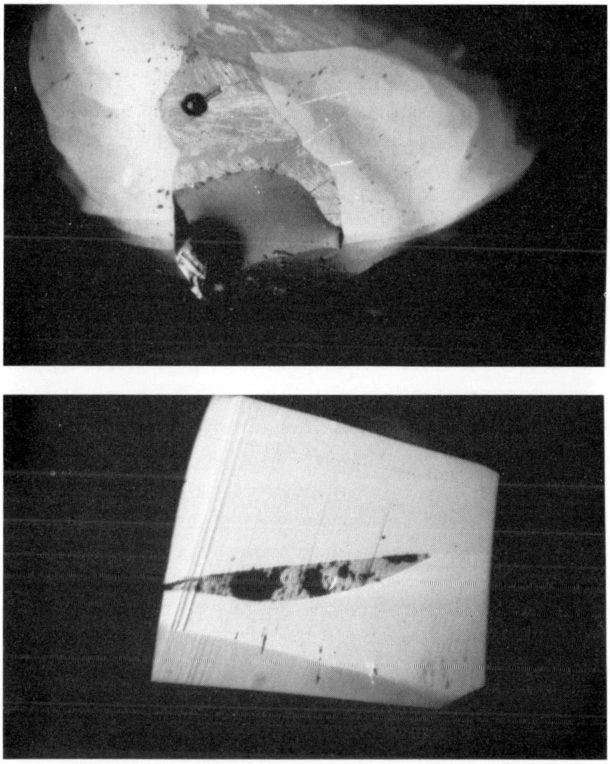

Figure 6.7 Surface-repaired (glass-infilled) rubies display the glass filling as areas of lower lustre compared with the surrounding corundum. Overhead lighting highlights the lower lustre of the glass. The black circular areas indicate gas bubbles in the glass which were cut through during cutting. (Photographs by the author)

Should a surface examination fail to reveal repaired areas, immersion in methylene iodide is necessary. The difference between the refractive indices of the glass filling and the gem causes repaired areas to stand out in bold relief at certain oblique angles of viewing. Particular attention should be given to the girdle, culet and pavilion, for it is these areas that most often possess cavities. All open cavities must also be checked, for if the glass does not fill them completely, as is often the case, they may go undetected in the initial dry examination.

The colourless nature of the glass filling is not normally apparent in surface-repaired gems. Only if a repaired area completely penetrates the girdle or culet can this be seen, and even in this event the gem must be immersed. Gas bubbles within the glass, however, should be visible during both dry (dark-field illumination) and wet (diffuse light-field illumination) examinations. In addition, many repaired areas exhibit spherical cavities formed as a result of included gas bubbles breaking the surface.

Since the original discovery was made, the author has encountered a number of repaired stones in which the repaired areas constitute only a tiny fraction of the entire gem, having virtually no effect on its weight or appearance. This fact raises questions with regard to the actual motive behind the treatment. The only possible explanation is either that the filling occurred accidentally during heat treatment or that these stones may have been part of a larger lot which was treated *en masse* while still in the rough state. Therefore, after treatment and cutting, the gems would show varying amounts of repaired surfaces, some more and others less. Rubies and sapphires are not generally heat treated after cutting, but ideally before cutting. This allows the cutter to make better use of the material, for by cutting the gem after heating the lapidary can remove flawed areas caused by the burning or, conversely, incorporate areas of fine colour and/or clarity which might be removed if the cutting was performed first. Finally, if the surface repair was applied to cut stones only, it would then be extremely difficult to explain how such large repaired areas (as have often been found) were produced, because it is not often that a cutter will leave large gaping cavities in prominent areas on a finished gem.

One final note on surface-repaired stones concerns the separation of repaired areas from included crystals which sometimes break the surface. Since foreign inclusions of a solid nature may also show gas bubbles and different lustre and relief when breaking the surface, confusion is possible. In cases such as this it is best simply to indicate the presence and extent of the filled areas, leaving open the question of whether or not it resulted from natural causes, was an accidental by-product of heat treatment, or resulted from the intentional filling of surface defects.

Irradiation

The irradiation of the corundum gems involves their exposure to one of three principal types of electromagnetic radiation: short-wave ultraviolet (253.7 nm, 5 eV), x-rays (10 000 eV) and γ-rays (1 000 000 eV). This bombardment causes, in suitable varieties, formation of colour centres and a subsequent change in colour. Although other types of radiation will produce similar results, it is those above that are most often employed with corundum.

The first documented experiments on the effects of the irradiation of corundum were those of Pough and Rogers in 1947. While studying the colour changes produced in gems when bombarded with x-rays from a new high-intensity unit, they were completely taken aback by the metamorphosis of Sri Lankan sapphires. After only 5–10 minutes exposure, the pale coloured or even completely colourless gems deepened to a rich amber hue. Less dramatic were the changes of more deeply coloured varieties. Blue sapphires turned a muddy green, but the light-lilac stones acquired a pinkish-orange colour reminiscent of that most desirable of sapphires, the padparadscha. Alas, the pipe dream was ended when it was discovered that the radiation-induced colours were unstable, fading rapidly upon exposure to either heat or light.

Studies of irradiation-induced colours in minerals have determined that the changes usually result from 'colour centres'. Irradiation can, in certain specimens, dislodge one of a pair of electrons from a suitable atom, ion, molecule, impurity or other defect to create a 'hole centre'. This electron is then trapped by a similar atom, ion, etc. to form an 'electron centre'. If both centres were originally neutral, the transfer of this electron produces a positive charge in the hole centre and a negative charge in the electron centre, leaving a single unpaired electron in each. Either or both unpaired electrons can now become excited by absorbing energy from white light, thus creating colour. When the hole centre does this, a 'hole colour centre' is present, and if the electron centre is responsible an 'electron colour centre' is present (Nassau, 1984). Currently, further study remains to be done before the exact nature of the colour centres in sapphire can be fully understood.

The colour centres of some minerals, for example amethyst and blue topaz, are deep enough to be completely stable to light. Others, as in Maxixe beryl, may fade because visible light or mild heating gives enough energy to free the trapped electron and return it to its original site. In the yellow sapphires from Sri Lanka, both the fading and non-fading type of colour centres can be present. The irradiation of pale-yellow or colourless sapphires from Sri Lanka generally produces an intensified yellow colour which fades rapidly upon exposure to light or heat. This is the fading colour centre. Reliable sources in Sri Lanka have also told of yellow sapphires which faded soon after being dug from the gem pits. Yet the colour of other deep-yellow stones examined by the author has remained intact even after long exposures to the heat and light of a 150 W spotlight. Apparently these naturally yellow sapphires owe part or all of their colour to non-fading colour centres produced by relatively low doses of radiation from natural sources over extended periods of geological time.

If heated to a sufficient temperature, the non-fading colour centres can be bleached just like the fading type, but more than 500°C is required, compared with 200°C for the unstable kind. In both cases the colour can be restored by irradiation if the colour centres have not been totally destroyed by overheating. Dr Kurt Nassau (1982), of Bell Laboratories, has reported cases where the stable colour centres were accidentally bleached by heating. At his suggestion the sapphires were irradiated, burning them a deep brownish yellow that then faded back to the original stable medium yellow. Thus, in this case the irradiation created not only the non-fading colour centre, but also the fading colour centre. It was this unstable colour centre that was bleached by the light.

Kurt Nassau and Kay Valente (1987) recently completed a detailed study of yellow to orange sapphires. The results of their study suggest that there exist seven different types of yellow/orange sapphires, as follows (typical sources are in parentheses):

Type 1: natural stable colour centre (Sri Lanka)
Type 2: naturally or laboratory-irradiated fading colour centre (Sri Lanka)

Type 3: natural iron-produced stable colour (Thailand, Australia, Tanzania)
Type 4: heat-treated stable colour (Sri Lanka)
Type 5: surface-diffused additive colour
Type 6: synthetic with impurity-caused colour (Verneuil, nickel doped)
Type 7: synthetic with laboratory-irradiated fading colour centre (Verneuil)

One unusual fact which they discovered was that in type 1 sapphires the colour could be altered by either heat or irradiation. Heat (1 hour at 150–600°C) produced a fading of the colour, while irradiation deepened it. However, in both cases the exposure of the stone to light (typically 2 days in the bright California sun, or longer under less bright conditions) caused the colour to return to its original stable state. The behaviour of both type 1 and type 2 yellow/orange sapphires is shown below (after Nassau and Valente, 1987):

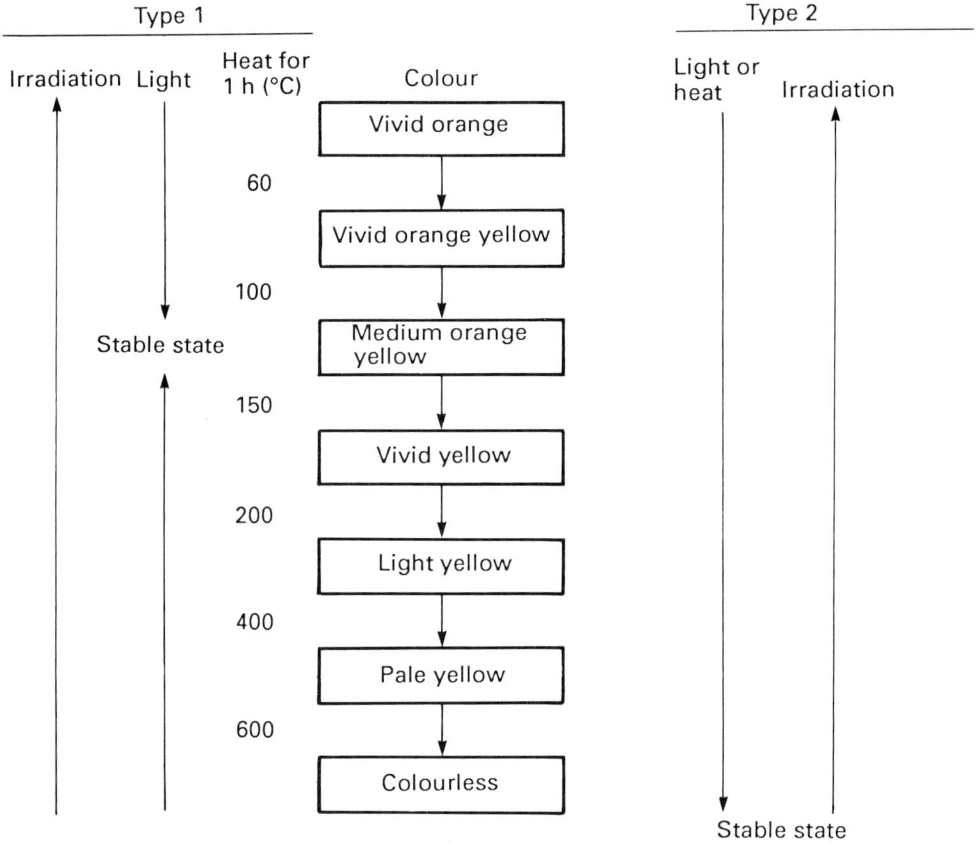

(a)

Of the various radiation sources commonly used for treating sapphires, by far the most effective is the γ-ray cell. This consists of a large lead shield surrounding a central cavity within which are rods, plates or pellets of a highly radioactive isotope, typically cobalt-60. Since γ-rays are 100 times more energetic than x-rays and 200 000 times more energetic than short-wave ultraviolet radiation, they can induce a deep golden-yellow colour into a pale Sri Lankan sapphire in a matter of minutes. Furthermore, the γ-ray-induced colour penetrates the gem from top to bottom, whereas that induced by x-rays or short-wave ultraviolet radiation is largely confined to the surface regions. Indeed, at Colombo Hospital, home of Sri Lanka's only cobalt-60 unit, a steady stream of customers can be found on any given day, their pockets bulging with 'patients' awaiting therapy.

Typical of the unpredictable manner in which mother nature hides her treasures, only sapphires from Sri Lanka are susceptible to irradiation. Gems found in other parts of the world apparently lack the necessary impurities or defects that produce the colour centres. Even amongst Sri Lankan sapphires, not all will respond. Best of all are the colourless or pale-yellow stones, which may turn a rich golden yellow. Gems of a pink colour, produced by traces of chromium, sometimes change to the pinkish-orange padparadscha hue. Robert Stevenson, a dealer from Bangkok, told once of being shown a truly extraordinary gem by a broker in Colombo – a star padparadscha! Naturally suspicious of such a rarity, he reached for his butane cigarette lighter and, with a quick flick, struck the flame. A rather drastic course of action, to be sure, but also quite telling, for it wasn't long before the expected metamorphosis occurred. Under the heat of the flame, this once lovely flower wilted like a dandelion in the midday sun, leaving the broker looking like a boy who had just found a lump of coal in his Christmas stocking. He picked up his fallen beauty and quickly exited, mumbling something about an appointment at the hospital.

Unfortunately, the 'fade test' is the only certain method of identifying radiation-induced colour in sapphires. It can be performed by gently heating the stone with a cigarette lighter or other similar tool, as above, but this runs the risk of damaging the gem through thermal shock. A safer technique involves placing the gem on a metal platform within 1 cm of a 150 W spotlight. Exposure to the heat and light of such a bulb will generally produce some fading in irradiated gems within 1 hour or less. The colour of untreated gems is generally not affected, while heat-treated specimens suffer a temporary darkening, as described previously. Contrary to some published reports, a fade test should not be attempted by exposure to ultraviolet light, as short-wave ultraviolet radiation, in particular, may deepen the colour.

When short-wave ultraviolet light is used to irradiate a specimen, only the surface regions are affected because of the less energetic nature of the medium. This imparts a colour which is but 'skin deep', causing the gem to appear slightly darker around the girdle. The same effect can be seen when x-rays are used, but to a lesser degree, because they achieve greater penetration. γ-rays are preferable to both for they can colour a stone from top to bottom, affecting not just the surface, but the entire piece.

Dyes and oils

Contrary to the many reports that have appeared recently in Western trade publications describing coloured oils for the dyeing of ruby and sapphire, these treatments are of little importance. The reason for this is that few cut rubies and sapphires possess the network of fine cracks necessary for the introduction of dyes and/or oils.

It is true that many lapidary supply shops in the gem-trading town of Chanthaburi, Thailand, sell a variety of local-brand 'ruby oils' and other gem elixirs guaranteed to put a smile on your ruby. These are not normally used with faceted stones, however. Instead, these ruby-red oils are applied to rough stones not so much to influence the colour, which in the case of Thai ruby is already even and intense, but to add sparkle and shine to the otherwise dull unpolished surfaces.

The application of red oil to rough ruby in Thailand is generally performed simply by pouring it on the stones at room temperature. Again, this is consistent with the aim to improve the surface lustre rather than to impart added colour. At the early morning rough markets held throughout Thailand's ruby-mining districts of Chanthaburi and Trat, it is not uncommon to see traders putting their stones upon the rack for a quick 'lube job', and bottles of ruby oil are to be found atop many of the tables, next to the small piles of uncut ruby. The oiling of rubies in this manner has very little to commend it, since the oil rarely penetrates the fractures and so is, for the most part, removed during cutting. If it is feared that the oil is affecting a gem's appearance, soaking the stone in alcohol, acetone or other suitable solvent should resolve any doubts about its true colour.

In rare cases rubies have been examined in which the oil did penetrate the fractures. This was clearly visible under magnification as dark-red stains in the cracks. Pale-coloured rubies and sapphires from India and Sri Lanka which are heavily included are, at times, dyed with a coloured oil to deepen the colour. Immersion in methylene iodide will uncover this fraud, as will examination under ultraviolet light if the dye fluoresces. One final point with regard to dyed stones concerns a heat-treated blue sapphire from Sri Lanka encountered by the author. The bottom one-third of the pavilion had been coated with a blue substance resembling ink. Although easily detectable, even with a 10× loupe, this crude fraud could fool the unwary, particularly if the stone were to be mounted. Regarding the detection of dyeing in heavily included stones, careful use of a hot point will often resolve the issue. Under magnification, when the hot point is applied close to a fracture at the stone's surface, the dye will be seen to move and even drip out.

On occasion, stones have been encountered which possess cracks containing coloured stains of apparently natural origin. At other times, the origin of the stains may be uncertain. Regardless of origin, if the stain does influence the colour of the gem, it is important that the buyer be informed, both of its existence and the extent to which the gem's colour is affected.

Disclosure of treatments – pros and cons

To tell or not to tell. That is the question which of late has been haunting many of those involved in the ruby and sapphire business. It is a matter of great importance indeed, with fuel for the fire being provided in the United States by legislation now pending in at least one state house to make disclosure of all treatments a requirement by law. Clearly, in the United States at least, the handwriting is on the wall. If the trade does not act quickly on the disclosure issue, the government will do it for them.

There is general agreement that non-permanent treatments such as irradiation, surface diffusion, surface repair, oiling and dyeing should be disclosed. However, many would draw the line at permanent treatments involving heat. Opponents to the disclosure of permanent gemstone treatments feel that it is unnecessary and will only confuse the customer, eventually hurting business. They believe that so long as a treatment is permanent there is nothing to worry about; the whole controversy is so much needless fuss generated by a group of consumer witch hunters.

Is it really, though? The most important factor in maintaining the value of precious stones is the rarity. To see that this is true we need only look to the lowly flame fusion synthetic corundum. A Verneuil-produced synthetic ruby is a ruby in every sense of the word, with the same hardness, durability, colour permanence and beauty as nature's own. However, because it is produced in almost limitless quantities, its cost is but pennies per carat. The major factor separating it in price from the natural stone is rarity, and should deposits of natural ruby be discovered which yield quantities similar to the synthetic production, we could expect that its price would also drop to a similar level. Fortunately for those who own or deal in natural rubies, this possibility is remote, at best.

Although many would deny it, the additional quantities of ruby and sapphire produced by treating inferior material also have an impact on supply and therefore price. However, since the amount of treatable rough is limited, this impact is also limited. The heat treatment of yellow and orange sapphires from Sri Lanka was not widely practised before 1981. Before this, a fine, naturally coloured, deep-yellow/orange sapphire from Sri Lanka of 5 carats fetched $400 a carat, or more, at the wholesale level. Today, the same stone would bring only a fraction of that amount because it must compete against heat-treated Sri Lankan sapphires. It is unfair competition, to say the least, for the treatment process yields deep-orangy-yellow colours with unnerving frequency, colours that were almost unheard of before 1981. If, however, a distinction were to be made between natural and treated gems, it could be sold for what it is: a rare, naturally coloured, deep-yellow Sri Lankan sapphire.

Many of those involved in treating or selling treated rubies and sapphires fear that prices for these stones would fall should they be sold with a 'colour-enhanced' label attached. This seems unlikely, as supplies of treatable rough are limited. Studies undertaken in the United States by the trade magazine *Jeweler's Circular–Keystone* have also shown that if consumers are offered a choice of either a naturally coloured sapphire or a heat-treated sapphire of similar quality but lower in price, many will opt for the treated stone. Thus, the trade need not fear a drop in the prices of heat-treated ruby and sapphire from the disclosure of such treatment. Instead, the entire gem and jewellery business will benefit from the increase in value of naturally coloured gems. Perhaps of even greater importance, by the total disclosure of gem treatment jewellers will no longer have to incur the wrath of a customer who learns that their new ruby or sapphire has more likely than not been heated to improve the colour. The increased confidence in jewellers brought about by the disclosure of treatments can only help boost sales in the long run.

There exist many persuasive arguments which are used by those who are against disclosure. Among the most often heard are:

1. Heat treatment merely duplicates nature, acting upon the potential already existing within the stone.
2. Had the stone remained in the ground longer, it could have changed all by itself.
3. The colours produced are permanent and will not fade with time.
4. Heat treatment is traditional and has been done for centuries.
5. In heat treating corundums nothing is added to the stones; therefore such treatments are acceptable.

In the opinion of the author, each of these arguments is flawed, but this hardly matters as they are but a smokescreen designed to deflect enquiry away from the true reason – money.

Regarding heat being a duplication of nature, well didn't Verneuil and Chatham duplicate nature in their corundum synthesis processes, by merely acting upon the potential of the alumina powder? Graphite has the potential to become diamond, but when man causes the change we do not consider it natural. As for the stone changing colour if it were to remain in the ground, well this ranks with the beliefs of the past that blue sapphires are merely unripened rubies. Any further comment would be pointless. Regarding the permanence of the colours produced, well so are the colours of irradiated blue diamonds, but we don't sell them for the same price as natural blue diamonds. As for heat treatment being traditional, having been done for centuries, I would reply that corruption in high places is also traditional. Should one then support it on that basis? Finally, with regard to the idea that nothing is added to the stone during heat treatment, this is also untrue. Oxygen is being added or removed by a diffusion mechanism, as shown in Figure 6.8. Without such a diffusion of oxygen into or out of the corundum lattice, many of the heat treatments currently used for ruby and sapphire would be impossible.

Members of the gem trade often discuss the impact which treatment disclosure will have on the trade. What about the consumer? We should not be discussing what is good for the trade but, instead, what is good for the consumer. What is good for the consumer is, in the end, good for the trade. We must put ourselves in

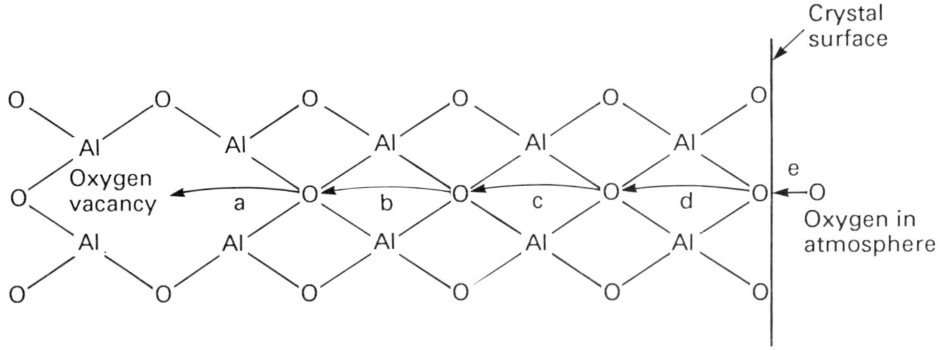

Figure 6.8 The diffusion of oxygen from the atmosphere to fill an oxygen vacancy in corundum. O, oxygen; Al, aluminium. (After Nassau, 1984)

the shoes of the consumer. What would we want to know about the stone if we were considering its purchase? I, for one, would certainly want to know if the colour and clarity were creations of nature, or if they were synthesized (created) by man.

I do want to stress that I personally see nothing wrong with the heat treatment itself, or any other treatment for that matter. I do believe, however, that it is wrong to sell a heat-treated stone as completely natural, with no mention of the treatment. Because the colour and clarity of a gemstone have such an important bearing on its quality, with a treated gem it is incorrect to imply, by the silence of non-disclosure, that it is a creation of nature. Man will always be able to produce by treatment rubies and sapphires more beautiful than untreated stones. If no distinction is made, we run the risk of ruining the ruby and sapphire market in way similar to that of natural pearls. Today, laypeople cannot appreciate genuine pearls because they are so used to seeing cultured pearls and have been told that they are essentially the same as the natural ones. When a natural pearl is shown to laypeople nowadays, they want to know why it is not round! Why isn't the surface smooth like normal (i.e. cultured) pearls? Why is there just one instead of an entire strand? A refusal to disclose treatments will lead laypeople to come to accept the treated stones as natural. As technology improves, the treatments will also improve, releasing larger and larger quantities of bigger and better stones onto the market, in effect pushing aside the more lacklustre genuine article. This is already happening. How can the products of nature possibly compete with man's cosmetic facelifts? Only by the disclosure of *all* treatments beyond ordinary cutting and polishing (which is obvious, even to the layperson, and is not the subject of deception) can the natural stone compete. In cases such as with aquamarine, where the treatment cannot be detected, but it is common knowledge that most stones are treated, the customer should be told that it is probably treated. Failure to act could mean the disappearance of natural stones from the market. After all, if, as those who oppose disclosure claim, there is nothing wrong with treating gems, then there should be nothing to fear by their disclosure to the public.

Bibliography

Abraham, J. S. D. (1982) Heat treating corundum: the Bangkok operation. *Gems and Gemology*, **18**, 79–82

Anonymous (no date) New gem locations and the present treatment of gemstones. *Journal of Gemmology*, 265–269

Carr, R. R. and Nisevich, S. D. (1975) Altering the appearance of corundum crystals. *US Patent* no. 3,897,529

Carr, R. R. and Nisevich, S. D. (1976) Altering the appearance of corundum crystals. *US Patent* no. 3,950,596

Carr, R. R. and Nisevich, S. D. (1977) Altering the appearance of corundum crystals. *US Patent* no. 4,039,726

Coldham, T. (1985) Sapphires from Australia. *Gems and Gemology,* **21**, 130–146

Crowningshield, R. and Nassau, K. (1981) The heat and diffusion treatment of natural and synthetic sapphires. *Journal of Gemmology,* **17**, 528–541

Crowningshield, R. and Nassau, K. (1982) The heat and diffusion treatment of natural and synthetic sapphires. *International Gemological Symposium*, pp. 101–109

Fjordgren, O. (1984) Heat-treatment and its effects on sapphires. *Retail Jeweller*

Fjordgren, O. (1986) Infrared detective. *Lapidary Journal,* **39**, no. 12, 39–41

Gübelin, E. J. (1983) Identification of the new synthetic and treated sapphires. *Journal of Gemmology,* **18**, 677–706

Gunawardene, M. (1984) Contribution towards the identification of treated corundums: heat and diffusion treated rubies. *Journal of Gemmology,* **19**, 298–310

Hänni, H. A. (1982) Characteristics of heat-treated and diffusion-treated corundums. *Swiss Watch and Jewelry Journal,* **5/82**, 573–577

Hideki, O. (1982) Improvement of color of sapphire and its identification. *Proceedings of the First International Colored Gemstones Conference and Gem and Jewellery Exhibition*, pp. 1–9

Hughes, R. W. (1984a) Surface repaired rubies. *The Australian Gemmologist,* **15**, 279–280

Hughes, R. W. (1984b) Surface repaired rubies – a new gem treatment. *Journal of Gemmology,* **19**, 384–386

Jobbins, E. A. (1971) Heat-treatment of pale blue sapphire from Malawi. *Journal of Gemmology,* **12**, 342–343

Kane, R. E. (1984) Natural rubies with glass-filled cavities. *Gems and Gemology,* **20**, 187–199

Keller, P. C. (1982) The Chanthaburi–Trat gem field, Thailand. *Gems and Gemology,* **18**, 186–196

Nassau, K. (1980) *Gems Made By Man*, Chilton, Radnor, PA

Nassau, K. (1981) Heat treating ruby and sapphire: technical aspects. *Gems and Gemology,* **17**, 121–131

Nassau, K. (1982) Color enhancement of gemstones: heat treatment of corundum. *International Gemological Symposium*, pp. 111–117

Nassau, K. (1983) *The Physics and Chemistry of Color*, Wiley, New York

Nassau, K. (1984a) *Gemstone Enhancement*, Butterworths, London

Nassau, K. (1984b) Heat treatment used on gemstone materials, parts I, II. *Lapidary Journal,* **38**, no. 1, 18–24; **38**, no. 2, 292–299

Nassau, K. (1984c) The early history of gemstone treatments. *Gems and Gemology,* **20**, 22–33

Nassau, K. (1985a) The early history of gemstone treatments, part II. *Lapidary Journal,* **38**, no. 12, 1532–1541

Nassau, K. (1985b) Miscellaneous gemstone treatments, parts I, II. *Lapidary Journal,* **39**, no. 2, 254–262; **39**, no. 3, 30–40

Nassau, K. (1985c) Gemstone enhancement. *Lapidary Journal,* **39**, no. 4, 16–22

Nassau, K. and Valente, K. (1987) The seven types of yellow sapphire and their stability to light. *Gems and Gemology,* **23**, 222–231

Pough, F. H. (1947) Experiments in x-ray irradiation of gem stones. *The American Mineralogist,* **32**, 31–43

Rossman, G. (1982) Irradiation of colored gemstones. *International Gemological Symposium*, pp. 94–99

Scarratt, K. (1983) Heat-treated sapphires. *Retail Jeweller,* **22**, no. 543, 16–17

Scarratt, K. (1984) Notes from the laboratory. *Journal of Gemmology,* **19**, 98–124

Scarratt, K. and Harding, R. R. (1984) Glass infilling of cavities in natural ruby. *Journal of Gemmology,* **19**, 293–297

Schmetzer, K., Bosshart, G. and Hanni, H. A. (1983) Naturally-colored and treated yellow and orange-brown sapphires. *Journal of Gemmology,* **18**, 607–622

Tennent, J. E. (1977) *Ceylon – An Account of the Island Physical, Historical and Topographical*, Tisara Prakasakayo, Dehiwala, Sri Lanka (reprint of the 1859 edn)

Voynick, S. M. (1985) *The Great American Sapphire*, Mountain Press, Missoula, MT

Yaverbaum, L. H. (1980) *Synthetic Gems Production Techniques*, Noyes Data Corp., Park Ridge, NJ

Chapter 7

Synthetic corundums and assembled stones: their manufacture and identification

History

The use of rocks and minerals as objects of adornment stretches back in time more than 40 000 years, predating even the mining of ores and coal. Chosen primarily for their vivid colours and rarity, gemstones have played a significant role in commerce and culture throughout history and have long been coveted as the most valuable of all of the earth's products.

Precious stones were of much greater importance to trade in ancient times than today. Due to the small carrying capacity of the available transportation in early times, it was necessary that the goods being carried should be compact and of high value. Thus, in the beginning, international commerce consisted largely of silks, spices, gems and other products of a similar nature (Ball, 1950).

One of the oldest trade routes to develop around the traffic in precious stones was that between the lapis lazuli mines of Afghanistan and the Egyptian empire. The rich blue hue of lapis lazuli made this especially popular in Egypt, and, as the only known source was so far distant, demand often exceeded available supplies. Moving in to take advantage of this situation, merchants began to sell steatite, a relatively worthless mineral that was glazed blue so as to resemble lapis lazuli (Elwell, 1979).

In the modern sense of the word, this glazed-blue steatite would be considered an imitation instead of a synthetic mineral, as only the appearance of the lapis lazuli was reproduced. It was, however, significant, for it was the first documented attempt by humans to manufacture the mineral product they needed and thus marked the birth of the synthetic gem industry.

Today, the term 'synthetic' is properly used only in describing a man-made mineral which duplicates exactly the chemical composition, crystal structure, properties and appearance of the gem it represents. It was not until the nineteenth century that man's knowledge of the chemical and structural make-up of minerals had advanced to a sufficient degree to allow for the true synthesis of gemstones. Corundum was among the first to be duplicated in the laboratory, in the year 1837, by a French chemist named Marc Gaudin. However, because of a slight difference in specific gravity between his product and the natural stone, Gaudin concluded that his experiments had failed. Finally, in 1870, after many more years of work, he abandoned his experiments, still unable to reach his goal (Nassau, 1980).

Gaudin's research may not have produced the desired result, but it did prompt others into taking up the challenge. One of these was Dr Edmond Fremy (1814–1894), professor of chemistry and head of the chemistry laboratory at the Museum of Natural History in Paris.

The technique employed by Fremy involved dissolving a mixture of alumina and potassium dichromate in a type of solvent known as a 'flux'. Through a chemical reaction, the flux causes this raw ruby mixture to dissolve at a temperature far below its normal melting point of 2050°C. Then, either by evaporation of the flux, or by slowly lowering the temperature, crystals of ruby precipitate out.

Fremy eventually settled upon a flux of potassium hydroxide and barium fluoride, which was dissolved together with the raw materials for ruby in a ceramic crucible at 1500°C. Resulting from this were rhombohedral ruby crystals of good clarity, in sizes up to 0.30 carats. Some of Fremy's synthetic rubies were actually mounted in jewellery (both rough and cut stones) while others were employed as watch bearings, but their small size prevented his work from being a commercial success. Plans were eventually drawn up for experiments aimed at producing larger stones, but Fremy died before these could be performed. In 1891, three years before his death, Fremy published a book, *Synthese du Rubis*, containing illustrations of his synthetic rubies and a summary of his research.

Throughout his work on the synthesis of ruby Fremy was ably assisted by a number of young men. Chief among these was Auguste Verneuil (1856–1913), the man credited with being the first to develop a commercially viable process for the synthesis of ruby (Nassau, 1980). The method which was developed by Verneuil and which bears his name was truly revolutionary. This is borne out by the fact that even some 90 years after its initial development, the Verneuil process remains largely unchanged and today is responsible for over 90% of all synthetic corundum produced worldwide.

Auguste Verneuil's interest in chemistry grew out of his employment in his father's photographic studio. In 1876, at just 17 years of age, he applied and was accepted for the position of laboratory assistant at the Paris Museum of Natural History, under Edmond Fremy. It was during this collaboration with Fremy that an incident reportedly took place which was to completely alter Verneuil's ideas about the synthesis of ruby. Prior to this, all of their experiments involved the growth of ruby within a flux bath. Then, in 1886, a colleague named P. M. E. Jannettaz supposedly showed Verneuil a sample of a new type of synthetic ruby which had been circulating in the market during the previous year. Dubbed 'Geneva rubies' after the Swiss city in which a priest was said to be producing them, their appearance in European gem markets initially caused panic amongst jewellers because, although they resembled and were represented as natural stones, examination by experts suggested that, in fact, they were synthetic (Nassau, 1980).

The first detailed study of the Geneva rubies was performed in 1886 by Professor F. Friedel of the Sorbonne. He found that these rubies possessed identical physical and optical properties to the natural stones, but also contained as inclusions dense swarms of what appeared to be gas bubbles. Thus, Friedel concluded that the Geneva rubies were actually synthetic, and on the basis of his report the French Trade Association ruled that they could not be represented as natural.

The Geneva ruby dealers, however, did not give up without a fight. Barred by law from selling their stones as natural, they protested that, in fact, the stones were manufactured by fusing together small chips of genuine ruby. This resulted in Geneva rubies acquiring the title of 'reconstructed rubies', a name which they were to carry until only quite recently.

In 1969, Dr Kurt Nassau of Bell Laboratories and Robert Crowningshield of the Gemological Institute of America published a report of their investigations into the mystery of the Geneva ruby (Nassau, 1980). Nassau and Crowningshield examined several Geneva ruby boules and concluded that they appeared to be grown from purified alumina powder, not crushed fragments of natural transparent ruby as claimed. The process as hypothesized involved three separate stages. First, a single gas torch was used to melt a small globule on the top of a cone of sintered ruby powder. This globule was broken off, inverted, and more material then grown on top of it. The final step, which was responsible for the largest amount of growth, involved dropping the feed material onto the boule from above, with two torches at either side providing the necessary heat for fusion.

Nassau and Crowningshield's study of Geneva rubies not only involved examination of actual Geneva boules, but also featured an attempt to produce them utilizing small chips of natural ruby from North Carolina. The result was an almost opaque mass filled with extremely large gas bubbles which bore no resemblance whatsoever to the transparent Geneva product. Thus, the conclusion was that there was virtually no chance that the Geneva rubies could have been 'reconstructed' using natural ruby fragments. Instead, the Geneva rubies appeared to be grown from a purified alumina powder in a manner somewhat similar to that which Verneuil was later to employ.

The evidence presented against the 'reconstruction' theory appeared to put an end to the Geneva ruby controversy. However, in 1980 the eminent British gemmologist, B. W. Anderson, published two articles in the British trade magazine *Retail Jeweller* which again put the matter in doubt (Anderson, 1980). Quoting from personal as well as published sources, Anderson told of a gentleman named Rudolph Oblatt who claimed to have produced Geneva rubies by the fusion of small pieces of 'ruby sand' from Burma. That Mr Oblatt's claim was no mere boast is given added weight by the fact that he provided a very detailed account of not only the Geneva process for the making of reconstructed rubies, but also the Verneuil process using purified alumina (Wodiska, 1909). Even more amazing was Oblatt's claim that it was none other than Verneuil himself who taught him how to manufacture rubies by the fusion of small chips of natural stones! According to Oblatt, after Verneuil confided in him he set up a laboratory but was forced to take in partners owing to a lack of funds. This allegedly resulted in the reconstructed rubies being produced in several different localities all across Europe.

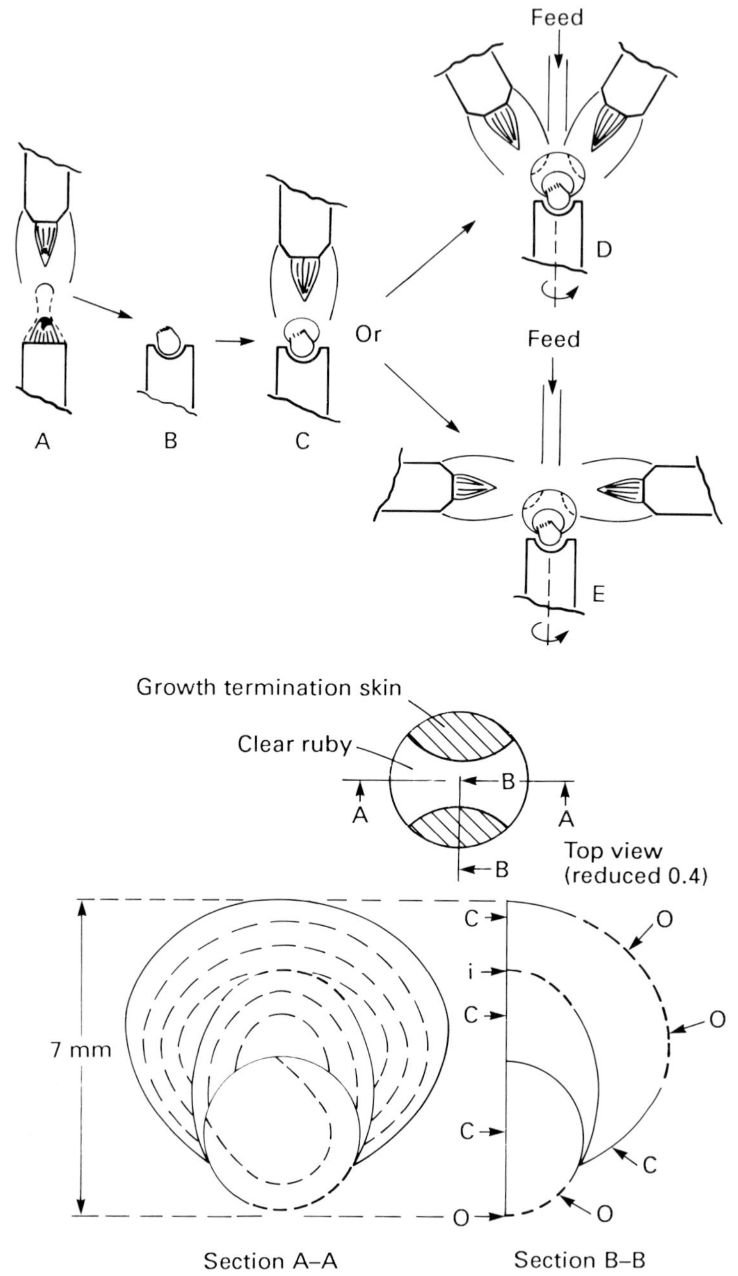

Figure 7.1 The believed manufacturing process and structure of the 'Geneva'-type synthetic ruby, an early precursor of the flame fusion stones of Verneuil. Full lines, clear interface; bold broken lines, growth termination skin; broken lines, growth striations (only some shown). Analysis locations: C, clear ruby; O, outside skin; i, inside skin. (After Nassau, 1980)

Was Rudolph Oblatt in fact the man behind the Geneva rubies, or was he simply attempting to grab headlines with his tale? Oblatt, in his description of the Geneva process, did state that small Burmese rubies were used, whereas Nassau used rubies from North Carolina. As Burmese rubies contain much less iron than those from North Carolina, and as the high iron content was one factor which led Nassau to disbelieve the reconstruction theory, further questions are raised. Unfortunately, because of the deaths of the participants in this drama, the world may never learn the truth of the Geneva ruby mystery.

Whether Verneuil was already working on the flame fusion growth of ruby, or whether the appearance of the Geneva stones started him in this direction, cannot be known with certainty. In any case, it is evident that about 1886 Verneuil began experiments in this direction (Nassau, 1980). By the year 1891 he had made sufficient headway to record the major details of the technique and deposit them in a sealed envelope with the Paris Academy of Science. Already Verneuil was able to grow crystals of sufficient size, but problems were encountered with the cracking of the boules. This shortcoming was soon taken care of, leading to a second sealed note being handed to the Academy in 1892, which was opened with the first in 1910 at Verneuil's request.

The first hint of what was to come occurred at the Paris World's Fair in 1900, where Verneuil's assistant, Marc Pacqier, displayed several of the new synthetic ruby crystals. However, no details of the process were released and it is not known whether or not Verneuil had approved of the showing. Finally, in 1902, Auguste Verneuil proudly announced to the world that he had succeeded in producing rubies in the laboratory which were the equal of nature's finest. Not only did his 'scientific' rubies duplicate exactly the appearance and properties of the genuine article, but, most importantly, they could be grown in sizes large enough for use in jewellery. Thus, at long last, the world had its first truly successful synthesis of a precious stone.

Two years later Verneuil published an extensive article giving the details of his technique. This threw the door wide open for anyone with the right amount of chemical and mechanical training to produce synthetic rubies. Several firms immediately started up work and by the year 1907 the annual production of synthetic ruby soared to over five million carats.

Melt growth processes

Flame fusion (Verneuil) process

Let us now take a closer look at Verneuil's revolutionary technique. In its most basic form the flame fusion process involves melting a mixture of alumina powder and a colouring agent in such a way that a single crystal of corundum forms. Deceptively simple at first glance, it is in actuality a highly complex chain of events which took years of research to refine and perfect.

The first step of the process is the purification of the raw materials. Corundum consists of chemically pure aluminium oxide (Al_2O_3) which is obtained from ammonium alum. Its solubility greatly increasing when heated, this alum is dissolved, and the solution brought to the boiling point. Then, after being filtered and cooled, the wet crystals are dried and roasted in furnaces at 1200°C, producing alumina. A fluffy white powder is obtained by gathering the alumina cake and processing it in vibrating sieves. Although light enough to be inhaled, incredibly it is this white powder which, when melted, turns into a glistening gem that is among the hardest of all known substances.

Colourless sapphire will result from a fusion of the pure alumina powder. However, if coloured gems are desired, small amounts of metallic oxides must be added to the alumina. A number of elements in various combinations have been used on an experimental basis over the years to produce differently coloured varieties of synthetic corundum. Those used most commonly by modern Verneuil manufacturers are compared with the colouring agents of natural corundum in Table 7.1.

TABLE 7.1 Common colouring agents of natural and Verneuil synthetic corundum

Variety	*Colouring agent(s)*	
	Verneuil synthetic[a]	*Natural*
Colourless sapphire	Pure	Pure
Ruby – ordinary type	Chromium	Chromium
Ruby – dark-red type	Chromium + Iron	Chromium + Iron
Pink sapphire	Chromium	Chromium
Blue sapphire	Iron + Titanium	Iron + Titanium
Purple and violet sapphire	Chromium + Iron + Titanium	Chromium + Iron + Titanium
Yellow sapphire	Nickel	Iron + colour centres in some types, or haematite plates
Orange sapphire	Nickel + Chromium	Iron + Chromium + colour centres in some types, or haematite plates
Green sapphire	Cobalt + Vandadium + Nickel	Iron
Colour-change sapphire	Vanadium	Chromium + Iron + Titanium (+ vanadium in some types)
Black star sapphire	–	Haematite plates

[a] Nassau (1980).

In addition to the dopants commonly used in the commercial Verneuil production of synthetic corundum gems, many other elements have been tried on an experimental basis.

Corundum is noted for its high melting point (2050°C) and this provided one of the major barriers in early attempts at its synthesis. Verneuil surmounted this difficulty by utilizing an oxyhydrogen blowpipe mounted vertically such that the flame is directed downwards onto a ceramic pedestal which can be lowered by means of a fine screw. This allows the growing boule's upper surface to be kept within the hottest part of the flame. The powder is placed in a metal basket, the base of which is a screen, and is hung just above the central tube of the blowpipe. A mechanical device causes a hammer to tap the basket gently at selected intervals, sprinkling the alumina powder through the flame. Once these grains reach an area of sufficient heat they will be fused. This occurs near the ceramic pedestal, positioned just below the area of fusion. As the alumina powder continues to fall, a small 'sinter cone' develops, steadily inching its way upwards towards the zone of fusion in the hottest part of the flame. Finally, the tip of the cone enters this zone and is liquefied. From this point on, all of the grains which fall onto this molten area will also be fused. Thus, the actual growth begins from a slender stem. It is vital that this neck be as small as possible, for if the contact area between the boule and its support is too wide, the finished crystal will shatter into tiny pieces when the flame is turned off.

Once the neck has formed, the boule grows upwards, steadily increasing in breadth as it rises into the hotter and wider portions of the flame. In order to increase the diameter of the boule, the powder feed rate is raised and the

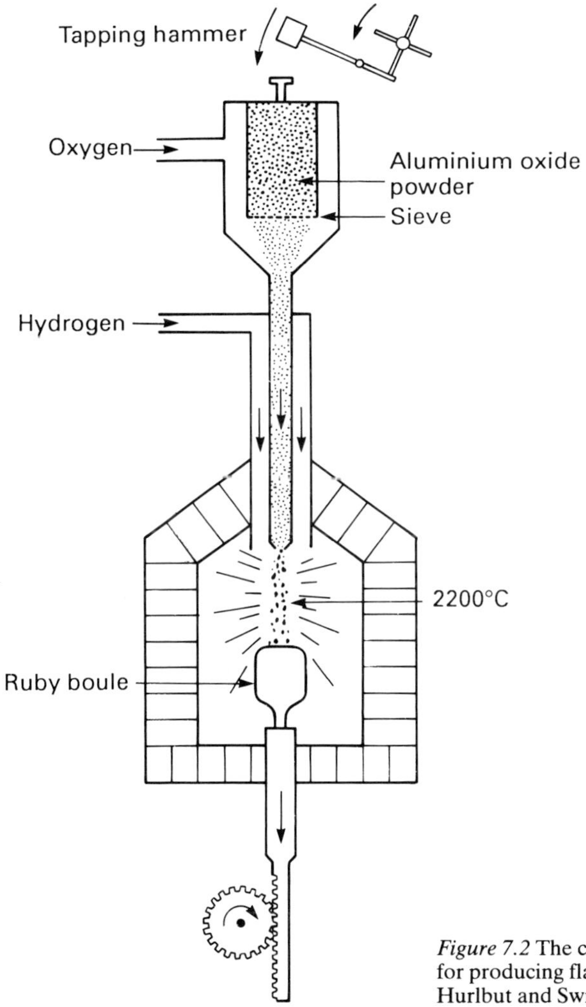

Figure 7.2 The construction of the Verneuil furnace used for producing flame fusion synthetic corundum. (After Hurlbut and Switzer, 1981)

temperature of the flame is also raised by increasing the flow of oxygen. Powder continues to cascade downwards, drawn through the flame by the force of the gas and onto the glowing crystal, which is incandescent with the heat. Thus the molten mass slowly grows upwards by means of thin layers deposited, one on top of the other, on the spherical upper surface. Its vertical growth is compensated for by lowering the pedestal at an identical rate, keeping the domed upper surface in the warmer regions of the flame at all times. Should the flame be too hot, a characteristic bubbling will appear, causing the inclusion of gas bubbles in the finished product. In order to avoid these defects the flame must be turned down slightly and the boule must be lowered to a slightly cooler region using the screw device. Monitoring of the boule to ensure its perfection is achieved visually through a small window built into the oven for this purpose. At last, when the crystal has grown to the desired size, the gases are abruptly shut off. This is the most critical stage, as fractures are now most likely to develop. If the boule has been grown correctly, is well centred and is heated evenly throughout, then it can be split vertically into two halves with a gentle tap of a hammer. Because the growth of a boule is accomplished in a matter of hours, considerable strain builds up within the finished product. Splitting the boule lengthwise along an incipient parting plane relieves this pressure which would otherwise cause irregular fractures. For certain industrial uses where whole boules are desired, the splitting can be avoided by 'annealing', or reheating the boules. Annealing to dissipate the strain is performed most frequently on colourless sapphires for use as cover glasses in watches, gauges, etc., but is not necessary for gem boules.

The term 'boule', which is used to describe crystals grown by the Verneuil process, is derived from the fact that the early crystals were somewhat spherical in shape and thus resembled the heavy balls used in the French game of the same name. Although modern Verneuil crystals bear a greater likeness to carrots than to balls, they are still referred to as boules.

The sizes of the early flame fusion crystals were generally small, averaging perhaps 15 carats in weight. A typical boule of 5–6 mm diameter might yield four or five cut stones of under 2 carats. However, improvements in the quality of the feed material and the furnace apparatus allow the growth of much larger crystals today. Modern boules can measure up to 9 cm in diameter and faceted stones of well over 100 carats are in existence, but these are extreme rarities. On average, most Verneuil boules produced today are roughly 6.5 cm in length and weigh between 150 and 200 carats. Because the furnace pedestal is lowered at the rate of 1 cm per hour, the boule's growth rate is the same. Thus, a 6.5 cm boule would require approximately 6.5 hours for its manufacture.

When the Verneuil synthetic rubies first entered the world's gem markets after 1904, fears were immediately raised regarding their distinction from natural stones. The situation was similar to that which prevailed in 1885–1886 with the advent of the Geneva rubies, but with important differences. Although the Geneva ruby did qualify as synthetic corundum, it fell short of the mark in several areas. Geneva boules were generally filled with fractures due to uneven cooling, and clouds of gas bubbles were often so numerous as to render the stones hazy. Furthermore, the colour was quite unnatural in appearance and so the market soon developed methods for their distinction, restoring public confidence and causing a recovery of prices for the genuine article.

In contrast to the Geneva product, Verneuil synthetic rubies were of a colour and clarity which equalled or even exceeded nature's finest specimens, and their appearance in the market created quite a stir. Large amounts of publicity were given to Verneuil's revolutionary rubies and their amazing likeness to genuine stones. This interest in the 'scientific' ruby, coupled with a severe financial crisis in the United States in 1907, caused demand for natural rubies in the important London market to fall. The years between 1907 and 1914 were particularly bleak for the natural ruby trade. Not only were prices and sales down in Europe and the USA, but also large quantities of synthetic stones were being shipped to Colombo, Calcutta, Rangoon and other Asian centres, eating into profits there as well. At last, in the face of falling revenues, the natural ruby industry began to act. In Colombo, the Ceylon Legislative Council imposed an import tax of 100 rupees ($33) per carat on almost all imitation and synthetic gemstones, which, in effect, barred their importation. With the aid of experts, the Burma Ruby Mines Company in 1912 prepared instruction charts for separating natural and synthetic rubies. These were then distributed free of charge to jewellers and gem dealers around the world (Kunz, 1913).

Eventually these tactics paid off. Keen competition amongst the various manufacturers of the synthetic stone flooded the market, and in doing so depressed its price. At the same time, demand for natural rubies increased as fears regarding their distinction were shown to be unjustified. Thus, the natural ruby soon regained its pre-eminent position in the eyes of the public.

It is interesting to note that recently similar fears have again been voiced regarding the ability of detection methods to keep pace with modern growth processes. Let us consider the answer that was given to this very same question by the German mineralogist Professor Dr Conrad Oebbeke in the late 1900s:

> Between the natural and the artificial precious stones, the material difference will always exist, that one is a natural, the other an artificial product. Up to the present time, I have not seen *a single artificial precious stone that could not be recognized as such.* The claim that the artificial stones are not to be distinguished from the natural gems, that they are absolutely free from defects, etc., according to my experience, is *not justifiable.* Even if it is possible to produce precious stones having the same crystallographic, physical, and chemical properties as the natural gems, they are nevertheless *not equal in value* to the natural product. No more so than an ever so carefully executed and deceptively similar copy of a work of art, a painting, a piece of sculpture, etc., can be called the original. The artificial products, made in the laboratory, are not formed under the same conditions as the natural article, and for this reason we may rest assured that, even should the present scientific methods of distinguishing the genuine from the artificial precious stones fail, further scientific investigation will reveal a method that will make the distinction possible. Interesting as may be the success thus far attained in the production of artificial precious stones, and while we may congratulate ourselves on the progress made in chemical techniques in this

direction, to the *connoisseur, these articles will always be artificial products* that can never deprive the natural stones of their value. On the contrary really beautiful natural precious stones will *only be the gainer*. The claim that synthetic stones will ever break the market for real precious stones, is, in my opinion, utterly unfounded.

(From Wodiska, 1909)

It has been more than 75 years since these brave words were spoken, and yet they still ring true.

Although the Verneuil synthetic rubies at that time represented a tremendous leap forward, amongst the experts who examined them there was never the slightest doubt as to their artificial origin. It was Verneuil himself who, in his 1904 publication, systematically described the characteristics by which they could be identified. Flame fusion boules are formed by the gradual crystallization from the bottom up of one layer on top of another. However, slight variations exist in the distribution of the colouring agent from one layer to the next, resulting in the formation of curved growth lines. These lines, or striae, follow the successive curved contours of the top surface and so are visible when looking in directions roughly perpendicular to the length of the boule. As the growth lines of natural corundum are always straight, curved lines represent positive proof of synthetic origin.

Gas bubbles are frequently found within the Verneuil stones and result from a bubbling or boiling of the top of the boule during growth. They are generally spherical, or 'tadpole shaped', with the swollen head of the bubble pointing away from the surface (perpendicular to the direction of the curved growth lines). Rotation of the growing crystal, which is done in the production of large boules, is sometimes employed in an effort to force gas bubbles out from the central core towards the surface.

As previously mentioned, the finished boule is usually split lengthwise along a parting plane to relieve the tremendous strain built up during its rapid growth. If the boule has developed properly and been allowed to cool evenly, the *c* axis will be parallel to its length. However, because of the added costs involved, in practice less care is taken and the *c* axis may be inclined or even perpendicular to the boule's length. In all cases, though, the *c* axis will lie within the plane in which the boule has split.

Crystallographic orientation can be controlled by using a preselected seed rod upon which the growth is initiated, with the growing boule conforming to the orientation of the seed. Three major growth directions are used and are designated 0°, 60° and 90° crystals, depending upon the angle between the *c* axis and the length of the boule: 0° boules appear roughly equidimensional in cross-section, while 60° boules are oval and 90° boules almost rectangular (Nassau, 1980). Among boules produced for use as gems, the 90° crystals are probably encountered less frequently than any other type, being found primarily in yellow and orange sapphires.

The crystallographic orientation of Verneuil synthetic corundum is of concern not only for industrial uses, but also to gemmologists. Natural rubies and sapphires are generally cut with the *c* axis (optic axis) close, or exactly perpendicular, to the

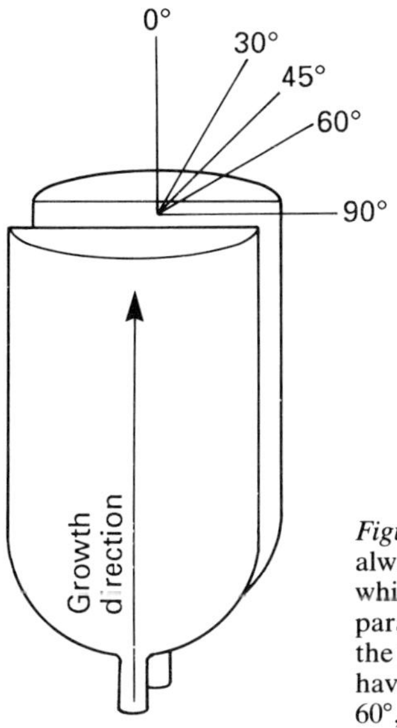

Figure 7.3 The optic axis (*c* axis) in Verneuil synthetic corundum is *not* always parallel to the boule's length, but always lies within the plane along which the boule is split. Boules of 0° are those in which the optic axis is parallel to the length, while in 90° boules the optic axis is at right angles to the length. The former are roughly circular in cross-section while the latter have flattened sides. Most boules have an orientation of between 30° and 60°, and so are oval in cross-section

table facet. This is done for two important reasons. First of all, due to the shape of most corundum gem rough, loss of weight during cutting is minimized with this orientation. Secondly, however, of even greater importance is the fact that in most varieties of corundum (natural or synthetic) the colour of the o-ray is more intense and beautiful than that of the e-ray. Thus, when lapidaries cut natural corundum they usually try to orientate the gem with the table facet, or cabochon base, perpendicular to the *c* axis so that the o-ray colour alone is seen, undiluted by the more insipid e-ray hue. In contrast, lapidaries are rarely concerned with obtaining the best colour when cutting synthetic corundum. Instead, convenience, speed and weight retention are the factors which dictate orientation. Because the boules are split lengthwise after removal from the furnace, the tendency is for the cutter to place the table facet parallel to the flat inner surface produced by the split. The *c* axis, which always lies in the plane of the split, is thus parallel to the table. As a result, synthetic gems so cut show a slight difference in colour compared with the natural ones, due to the dilution of the o-ray by the weaker e-ray. Examination of the stone with a dichroscope clearly reveals this difference, with the synthetic gem displaying strong dichroism through the table, while the natural one shows little or no dichroism in this direction. It must be kept in mind, of course, that other factors may cause both natural and synthetic stones to be oriented differently. However, the experience of the author has shown that at least 70–80% of all natural and Verneuil synthetic rubies and sapphires are cut as described above. Moreover, in the case of large (above 10 carats) synthetic gems this increases to perhaps 90% or more.

After 1904, with the successful synthesis of ruby now behind him, Verneuil focused his attentions on yet another corundum gem, blue sapphire. The synthesis of blue sapphire came late in 1909, after chemical analysis by Verneuil of natural

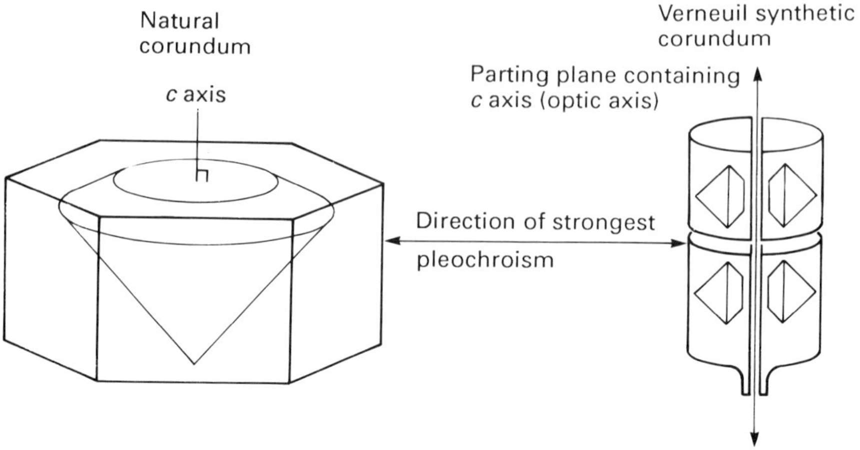

Figure 7.4 Natural corundum is usually (but not always) oriented during cutting with the *c* axis at 90° to the table facet. This generally produces the most intense (i.e. best) colour and greatest yield. Verneuil synthetic corundum boules, however, are split down the centre along an incipient parting plane (to relieve tension) before cutting. As colour is not of great importance in synthetic corundum (it generally has good colour no matter how it is oriented), the boules are then cut to achieve greatest weight retention. This is done by slicing them into small chunks across the width and then placing the table facet parallel to the largest flat surface (parallel to the plane of the split). As the *c* axis of the Verneuil boule always lies in the plane of the split, this means that the *c* axis will then generally lie parallel to the table of the faceted stone. This means that strong pleochroism will usually be seen at 90° to the table facet in the Verneuil synthetic corundum, while the same direction (perpendicular to the table) in most natural corundums generally shows no pleochroism. Of course this is true only if the stones are cut in the above orientations

stones revealed traces of iron and titanium in all blue sapphires. With the arrival of this key piece of information, it was not long before blue sapphire was added to the list of successful synthetic gemstones, and quantities of the new material soon began pouring out of synthetic ovens around the world.

In comparison with ruby, the growth of blue sapphire presents a much greater challenge, for there is a tendency to develop a violet coloration unless strict control is maintained over the flame temperature and boule position. As with natural sapphires, the blue colour of the man-made stone is more irregular than that in ruby. Thus, the Verneuil products often show a colourless central core with the blue colour being restricted to the outer areas close to the surface. In recent years, lapidaries have taken advantage of this fact to produce stones which resemble the strongly zoned heat-treated Sri Lankan blue sapphires now so prevalent in the marketplace. The uneven colour distribution in the synthetic stone apparently arises from the volatile nature of the iron, most of which burns off in the intense heat of the furnace. Enough iron remains behind to influence the boule's colour, but not the spectrum, as the synthetic stones lack the 451.5, 460, 470 nm absorption complex that is the hallmark of so many natural sapphires.

Modern Verneuil synthetic blue sapphires are produced in a wide range of different colour intensities, of which the zoned type is only one example. These vary from near colourless to the deepest blue, with dense concentrations of gas bubbles often filling the interior of the latter. Rubies are also made in a range of tones. This is not just to allow stones of different shades to be cut, but also to allow stones of different *size* to be matched in colour, for if two gems which differ greatly in size are cut from the same boule, the larger of the two will have a deeper colour. Hrand Djevahirdjian SA, of Monthey, Switzerland, one of the world's largest manufacturers, produces ten different shades of ruby and pink sapphire and six different shades of blue sapphire. Altogether, a total of 32 different colours are made, ranging from colourless to pink, red, orange, yellow, green, blue, violet and purple, as well as a colour-change type. In addition to the wide range of single-colour gems, bicolour and even tricolour stones have also been made in the past.

Modern Verneuil manufacturing plants are capable of producing enormous amounts of synthetic corundum; the Djeva facility alone has a capacity of 60 000 kg per year. Most of this material is slated for industrial use, however, with only a part being cut for jewellery. Due to the enormous quantities involved, prices for the rough material average but a few cents per carat. Blue sapphires cost up to twice as much as rubies, while the green synthetic sapphires fetch ten times more.

Synthetic star corundum

The Verneuil process is used not only for the manufacture of transparent synthetic corundum, but also for star rubies and sapphires. Production of asterated corundum was first achieved in 1947 by the Linde Division of the Union Carbide Corporation in the United States. Linde later received a patent for the process in 1949 (Pough, 1966).

Asterism in natural corundum is caused by the reflection of light from sets of microscopic inclusions which intersect in three directions at 120° within the basal plane. Most of the silk consists of the mineral rutile (TiO_2), although haematite (Fe_2O_3), ilmenite ($FeTiO_3$) and magnetite (Fe_3O_4) have also been identified. It forms parallel to the prism faces of the host corundum by a process known as exsolution, the unmixing of a solid solution.

Inclusions of rutile were identified in asterated corundum as early as 1878 and the principles of exsolution were also well known at an early date. It was only after a chance observation in 1947, however, that star rubies and sapphires were synthesized. While experimenting with titanium as a colouring agent in synthetic corundum, Linde production chief J. N. Burdick noticed a trace of silk in one of the boules. Appreciating its significance, he then began a series of experiments which succeeded in developing enough silk to produce a star. The process involves the addition of 0.1–0.3% of titanium oxide (TiO_2) to the feed powder, over and above the normal percentages of colouring agents. Then the boule is grown in the usual fashion, with the titanium entering into solid solution with the alumina. After cooling, it is reheated to between 1100 and 1500°C for between 72 hours and 2 weeks, or more, a process termed annealing. The maintenance of this temperature for a relatively lengthy period of time allows the titanium molecules to break out of solid solution and form needles within the synthetic corundum host. Thus, the silk forms and a star is born.

Reports on the exact identity of the needles, or silk, in synthetic star corundum differ. Some suggest that they consist of rutile (TiO_2), while others state that aluminium titanate (Al_2TiO_5) is formed instead, a result of the titanium reacting with aluminium.

Linde's early production suffered from the tendency of the titanium to move to the outside surface of the growing boule, creating a star in which the rays stretched only halfway down the sides of the cabochon, fading away near the girdle. Other stones showed incomplete stars and transparent areas, due to the uneven distribution of the silk. The problem was eventually solved by continuously altering the flame temperature as the boule grows, thus allowing one layer to solidify before the next is deposited. The result is a boule containing layers of a uniform distribution of titanium, alternating with layers of uneven distribution. As the layers are only 0.10 mm thick, the star appears continuous to the eye and the stone can even be recut without fear of losing any part of it.

One disadvantage of the modification in the temperature of the torch is that the boules must be kept smaller in size. The largest cut stone ever produced by Linde was a synthetic star ruby of 109 carats, which was donated to the American Museum of Natural History.

Linde began selling synthetic star rubies in September 1947 at $30 per carat, and several years later they introduced synthetic blue star sapphires into the market. Only cut stones were sold, with cutting being performed automatically by Linde. In addition to star rubies and blue sapphires, small quantities of white, grey, purple, green, pink, yellow, brown and black stars were made.

During the early years of production, Linde enjoyed a monopoly on the manufacture of synthetic star corundum. However, this was short lived. Sometime before 1957, the Wiedes Carbidwerke of Freyung, West Germany, began selling synthetic star rubies and sapphires produced by a slight variation on the Verneuil method used by Linde. The German stars were termed 'Stars of Freyung' and were more transparent and natural appearing than Linde's stars.

Synthetic star corundum has also been produced in Israel, Japan, Switzerland and the USSR. Currently, synthetic star corundum is produced solely by the Verneuil process. Linde did receive a patent for the manufacture of synthetic star corundum by the Czochralski (pulling) process, but these stones were never marketed.

Czochralski (pulling) process

The vast majority of all synthetic rubies and sapphires used in the jewellery trade are grown by the Verneuil process. However, certain industrial applications require crystals of greater perfection than that attainable by the flame fusion technique, thus leading to the growth of corundum by other methods. One such method is the Czochralski process.

The Czochralski process of pulling from the melt achieves a greater degree of perfection than almost any other. Although the basic apparatus was first developed in 1918 by J. Czochralski, it was not until the advent of lasers in the early 1960s that it was applied to the growth of rubies on a large scale. Today, in addition to ruby, pink, colourless, blue and orange sapphires are also manufactured by the pulling techniques for use in both jewellery and industry.

The principle of Czochralski growth is rather simple, but sophisticated apparatus is required. In the initial stage, aluminium oxide plus traces of the proper colorant (if a coloured crystal is desired) are melted in an iridium crucible by means of a radio frequency (RF) coil heater. Then, a small corundum seed crystal attached to the end of a rod in a specific crystallographic orientation is lowered so that it just touches the surface of the melt. Accurate temperature control is critical, for if the alumina is too hot the seed will melt. Pulling now begins by a simultaneous rotation and slow retraction of the seed. As it is withdrawn from the melt, surface tension causes some of the molten alumina to cling to the seed, where it solidifies. In addition, heat conduction allows the seed to extend a slight distance into the melt without being dissolved itself. Corundum is then added continually to the end of the rod as it is slowly lifted out from the crucible. Thus, a single cylindrical crystal up to 11 cm in diameter and 63.5 cm in length can be obtained.

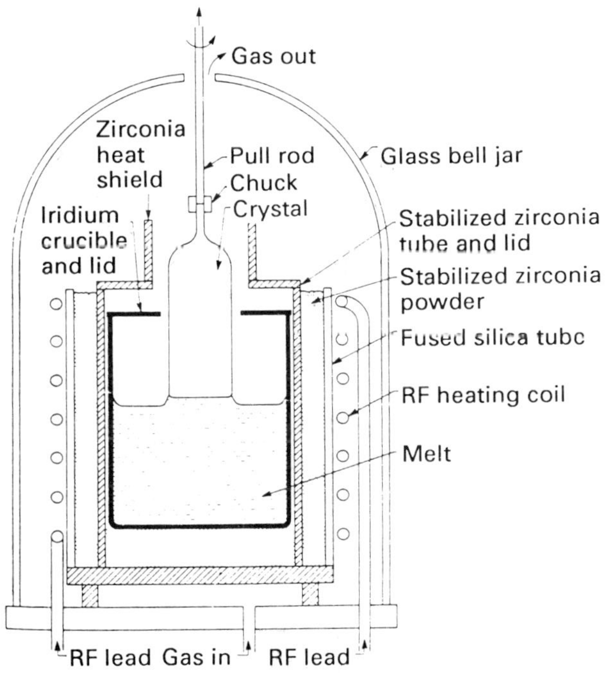

Figure 7.5 The typical apparatus used for growing synthetic corundum by the Czochralski ('pulling from the melt') process. (After Nassau, 1980)

Pulling rates for Czochralski crystals range from 6 to 25 mm per hour, with larger pieces utilizing the slower rates. It is of great importance that the diameter of the growing crystal be kept uniform because sudden changes tend to cause the inclusion of gas bubbles and other imperfections. Various methods are used to control the

Figure 7.6 A large single crystal of synthetic ruby grown by the Czochralski (pulling) process. As a result of the curved surfaces, stones cut from such crystals will often display curved growth lines. (Courtesy of the Gemological Institute of America)

temperature and growth, such as continuously weighing the crucible contents or even weighing the growing crystal itself (Nassau, 1980).

Specimens of Czochralski-grown synthetic corundum gemstones have in the past been primarily cut from crystals originally intended for use in industry. These consisted or rubies and pink sapphires made for lasers, and colourless sapphires for use as watch crystals, among other things. However, beginning in 1980, the Inamori company of Japan began selling faceted Czochralski-grown synthetic rubies, either as loose stones or mounted in their own line of luxury jewellery. The Inamori product is far more expensive than that from the Verneuil furnace and reflects the company's marketing strategy of creating an exclusive image for their line. In 1980, wholesale prices for the Inamori ruby ranged from $190.00 per carat for 0.20–0.50 carat stones to $343.00 per carat for stones of 4.00 carats or more, with retail prices being double wholesale prices. Besides ruby, Inamori also markets blue and orange sapphires (Inamori Jewelry Division, 1980).

Floating zone process

Among the latest of the various growth techniques to be applied to the synthesis of gem-quality corundum is the floating zone process. It is a variation on a crystal growth, or purification, method called zone refining and has been adapted for the manufacture of gems by the Japanese company, Suwa Seikosha, maker of Seiko timepieces (Scarratt, 1984).

The floating zone process makes use of a moving temperature gradient, or zone of steadily decreasing temperature. In the manufacture of synthetic ruby, a sintered rod of aluminium and chromic oxides is rotated vertically between two chucks. To begin the growth, the lower end is melted with an infrared radiation convergence heater to form a molten zone. Then, by slowly lowering the rod past the heater, this molten zone is made to travel along the rod, supported by surface tension. In doing so, any impurities in the material tend to remain within the molten zone, thus leaving a purified crystal in its wake. Once the opposite end of the rod is reached, the heater is turned off and the rod is returned to the original starting point, where

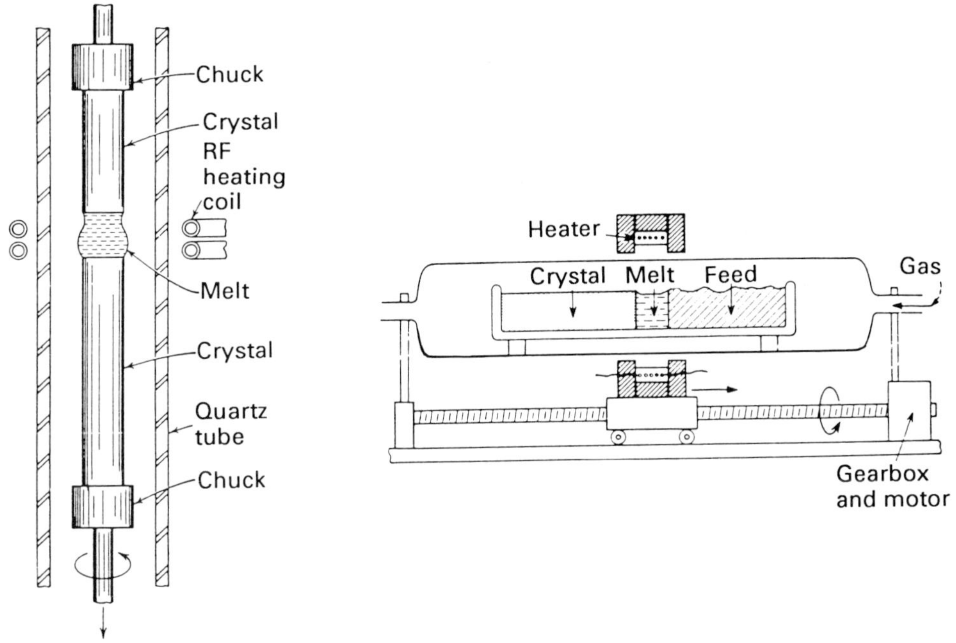

Figure 7.7 The apparatus used in the floating zone (zone refining) growth of synthetic corundum. Currently Seiko of Japan is the only commercial manufacturer of synthetic corundum by this process. (After Nassau, 1980)

it can then begin another pass. Ideally, after several passes have been made a single crystal of great purity has formed.

It is only quite recently that the floating zone process has been employed in the synthesis of corundum for gem use. Since 1979, when Seiko made their first quartz watch, they have been attempting to develop a better crystal. These experiments eventually led to the synthesis and manufacture of corundum by the Division which performed them, the Matsushima Kyogo Co. Ltd, who also make the quartz cover crystals used in Seiko watches. Ruby, blue and orange sapphires are presently (1984) being produced at the rate of 1000 carats per month (Scarratt, 1984). However, it is hoped that production will increase up to 5000 carats within the next few years. Stone sizes are mostly less than one carat cut, to keep the prices of the finished jewellery down (Brown, 1985).

The Seiko synthetic corundum is not being sold in the rough. Instead, it is available only in finished jewellery under the Bijoreve trademark. Gemmologists who have examined the rough have stated that, in shape, it appears much like the Verneuil material.

Solution growth processes

Hydrothermal process

Included among the many different techniques by which corundum has been synthesized is the hydrothermal transport process. Otherwise used primarily for the synthesis of quartz, this process differs from the methods discussed so far in that growth is achieved via a solution rather than a melt. However, due to the difficulties of growing corundum by this method and the high costs involved, to date it has been applied only on an experimental basis.

Corundum is completely insoluble in water at room temperature and only slightly soluble at the boiling point of water (100°C). However, if a pressurized vessel is used and the temperature is further increased, its solubility rises dramatically (Butcher and White, 1964). This is the principle of the hydrothermal process, the method that, more than any other, comes closest to duplicating the natural growth of corundum crystals within the earth.

In the growth of ruby the nutrient material, consisting of either corundum (Al_2O_3) or gibbsite ($Al(OH)_3$), is placed in the bottom of a high-pressure vessel known as an autoclave. Autoclaves are typically made of thick steel and incorporate sophisticated closures to withstand pressures which average approximately 3000 atmospheres. So great, in fact, are the pressures involved that explosions are a very real possibility, hence the origin of the term 'bomb', a name by which the autoclave is sometimes known.

The synthesis of ruby requires that the autoclave be lined with a suitable metal, such as silver or platinum, for otherwise the iron from the container's walls will colour the crystals green. Although early hydrothermal growth was achieved using water alone, it was later found that the solubility could be further increased by the addition of a mineralizer, such as potassium or rubidium carbonate, which creates an alkaline solution.

Seed crystals of either natural or synthetic (Verneuil) corundum are hung from a silver frame at the top of the vessel and, once sealed, the autoclave is then heated on the base. While the temperature at the base is approximately 580°C, in the upper regions (where the seeds are located) it is only 540°C, creating the temperature gradient which is the key to the hydrothermal transport process. Because of the higher temperatures at the bottom, a saturated solution of corundum forms in the mineralized liquid. Although the dissolved corundum increases the density of the solution, this is more than offset by the higher temperatures, which have the opposite effect. Thus, as a result of its lower density the saturated solution rises from the bottom in what is known as a convection current. Once it reaches the cooler regions near the top, it can no longer hold all of the dissolved corundum in solution, and so the excess material crystallizes out on the strategically placed seed crystals. Then the solution is carried back down by convection and the process repeats itself again and again until the corundum nutrient is depleted; hence the origin of the term 'hydrothermal transport'.

The growth of ruby by the hydrothermal process is quite slow compared with other techniques. Fastest growth takes place on the basal pinacoid at the rate of 0.05 to 0.25 mm (0.002 to 0.01 inches) per day with linear growth along the *c* axis rarely exceeding 1.00 mm over a period of 14 days (Butcher and White, 1964). Although it is estimated that the crystal growth rates and qualities could be improved upon, due to the high start-up costs involved and the abundance of comparatively inexpensive corundum grown by other methods, up to now there has been little incentive for such an undertaking.

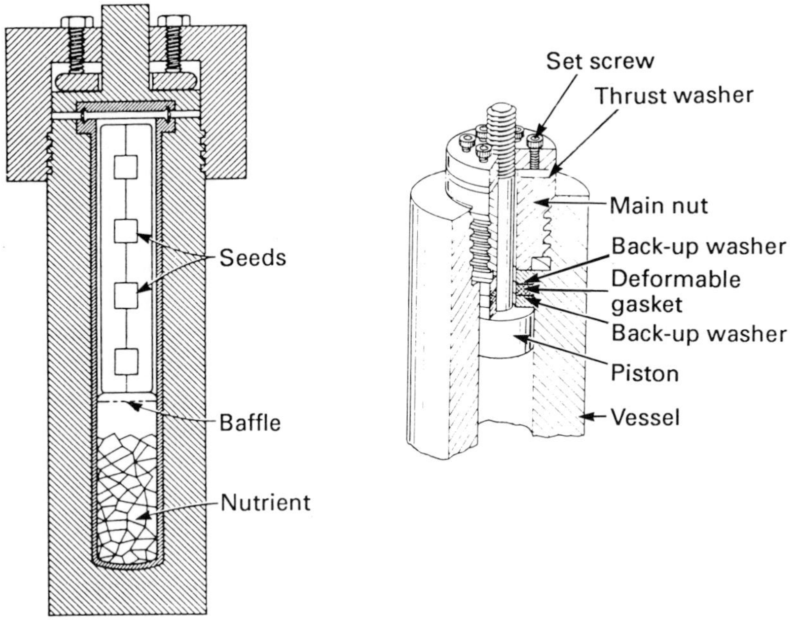

Figure 7.8 The autoclave and other apparatus used in the hydrothermal growth of synthetic corundum. (After Nassau, 1980)

In 1959, Carroll F. Chatham (1914–1983) of San Francisco began marketing synthetic rubies produced by an undisclosed process (Reinecke, 1959). Several different reports in the gemmological literature at that time suggested that the stone had been produced by the hydrothermal method (Gübelin, 1961a,b). This assumption was based largely on the presence of seed crystals and other inclusions which resembled somewhat the published descriptions of experimental hydrothermal rubies grown in 1958 at Bell Telephone Laboratories. However, recent information provided to the author by Thomas H. Chatham, son of Carroll Chatham and current president of Chatham Created Gems, states that the actual process used was, and continues to be, a type of flux growth.

Flux process

It is the production of synthetic corundum by the flux growth process which, more than any other type, gives gemmologists the greatest cause for concern. Unlike the Verneuil product with its distinctive curved growth lines and gas bubbles, flux-grown gems possess inclusions that cleverly mimic those of the natural stone. Only through meticulous study and a continuous honing of their diagnostic swords can gemmologists successfully unmask these skilful reproductions.

Flux-grown synthetic corundum gems were first manufactured by Fremy in the latter part of the nineteenth century, although the process was later abandoned in favour of Verneuil's less complex approach. It was only in the late 1950s, with the use of ruby in lasers, that researchers again began to explore the potential of flux growth. Carroll Chatham, who at the age of 16 pioneered the commercial synthesis of emerald, was the first to achieve a breakthrough. In 1959, after seven years of experiments, he introduced his 'cultured' rubies into the market. Other firms were later to follow, and at the present time at least five major manufacturers are involved in the commercial flux growth of corundum for gem use. Beyond saying that the flux technique is being used, none of these firms have released any detailed information regarding the method of growth. However, the scientific literature does contain several descriptions of the growth of corundum in a flux environment and the following details have been taken from these reports.

The term 'flux' refers to a material which possesses the ability to lower the melting point of another substance with which it is mixed. Just as the melting point of sugar is lowered when it is mixed with water, the melting point of a gem mineral can be dramatically reduced when it is combined with the appropriate flux. However, water is not considered a flux in the true sense of the word, for unlike water a flux dissolves a substance through a chemical reaction (Chatham, 1982). A solid at room temperature, a flux becomes a powerful solvent only after it has melted.

Commercial manufacturers do not normally release details about their growth processes or the particular flux being used. However, chemical analysis of the stones has suggested that the following fluxes (Table 7.2) are possibly being used in the commercial flux synthesis of corundumm gems (Schmetzer, 1985).

TABLE 7.2

Manufacturer	*Product*	*Flux*
Chatham	Synthetic ruby	Li_2O–MoO_3–PbF_2 and/or PbO_2
Kashan	Synthetic ruby	Na_3AlF_6
Knischka	Synthetic ruby	Li_2O–WO_3–PbF_2 and/or PbO_2
Ramaura	Synthetic ruby	Bi_2O_3–La_2O_3–PbF_2 and/or PbO_2

To grow corundum by the flux process a saturated solution is prepared by dissolving the nutrient material (Al_2O_3 + colouring agent) into the flux at a temperature just above the saturation point. Rotation of the crucible is often employed during this stage to ensure that the contents are well mixed. Once a complete solution has formed, the actual growth is initiated by gradually lowering the temperature through a range in which corundum will precipitate.

Although it is not mandatory, manufacturers often find it advantageous to keep one region of the crucible slightly cooler, thus causing convection currents to form, as with the hydrothermal transport technique. Growth usually begins in this cooler area, either by spontaneous nucleation, or by growth on seed crystals specifically inserted for this purpose. When the crystals have reached the desired size, they are removed by pouring off the still molten flux, or by draining it out through a hole punched through the red-hot bottom of the crucible. An alternative method involves cooling the crucible until the entire mass has solidified and then freeing the crystals with boiling acid which dissolves the flux. However, as this may require several weeks to accomplish, the former methods are usually preferred (Nassau, 1980).

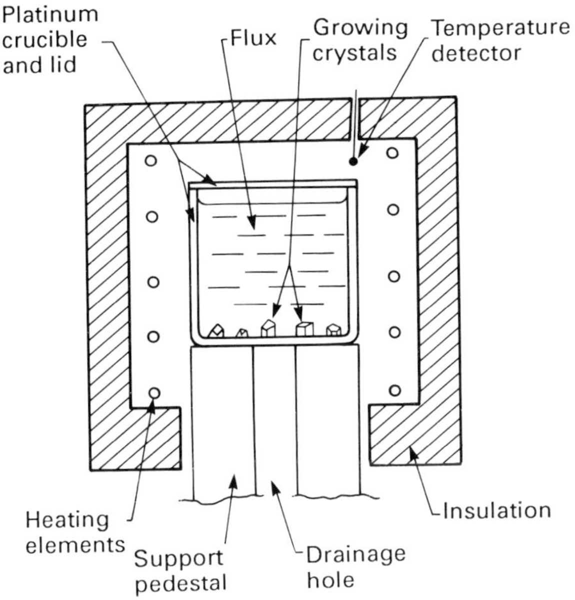

Figure 7.9 The construction of a typical furnace used in the flux growth of synthetic corundum. (After Nassau, 1980)

The tendency of ruby to grow as thin plates from a flux solution is one of the biggest obstacles in the production of usable gem material. Higher growth temperatures seem to help the situation, as does the addition of approximately 0.5% of lanthanum oxide to the solution (Elwell, 1979).

The use of seed crystals is another method of influencing the habit of crystals grown from high-temperature solutions, because seed crystals allow better control over the growth rate, as well as influencing the crystals' perfection and orientation. Growth rates are directly proportional to the surface area of the seed, and its initial orientation also affects that of the finished crystal. In addition, seeding helps to promote the growth of fewer and larger crystals, rather than multitudes of tiny ones. Thus, by using seed crystals, a manufacturer is able to produce larger crystals not only of more desirable shape for cutting, but also in a shorter span of time.

Despite these advantages, some producers have found it advantageous to forego the use of seed crystals and instead allow growth via spontaneous nucleation. This is because, although the growth is less controlled, resulting in crystals of a variety of shapes and sizes, it allows development of extremely clean tabular crystals, the shape most preferred for faceting. If grown on seeds, tabular crystals are usually far more heavily included (Kane, 1983).

The most common internal features of flux-grown synthetic corundum are inclusions of the flux from which the stone formed. These inclusions often bear a strong resemblance to the liquid-filled fingerprints and feathers of natural gems and so can create problems for even the experienced gemmologist.

Flux inclusions consist of solidified bits of the once-molten flux which was trapped as the crystal grew. They are usually in the form of a singly refractive glass and so superficially may appear to be liquid. This resemblance to liquid inclusions is further increased due to the frequent presence of what appears to be gas bubbles within the flux (Burch, 1984). However, because of the great difference between the refractive indices of the glass and surrounding corundum host, the flux will generally stand out in high relief. Particularly with larger inclusions, the flux may display flat crystal faces that are actually flux-filled negative crystals. Moreover, the surfaces are often crossed by irregular tension cracks, especially when solidification has started at one end of the inclusion (Flanigen *et al.*, 1967).

There are a number of different ways in which flux inclusions form, all of which are related to a single factor – overly rapid growth of the crystal. When the pace of growth proceeds too quickly, supersaturation of the flux solution decreases at the centre of the crystal face. This results in growth primarily along the edges and corners, creating a dendrite-like series of narrow channels. Instead of a flat planar surface, the face develops small depressions, or furrows, which imprison traces of the undigested growth solution. It is in the above manner that the thick flux globules and many of the parallel flux rods have their genesis (O'Donoghue, 1983).

Rapid growth is also responsible for the inclusion of thin curving flux fingerprints and feathers. As a ruby or sapphire grows, stress and strain within it build if the pace is too quick. This pressure is released by the formation of stress fractures in random directions or, more commonly, along directions roughly parallel to the major crystallographic faces, such as the rhombohedron, prism, pyramid and/or basal faces. Growth solutions now rush into the fractures, healing them shut, and, in doing so, trap small pockets of undigested residual flux.

This process by which thin drops of flux are trapped within healed fractures is identical to that which produces fingerprints and feathers in genuine stones. However, differences in the nature of the filling material and the relative development of these heated fractures allow a distinction to be made. Fingerprints and feathers within natural corundum contain liquid instead of the solidified flux glass of the synthetic. This liquid is generally more transparent and rarely, if ever, shows the irregular surface crazing of flux inclusions. In addition, the growth details within natural fingerprints are much more intricate as a result of a slower growth rate and lower viscosity of the liquid. Between large liquid-filled channels there exist smaller ones, forming what may be termed a 'fingerprint within a fingerprint'. With illumination at an oblique angle from above, vivid interference colours reflect from tiny growth steps and terraces. In contrast to the fantastic complexity of these thin liquid films, the greater viscosity of the flux and the faster growth rate produce much more coarse, less intricate, patterns. Interference colours may be seen on occasion, but the wonderful detail of the natural stone is, for the most part, absent in the flux synthetic.

Although mother nature possesses far greater patience when growing her crystals, the growth rates for the flux stones are still much slower than for most other types of synthetic corundum. Growing periods for flux rubies range from six to eight weeks for the Ramaura product (J. Osmer, personal communication) to as much as ten to twelve months with the Chatham process; with the Verneuil, Czochralski or floating zone methods, crystals large enough for cutting can be produced in a matter of hours. The slower growth of flux stones allows for the formation of angular flat faces similar to those on natural crystals. As a result, the curved growth lines which are so important for identifying the melt-grown synthetics will not be found in the flux stones, except when flame fusion seed crystals are used. Instead, they display straight bands which run parallel to the crystal faces and meet at angles, just like in the natural stone.

To date, five different manufacturers have successfully synthesized corundum on a commercial scale using the flux process. They are listed below with the place of origin in chronological order:

1. Chatham — San Francisco, California (USA)
   Varieties: Ruby (first crystal 1952, first marketed 1959), blue sapphire (first crystal 1972), orange sapphire (first crystal 1978), yellow sapphire (first marketed 1986).
2. Kashan — Austin, Texas (USA)
   Varieties: Ruby, pink sapphire (first crystals, Ardon Industries, 1964, first marketed 1968, Kashan Laboratories formed 1978).
3. Knischka — Austria
   Varieties: Ruby (first marketed 1980).
4. Ramaura — Los Angeles, California (USA)
   Varieties: Ruby (first marketed 1983).
5. Lechleitner — Innsbruck (Austria)
   Varieties: Ruby, blue, orange, pink, yellow, green, colour-change, colourless sapphire (first crystals 1983).

Figure 7.10 An early photograph of Carroll Chatham in his laboratory. (Courtesy of Chatham Created Gems)

Carroll Chatham was the first to produce and market flux-grown synthetic corundum and his company, Chatham Created Gems, is perhaps the most successful of the five. Best known for his pioneering efforts in the synthesis of emerald, Chatham was a brilliant scientist who did much to advance the synthetic gem industry. His dedication to his work eventually cost him his life when, in 1983, he passed away as a result of the cumulative effects of exposure to beryllium compounds and toxic gases. Today, his two sons, John and Tom, are in charge of his company.

Chatham's initial production of synthetic corundum was limited to ruby, with blue sapphire following in 1972, orange sapphire in 1978 and yellow sapphire in 1986. The ruby and the orange and yellow sapphires are currently grown in crystals ranging from 1 to 1500 carats each. This allows the faceting of relatively clean stones of up to 10 carats. Even though similar-sized blue sapphires have been grown, problems with the colour distribution and clarity have prevented its sale, except as small crystal groups which are not suitable for cutting. Unconfirmed reports state that Chatham has now abandoned the flux growth of blue sapphires, and new production of blue sapphires is made by the Czochralski process.

In Chatham's approach, growth takes place on seed crystals of either natural or synthetic corundum suspended from platinum wires. The crucible is also apparently made of, or lined with, platinum, for the stones frequently display included platinum plates and needles, resulting from the partial dissolution of the crucible walls. The use of another metal such as iridium for the crucible could eliminate these inclusions, but this change would certainly sadden gemmologists because these shiny bits of platinum provide one of the surest methods of distinction from the natural stone.

Beginning in 1963, work was started on the flux growth of synthetic ruby by scientists of the American oil company Ardon Associates (Everhart, 1985). Out of these experiments was born the Kashan synthetic ruby. At the prompting of the head of Ardon Associates, who was a gem enthusiast, Dr Trueheart Brown and two other Ardon scientists attempted the synthesis of ruby. The research was later carried on privately and, by 1968, Brown and his two colleagues had succeeded in producing commercial quantities of gem-quality synthetic ruby, which they sold under the name of Kashan.

Although the Kashan product was of high quality, it did not achieve widespread recognition until the latter part of the 1970s. Prior to this, Burmese rubies were readily available in the natural gem market, but by the mid to late 1970s they had become increasingly scarce as a result of political problems. This vacuum was quickly filled by Thailand, which soon developed into the world's biggest supplier of facet-quality natural ruby. Renewed interest in the darker-toned Thai rubies also resulted in greater demand for the Kashan flux-grown synthetic rubies because of their marked similarity in colour. An additional boost was provided by rumours sweeping the trade to the effect that several Kashan synthetic rubies submitted to major gemmological laboratories had been certified as natural. Although these reports were never thoroughly investigated or proven, they did succeed in thrusting Kashan rubies further into the spotlight.

In hindsight, the event which was probably most responsible for the shift in

Kashan's fortunes was when Aris 'Bob' Mallas, a gem dealer and newsletter publisher, joined Kashan and took charge of the marketing and cutting. Shortly thereafter, a new company, Kashan Inc., was formed in 1978, and Mallas promoted the stones vigorously. The effect was almost immediate. Before Mallas became involved with Brown and his colleagues in the mid 1970s, the business was largely a hobby and their rubies relatively unknown. However, just five years after Mallas joined the company, the Kashan name had become a household word amongst those involved in the gem business.

Kashan stones have over the years acquired a reputation for being extremely difficult to identify, due in no small part to the marketing tactics of the company. Evidently it was felt that sales would increase if the buying public believed that the Kashan product could not be separated from the natural stone, for the difficulties in making such a separation were repeatedly stressed in their promotional literature. This attitude is vividly illustrated by the following quotations from a 1975 newsletter written by the future Kashan president Aris A. Mallas and reprinted for a Kashan sales brochure:

> Kashans are not synthetic, which is really like glass – a haphazard arrangement of atoms. Kashans are like natural crystals which is a precise geometric pattern created billions of times over.

Mallas continues

> . . . Moreover, I've had them (Kashans) examined by experts who tell me what fine (natural) rubies they are. Kashans are so remarkable they fool these experts – and I don't mean your friendly jeweler down the street, since almost anyone can fool him on gems other than diamonds – I mean top experts.

Contrary to the above statements, synthetic gemstones are not like glass and Kashan rubies are most certainly synthetic. In addition, the Kashan product very definitely can be separated from that of nature, provided one becomes familiar with the types of inclusions found in each.

In terms of visual appearance and colour, Kashan synthetic rubies generally bear a much stronger resemblance to natural rubies from Thailand than Burma. They are manufactured in a broad range of reds, from pure hues to orangy reds and purplish reds, including dark blackish reds. Kashan also produces synthetic pink sapphires of light, medium and deep tones and intensities. One peculiar characteristic of Kashan synthetic rubies (particularly the deeper red stones) is that they often display a slight orangy tint, somewhat reminiscent of natural red spinels. This is due to an overly yellowish extraordinary ray which, according to one report, results from a high titanium content.

Rough Kashan crystals range in size from 10 to 45 grams or more and are tabular formations which show some development of the rhombohedron faces. They are sawed in half before sale, with a further 'v'-shaped cut being made in the centre, apparently to detach the seed crystal. Non-seeded crystal clusters, or rosettes, are also marketed for use in jewellery, being too heavily included or too thin for faceting.

Figure 7.11 Rough rubies (flux grown) produced by Kashan. Note the 'v'-shaped cuts at the centre of each stone where the seed crystal was originally located. (Photograph by AIGS, Bangkok)

The saga of the Kashan synthetic ruby took an interesting turn recently when, in an article in the American trade magazine *National Jeweler* (1 February 1985), it was revealed that Kashan Inc. had filed for bankruptcy in October 1984 (Everhart, 1985). According to the account, Dr Trueheart and Nancy Brown, producers of the Kashan ruby and major stockholders of Kashan Laboratories, accused Kashan Inc. (marketing arm of Kashan Laboratories) president Aris 'Bob' Mallas of management irregularities and attempts to mislead court-ordered investigations into the company. The Browns' efforts at obtaining an accounting of Kashan Inc.'s assets allegedly caused Mallas to threaten to 'throw the company into bankruptcy' if they pushed too hard (Everhart, 1985). Evidently the Browns did just that, for Mallas made good his threat. Thus, the future prospects of the Kashan ruby appear anything but rosy. However, before other manufacturers break out the party hats and booze, they should keep in mind that it is highly likely that the stocks of unsold material will still find their way into the world's gem markets, with or without the help of Kashan Inc.

In 1980, Dr Edward Gübelin, the famous Swiss gemmologist, introduced the world to a new type of flux-grown synthetic ruby from Austria (Gübelin, 1982a,b). Produced by Professor Paul Otto Knischka, the ҀK (P.K.) ruby, as it is known, is distinctive in terms of its unique suite of inclusions and unusual crystal habits.

To date, Knischka has grown only ruby which, in terms of colour, tends towards a purplish red. Although reports have circulated that plans are afoot to market the stones commercially, present production is extremely limited, with one source estimating it to be barely 100 carats per year.

Knischka crystals differ from most other natural and synthetic rubies in the large number of faces they display. Among natural rubies it is rare to find a crystal which shows even 20 different faces. However, this new synthetic ruby has been grown with up to 40 faces on occasion. Their crystal habits are also unusual, in that some display a pseudo-isometric shape, while others have grown as hexagonal bipyramids.

Professor Knischka has recently raised the temperature of quite a few gemmologists by sending out promotional literature which describes his rubies as natural. The term natural should be restricted to those gemstones which are wholly formed by nature. This claim was based on an alternative flux-grown product offered by Knischka which used powdered natural ruby as the source material.

One of the recent flux-grown rubies to enter the marketplace is that of California crystal grower Ms Judith Osmer. First appearing in 1983, the Ramaura synthetic ruby, as it is known, is grown in various shades of red, including pure red, purplish red, orangy red and pinkish red, all of medium to high intensity (Kane, 1983). Ms Osmer has also produced blue sapphires, but only on an experimental basis. One notable feature of the rubies is their remarkable degree of transparency and clarity. These are said to be due to the spontaneous nucleation of Ramaura crystals, which reduces defects, such as impurities and dislocations, that can affect the passage of light.

Another interesting feature of the Ramaura synthetic rubies concerns their reaction to ultraviolet light. In an attempt to ease identification of the stones, the manufacturer has added traces of a rare earth element to the flux mixture so as to induce a characteristic orangy-yellow fluorescence. This is notable because, to the best of the author's knowledge, it represents the first documented case in which a manufacturer has deliberately attempted to build an identifying feature into its product. However, it would be wise to consider the following before you put your microscope up for sale. Even if this were to be done with all synthetic stones, it could not provide the panacea that many hope for because gemmologists would still have to guard against a dishonest manufacturer making up a batch of stones in which the dopant was omitted. In any case, the attempt has not been completely successful. Unfortunately, the dopant seems to be absorbed only into the surface regions of the crystals, resulting in the removal of the fluorescent areas during the cutting process.

The latest flux-grown synthetic corundum to appear in the trade is that of Austrian crystal grower Johann Lechleitner (Kane, 1985). Best known for his work with synthetic emerald, Lechleitner has only recently turned his attention to the growth of corundum. Unlike his emeralds, which were produced hydrothermally, the new synthetic corundums are grown from a flux bath, consisting of flux growth on seeds of Verneuil manufacture or on seeds of natural corundum.

According to a recently published account, Lechleitner first attempted the growth of corundum in late 1983. The production consists largely of ruby and blue sapphire, but also includes orange, pink, green, yellow, colourless and colour-change varieties. Faceted stones examined to date have ranged from 0.50 to 1.50 carats and appear relatively clean to the naked eye, except for a slight cloudiness. Magnification revealed this to be due to the presence of residual flux inclusions distributed in the typical wispy patterns which so often characterize flux-grown synthetic gemstones. Additionally, each stone was found to contain curved growth lines in the centre, indicating that a flame fusion seed crystal was used to initiate the flux growth. This, in itself, was not unusual. What was unexpected was the lack of any visible separation between the seed and flux overgrowth, making it impossible to discern the exact size of the seed and the thickness of the overgrowth. Small seed crystals could have been used to provide a method of nucleating and controlling growth, or larger seeds could have been overgrown with the aim of adding bulk, so that larger stones could be produced.

Unfortunately, instead of a clear division between each part, the flux veils penetrated deep into the gem, indicating that the seed may have been partially dissolved during the overgrowth stage (or they may simply represent fractures healed shut by the flux during the growth process).

The seed crystal used by Lechleitner may not necessarily be of the same colour as the overgrown material. Several different combinations have been produced, including flux pink sapphire over Verneuil colourless sapphire, flux ruby over Verneuil ruby, and even flux ruby over natural corundum. To date, most overgrowths are approximately 1 mm in thickness, which requires 10 to 20 days to produce. However, on occasion, covers as thick as 2 mm have been grown. According to Lechleitner, his goal is eventually to obtain thick enough covers so that gems can be cut from just the overgrowth itself, thus excluding the seed completely.

Basic identification techniques

The separation of natural corundum from other gem species, natural or synthetic, is not a great problem for the gemmologist. Differences in chemical composition and crystal structure produce differences in the characteristic physical and optical properties. Thus, in almost all cases, simply determining the optical character and refractive index is enough to allow separation between corundum and other species such as spinel, synthetic spinel, chrysoberyl, garnet, etc. Separating natural and synthetic corundum, however, is another matter entirely.

Synthetic minerals, by definition, closely duplicate the chemical and structural nature of their natural counterparts. As a result, the physical and optical properties are, with few exceptions, of little use in their distinction. This is particularly true with the different types of synthetic corundum, whose identification is almost totally based upon recognition of inclusions under the microscope. Unlike other common gem-testing instruments, the microscope does not yield quantitative information about the gem. Instead, its information is extremely subjective in nature and open to different interpretations by different individuals. More than any other instrument, its usefulness depends upon the skill and experience of the operator: skill in manipulating the stone and adjusting the lighting, and experience in distinguishing between natural inclusions and those of a similar appearance in the synthetic material.

Use of the microscope, however, allows valuable hands-on experience to be gained in working with the material in question. This includes not only the synthetic stone but, of equal importance, the natural material as well. Today, synthetic corundum displays such a bewildering variety of inclusions that it is necessary to examine hundreds of pieces of both synthetic and natural material before they can be identified with 100% accuracy. Only by becoming totally familiar with the range of possible inclusions in each can the gemmologist make the distinction between the works of nature and those of its imitator, man. This search for truth carries with it an awesome responsibility; but this is, and always has been, the challenge of gemmology.

Flame fusion (Verneuil)

Verneuil-produced synthetic corundum is encountered with much greater frequency than all other types combined. Given this fact, it is surprising that so few gemmologists take the time to study this product carefully. Even though the flux-grown synthetic stones offer more of a challenge, the huge quantities of flame fusion material in the market make it imperative that it is not taken too lightly.

The most important diagnostic feature of Verneuil synthetic corundum is the presence of curved growth lines, which result from minute differences in the distribution of the colouring agent(s) as the boule grows. With the growth taking place vertically, as layer after layer forms on top of one another, these lines will follow the contours of the domed top surface of the boule. Thus, in order to see them we must look in directions roughly perpendicular to the long axis of the boule.

The curved growth lines seen in synthetic ruby and colour-change sapphire are typically narrow striations which resemble the grooves of a phonograph record and are termed curved striae. In contrast, those found in other colours are usually broad arcing bands of colour, and so are termed curved colour banding. A distinction is made between the two different types because curved striae in stones of a greyish-green colour (in daylight) would suggest a synthetic colour-change sapphire.

Figure 7.12 Curved striae in a Verneuil-grown synthetic ruby. (Photograph by the author)

Locating curved growth lines in cut stones sometimes presents difficulties, especially when working with pale-coloured specimens or with the yellow, orange and pink varieties. The illumination of the specimen is extremely important but unfortunately often overlooked. In general, narrow growth lines, such as curved striae, will be resolved best under dark-field illumination, or light-field illumination using the 'shadowing' technique. By using transmitted light with an opaque light shield, such as a business card, inserted parallel to the growth lines between the light and the stone, fine details that might otherwise be missed are revealed. This is the shadowing technique.

In contrast to curved striae, broad growth lines (like the curved colour banding of

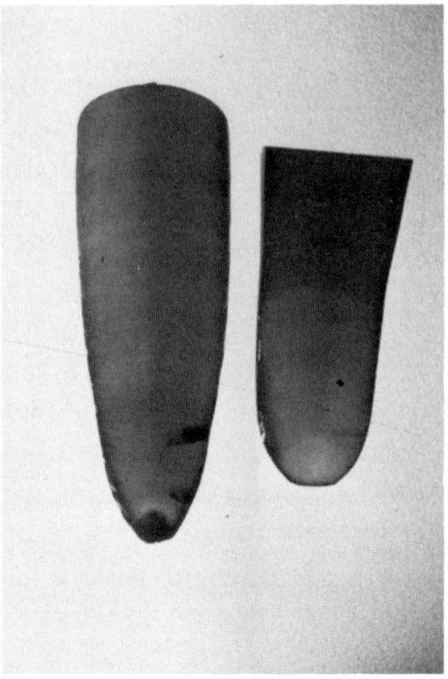

Figure 7.13 Curved colour banding seen in Verneuil synthetic blue sapphire following the curved upper surface of the boule (at left). (Courtesy of the Gemological Institute of America)

synthetic blue sapphire) are usually best observed by using dark-field or diffused light-field illumination. The latter can be obtained by the insertion of a translucent (frosted) white glass or plastic filter between the light source and the stone. Keep in mind that the only way to discover which illumination method is ideal for any given stone is to try them all. If one technique doesn't work, try another! It is also vital that the stone be examined from as many different directions as possible. If you don't see the lines in one position, then try another. This fact cannot be emphasized enough.

One of the chief difficulties of locating the growth lines in cut stones is reflection from the facets, making it nearly impossible to see into the stone from certain directions. This problem can be overcome by examining the stone in many directions or when it is immersed in pure methylene iodide. Immersion is extremely useful for locating both narrow and broad growth lines, and is often essential if the lines are to be seen in the pale yellow, orange, pink, blue and green varieties.

When using the immersion cell, the method of illumination is the same as if the specimen were being examined dry. Most pale-coloured specimens feature broad growth lines and so diffused light-field illumination is preferred. In 1981, the author discovered that the ability to locate growth lines in the difficult pale yellow and orange varieties can be tremendously enhanced by the use of a transparent blue colour-balancing filter, such as the #80A or 82A glass filters used with cameras, or simply frosted blue plastic. Because of the yellow bias of most tungsten microscope bulbs and the yellow colour of methylene iodide, it can be particularly difficult to locate the colour bands in yellow and orange stones. However, by inserting a blue filter under the immersion cell, or if the filter is of good optical quality, just beneath the objective lens of the microscope, the yellowish cast can be eliminated. This makes it possible to resolve the growth lines in even the most difficult yellow stones, provided the stone is patiently examined in all directions. Dry examinations are also improved by use of the blue filter, making it a tool that no gemmologist should be without. Yellow-green filters also make it easier to locate growth lines in rubies. Experiments have shown that the filter colour should match the absorption maxima (wavelengths of maximum absorption) of the stone.

There are other tricks which can also help in locating the colour zoning of difficult stones. One involves observation of the gem's ultraviolet fluorescence (if any) while under magnification and is especially useful for synthetic colourless sapphire. With low-power magnification (2×–6×), fine growth details such as curved banding can often be pin-pointed, which would otherwise be missed. This technique makes it possible to locate curved banding even in synthetic colourless sapphire. However, it should be mentioned that short-wave ultraviolet radiation is harmful to the eyes and so protective glasses which block out these rays must be worn when this test is performed.

Gas bubbles are another common feature of Verneuil synthetic corundum and are of great diagnostic value. They are found in a wide variety of shapes and sizes, but always appear in high relief because of the difference in refractive index between the gas and the surrounding stone. The concentration of gas bubbles may vary widely from one specimen to another. Some stones exhibit large numbers, while in others they appear only infrequently or not at all. Typically, gas bubbles are most common in dark-blue stones and least common in the yellow varieties.

The best type of illumination for the observation of gas bubbles is dark-field or, better still, fibre optic lighting. Not all fibre optic illuminators are created equal, however. In selecting one for gemmological use, care should be taken to ensure that the light source is strong enough and the lightguide large enough in diameter. Ideal sources are those utilizing a 150 W quartz halogen projector bulb with a built-in reflector, together with a flexible lightguide of 3/8 to 1/2 inch diameter. With a good fibre optic illuminator, bubbles that would be nearly invisible under ordinary dark-field illumination can be easily spotted as bright spots on a dark background. Large examples may resemble doughnuts, due to the manner in which light is reflected from the rounded surfaces. Spherical shapes are most common, with triangular or 'tadpole' shapes also being seen where the 'head' of the bubble points away from the top boule surface, perpendicular to the curved growth lines. Rarely, irregular worm-like patterns which resemble the fingerprints of natural stones may be encountered. These are among the most deceptive features found in the flame fusion synthetic stones and are believed to result from improper calcination of the raw alumina powder (Gübelin, 1953). Separation of gas-filled inclusions from the predominantly liquid-filled forms in natural corundum can be achieved by immersion with light-field illumination, since the gaseous inclusions appear dark while the liquid, in contrast, is transparent and light.

Gas bubble clouds generally follow the curved growth lines and thus reveal the direction of those lines. Within the clouds, some bubbles may be found which are elongated. Because this elongation is usually against the growth structure (perpendicular to the growth lines), their patterns can provide an important clue to the direction of the curved colour banding or striae.

All other inclusions seen in Verneuil synthetic stones are of minor importance compared with the curved growth lines and gas bubbles. Stones cut from the very base of the boule may include part of the seed rod upon which the growth is

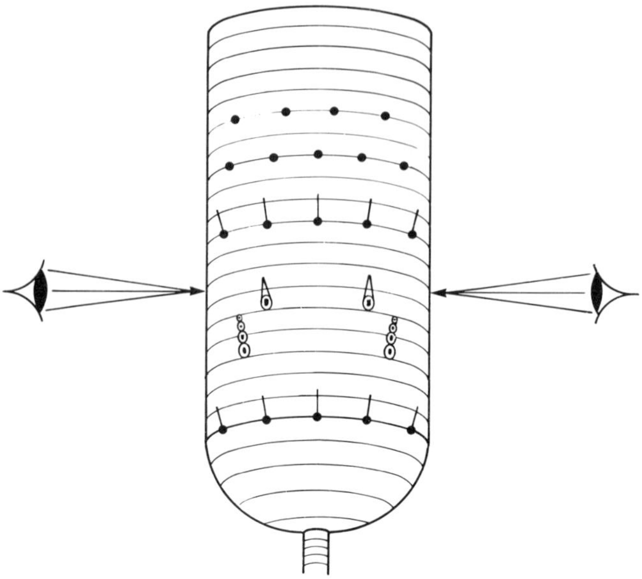

Figure 7.14 The distribution of curved growth lines and gas bubbles in Verneuil synthetic corundum. Curved growth lines are seen by looking at roughly 90° to the boule's length. Gas bubbles usually occur in layers which follow the curved growth lines. When the bubbles are elongated, this elongation is usually at right angles to the direction of the curved growth lines

sometimes initiated. Should the colour of the rod be different from that of the stone, the inclusions will show up clearly under magnification. In addition, the interface between the seed rod and the stone often shows a frosted appearance, making the seed readily visible.

It was long believed that oriented planar growth phenomena occurred solely in natural corundum. However, sometime about 1920, Mr Sandmeier, a Swiss gemmologist, observed an unusual banding in a Verneuil synthetic stone. In 1952 his findings were confirmed by Mr W. Plato of West Germany. These 'Sandmeier–Plato' striations, which are detected only with great difficulty, consist of slip or glide lines which form as a result of the rapid cooling of the boule and the relaxation of residual stresses when it is split. Slip lines represent the concentration of dislocations along crystallographic planes (Belyaev, 1980).

Plato lines can be found in each of the transparent varieties of flame fusion synthetic corundum, but only a small percentage of stones actually display them. To observe Plato lines, the stone must be examined under magnification between crossed polaroids. It is also necessary that the stone be immersed in methylene iodide and viewed parallel to the optic axis. Once the optic axis is correctly positioned, the lines will appear in single, double or triple sets which intersect at 60° or 120° angles. Irregular shadows similar to the anomalous double refraction effect of glass are also seen between the lines and result from the strain built up during the boule's formation. These are most notable in areas corresponding to the outer edges of the boule (Eppler, 1964).

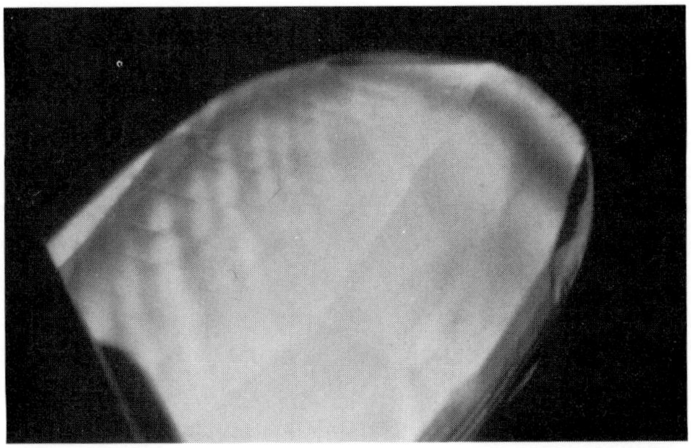

Figure 7.15 'Plato lines' seen in a Verneuil synthetic yellow sapphire parallel to the *c* axis while immersed in methylene iodide between crossed polars. The Plato lines are a form of repeated twinning parallel to the faces of the hexagonal prism. (Photograph by AIGS, Bangkok)

Plato lines are normally used only as a last resort, when all other tests have failed to identify the stone conclusively. This is because of the rather complicated set-up required to locate them and their complete absence in many Verneuil stones. However, if present, they do provide positive evidence of synthetic origin. Although the repeated twinning lamellae of natural corundum may somewhat resemble Plato lines, those in the natural material do not run parallel to the optic axis. The only oriented lines found parallel to the optic axis in natural rubies and sapphires are colour bands; these are dark between crossed polars and lack the irregular shadows of the true Plato lines.

Repeated twinning is normally considered to be a diagnostic feature of natural corundum. Few people are aware, however, that in certain rare instances twinning identical in orientation and appearance to the natural one has been found in Verneuil stones (Eppler, 1964). The twinning of flame fusion stones consists of one or more parallel planes oriented along the faces of the primitive rhombohedron. Generally the planes are few in number and are found in one direction only, but in rare cases may intersect at 87° and 93° angles. It must be stressed that these are not the Plato lines just discussed, but instead are planes oriented exactly like the repeated twinning of natural stones. In addition to the similarity in orientation, the repeated twinning of the synthetic stone also displays the same appearance as the natural one between crossed polars, showing interference fringes. Mechanical deformation and rapid cooling are believed to be the cause of these deceptive twinning planes, which occur most often in the blue, yellow and orange varieties. Even in these stones repeated twinning is by no means a commonly encountered feature, but we should be aware of the possibility.The presence of twinning planes in the Verneuil stones may come as an unpleasant surprise to many, but the curved growth lines and gas bubbles will still allow a positive identification to be made.

In recent years, a small number of flame fusion stones containing extremely realistic fingerprints and feathers have appeared in the world's gem markets. These stones possess all of the typical features of the Verneuil process and so are identifiable as such. Yet, in addition, they contain liquid fingerprints and feathers which, by themselves, are virtually indistinguishable from those found in natural rubies and sapphires. Much speculation has occurred on the possible causes of these inclusions, with some gemmologists suspecting that they have been deliberately induced by cracking the stones and subsequently healing these fractures at high temperatures. Whether these 'induced fingerprints', as they are termed, result from accidents during growth or from deliberate attempts to deceive still remains to be proved. However, those observed so far always meet the surface of the stone, and thus only fingerprints and feathers which lie completely within a stone can still be considered as definite indications of natural origin.

Another unusual type of Verneuil synthetic is that which is termed 'quench-crackled'. These are stones that have been heated and then, while still hot, plunged into a much cooler liquid. This produces fractures which cause the stones to resemble more closely the natural product. Synthetic ruby is most often treated in this way, with the fractures forming a characteristic 'honeycomb' pattern. Occasionally oils or other liquids are forced into the cracks in an effort to simulate the fingerprints and feathers of natural stones. Quench-crackled Verneuil stones can still be identified, of course, by the curved growth lines and gas bubbles.

The spectroscope

The spectroscope is an important diagnostic tool, for, with the exception of the microscope, it is the only basic gem-testing instrument that can provide the positive means to separate natural and synthetic corundum.

Natural and synthetic ruby are both coloured by traces of chromium and, as a result, in most cases they show identical absorption lines. However, for sapphires this is fortunately not the case. The blue colour of natural sapphires has been shown to be caused by small amounts of iron and titanium. These produce an absorption complex of three lines at 451.5, 460 and 470 nm, which collectively are termed 'iron lines'. On occasion, only the 451.5 nm line can be seen and in heat-treated stones this too may be absent, especially in those from Sri Lanka. Verneuil synthetic blue sapphires also owe their colour to the iron–titanium mixture, but because of the greater volatility of iron, the iron absorption lines are not seen. In very rare instances a weak poorly defined line at about 450 nm has been reported in certain synthetic sapphires of a medium- to deep-blue colour (Anderson, 1980). However, this is exceptional, for amongst the hundreds of specimens which the author has examined, only a few pieces have exhibited this line. Caution is essential, for a weak 451.5 nm line has also been observed in Chatham flux-grown synthetic blue sapphire (Kane, 1982). Although the complete series of iron lines can be taken as positive proof of natural origin, a weak line by itself at 451.5 nm should not. Given their complete absence, the stone in question could be either natural or synthetic.

Verneuil synthetic blue sapphires may have even weaker bands on either side of that at 451.5 nm which are nearly impossible to see. A vague blur is seen, or 'sensed', at about 490 nm, just where the green ends and the blue begins. The other is an even weaker blur in the violet at about 428 nm.

In addition to the above, deep-blue sapphires of either natural or synthetic origin may show a band several tens of nanometres wide centred at about 560 nm. This is due to the ordinary ray and can be intensified slightly by rotating a polaroid plate over the eyepiece to eliminate the extraordinary ray. Although quite strong at times, the 560 nm band has no diagnostic value because it is seen in both natural and synthetic sapphires (Anderson, 1980).

Iron also imparts the colour, in whole or in part, to most natural yellow, orange and green sapphires, so they too may show the iron lines at 451.5, 460 and 470 nm. Again, Sri Lankan stones can show only a weak 451.5 nm line, which is reduced or removed completely by heat treatment. Iron is not used as a colouring agent in Verneuil synthetic sapphires for these colours. Instead, the synthetic yellow sapphire is coloured by nickel, the orange by nickel and chromium, and the green by cobalt plus nickel and vanadium. In addition to the absence of iron lines, synthetic green sapphires have on occasion shown a line due to vanadium at 475 nm, or one at 528 nm, and another at 688 nm which may possibly be due to nickel.

Although iron is not used to colour synthetic yellow and orange sapphires made by the Verneuil process, a very weak, ill-defined line at 455 nm has also been reported on rare occasions (Anderson, 1980). In the deeper yellow and orangy-yellow synthetic sapphires, a complete cut-off from about 450 nm to the end of the violet may be seen. This is also seen in some natural sapphires and so is of no diagnostic value.

On occasion, the 451.5 nm complex has been detected in iron-rich purplish-red rubies, particularly those from Thailand. These stones will also show the normal chromium spectrum, but since the iron lines have yet to be found in synthetic rubies, their presence can be taken as proof of natural origin.

The final instance in which the spectroscope is diagnostic is for the Verneuil synthetic colour-change sapphire. Here vanadium is utilized as the colouring agent and it shows up in the form of features that somewhat resemble those of ruby (natural and synthetic), with a fluorescent line close to 690 nm and a broad absorption band centred at about 580 nm. However, instead of three lines in the blue as shown by ruby, the synthetic colour-change sapphire shows a single prominent line at 475 nm. This line has only very rarely been seen in natural corundum and thus is a strong indication of a synthetic stone.

Fluorescence

The use of ultraviolet fluorescence for the separation of natural and Verneuil synthetic corundum is perhaps the most overrated of all tests. At best its usefulness is extremely limited and under no circumstances should it be considered anything other than a weak indication of a stone's identity. Due to the overlap in the reactions of the natural and the flame fusion synthetic stone, great care is needed if mistakes are to be avoided.

To simplify matters somewhat, the various fluorescent reactions of Verneuil synthetic corundum have been listed in Table 7.3. These serve only as a general guide, however, and the actual fluorescence observed may differ considerably from the reactions listed.

TABLE 7.3

Verneuil synthetic variety	*Ultraviolet fluorescence*	
	Long wave (366 nm)	*Short wave* (253.7 nm)
Ruby	Moderate to very strong red to orange red; usually very strong, except in dark-toned stones	Weak to very strong red to orange red; usually very strong except in dark-toned stones
Pink sapphire	Moderate to very strong red, orange red, pink or pink red; usually very strong	Moderate to very strong red, orange red, pink or pink red; usually very strong
Blue sapphire	Generally inert; stones containing Cr may show weak to strong red to orange red	Inert to strong chalky blue to chalky green; it is the colourless areas which fluoresce while the blue bands are inert; stones containing Cr may show weak to strong red, orange red or pink red
Green sapphire	Weak to moderate orange-red, orange or red	Weak to moderate dull orange to brownish red
Orange sapphire	Inert to strong red, orange red or orange	Inert to strong red, orange red or orange
Yellow sapphire	Inert to moderate red to orange red	Inert to moderate red to orange red
Colourless sapphire	Inert to weak bluish white or green	Weak to strong white, or chalky pale blue
Violet and purple sapphire	Weak to strong red to orange red	Weak to strong red to orange red
Colour change sapphire	Weak to strong orange	Weak to strong orange, chalky blue or green

Fluorescence may reveal curved growth lines in some varieties (particularly synthetic colourless and blue sapphires) when examined under magnification. However, it is extremely important that the eyes be properly shielded from the short-wave radiation when this test is performed.

Synthetic star corundum

Synthetic star rubies and sapphires can in almost all cases be easily identified by their characteristic appearance and inclusions. Visually, the synthetic product appears 'too good', with most stones possessing an extremely sharp star due to the fine nature of the included silk. In addition, the synthetic stone is often less transparent than the natural one, bordering on opaque in the so-called 'porcelain' quality. Exceptions to the above are the early Linde and some recent synthetic stones, which are often more transparent overall and may show incomplete stars owing to an uneven distribution of silk.

In contrast to natural star corundums, the synthetic star stones are usually characterized by absolutely flat polished bases having little or no excess weight below the girdle. As an additional aid to their identification, many of the original Linde star stones feature an engraved 'ℓ' (standing for 'Linde') on the base, while the modern synthetic ones show a partial or complete 'bull's-eye' circular zoning pattern on the bottom surface.

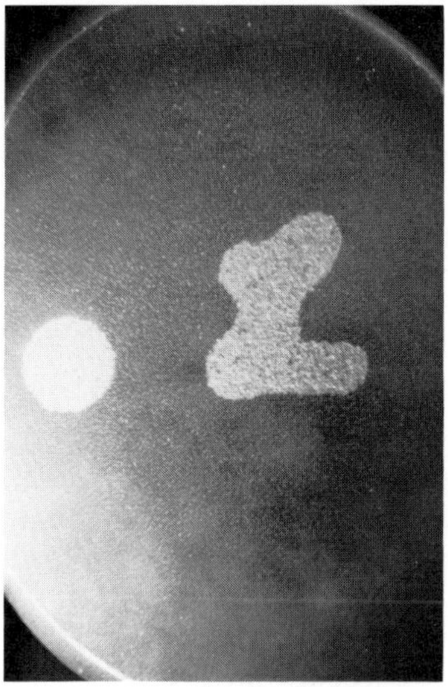

Figure 7.16 The engraved 'L' on the back of a Verneuil-grown synthetic star sapphire made by Linde. Also visible are numerous tiny gas bubbles. (Photograph by the author)

Positive means of identifying synthetic star stones are afforded by their characteristic inclusions. Gas bubbles are extremely common and will rarely, if ever, be absent entirely. However, due to the usual lack of transparency of the synthetic, they may not be seen clearly under dark-field illumination. Instead, when the nearly opaque 'porcelain'-grade stones are examined, the gas bubbles often show up better under light-field illumination, with the iris diaphragm open just enough to allow a small circle of light to pass through the stone near the girdle. Alternatively, the gas bubbles also show up clearly if the stone is illuminated from the side or edge with a fibre optic lightguide.

Other than gas bubbles, the most common and diagnostic inclusions in the synthetic star stones are the curved growth lines. Here again differences are seen between the Linde and modern products. Linde synthetic star rubies and sapphires show widely spaced easily visible curved growth lines at 90° to the optic axis. These differ greatly from the tightly packed curved striae of transparent Verneuil synthetic rubies. The modern star stones not only show curved lines roughly parallel to the cabochon base, but also display concentric bands which form a complete or partial bull's-eye pattern on the base of the cabochon itself. Often a circular completely transparent space surrounded by gas bubbles is seen. This is not found in Linde stones. In addition to the above, the modern product possesses a layered distribution of silk and bubbles. When the stone is viewed parallel to the optic axis under magnification, first silk comes into view. Then, as the focus is lowered, a layer of bubbles is seen, then silk again, and so on throughout the stone.

The silk of synthetic star corundum is generally much smaller in size and more evenly distributed than in the natural stone. That of natural star rubies and sapphires tends to be found in uneven strongly zoned areas easily visible under low-power magnification (10×–30×), whereas a minimum of 50× magnification is often required to resolve the silk in the synthetic. Due to differences in their respective growth rates, the needles in Linde synthetic star stones are usually longer and more slender, while the modern product contains shorter more coarse silk of uneven distribution. Differences in the size, shape and distribution of the needles cause differences in the appearance of the star. Linde star stones are very prominent and narrow as they contain long thin needles, while the modern synthetic star stones are wider and more diffuse, due to the short thick needles (Eppler, 1958). These factors, coupled with the greater transparency, result in a more natural appearance in the modern star stones.

Czochralski (pulling)

Czochralski-grown synthetic corundum is rarely seen in the jewellery trade, Inamori being the only manufacturer to grow it specifically for gem use. Although the following applies in particular to the Inamori product, Czochralski-produced synthetic corundum of other manufacturers will also show similar features.

Under the microscope Czochralski stones are often very clean and to the naked eye bear a remarkable similarity to gems of the Verneuil process. Careful examination, however, will reveal features which allow a positive distinction from both the natural and Verneuil synthetic stones. Foremost are gas bubbles, faint curved growth lines and, in the Inamori stones, nearly parallel rows of minute particles which resemble the 'rain-like' flux inclusions found in Kashan synthetic rubies. It is possible that these finely disseminated inclusions may consist of residual traces of crystallizers (flux) used to facilitate the growth. Platinum needles, crystals and plates have also been seen in the Inamori product, but are generally smaller and much less common than those of the Chatham flux-grown synthetic corundum.

The growth striae of Inamori rubies are tightly curved and so faint that their observation may require the use of immersion and shadowing. They can normally be seen from several different directions, with colour swirls and 'cigarette-smoke'-type veiling seen running across the striae. In addition to the above, the

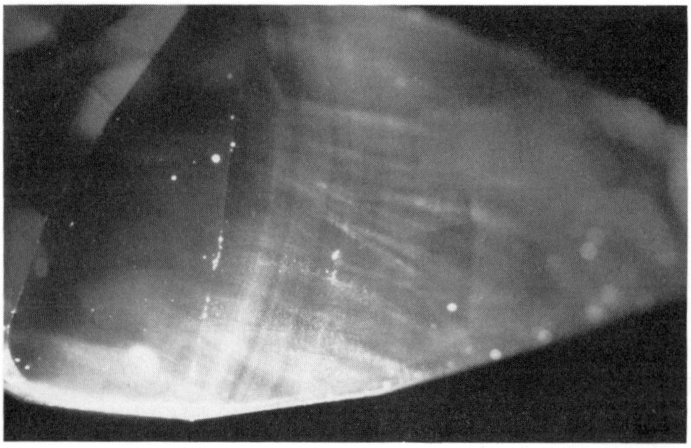

Figure 7.17 Curved striae and 'rain'-like particles (at roughly 90° to the striae) in an Inamori synthetic ruby grown by the Czochralski (pulling) process. (Photograph by the author)

manufacturer of Inamori stones states that feathers which resemble a 'bundle of twigs' have also been found. This may be in reference to the rain-like particles (Inamori Jewelry Division, 1980).

TABLE 7.4

Czochralski synthetic variety	*Ultraviolet fluorescence*	
	Long wave (366 nm)	*Short wave* (253.7 nm)
Inamori synthetic ruby	Strong to extremely strong red or orangy red; usually extremely strong	Strong to extremely strong red or orangy red; usually extremely strong
Inamori synthetic blue sapphire	Unknown	Unknown
Inamori synthetic orange sapphire	Strong orange red	Weak pinkish orange

Under long-and short-wave ultraviolet light (Table 7.4), the Inamori synthetic rubies show an intense red glow which is unequalled in strength by almost any other ruby, natural or synthetic. It is this fluorescence that is, by and large, responsible for the beautiful crimson colour of the Inamori product. The fluorescence of the Inamori synthetic orange sapphire is generally a weaker version of its red cousin, while that of the Inamori blue sapphire has not been reported on at the time of writing.

Floating zone

To date, Seiko is the only company to utilize the floating zone process for the growth of synthetic corundum intended for gem use. However, due to the fact that the Seiko synthetic corundum has only recently been released into the market, little information is available regarding its gemmological characteristics.

Identification of the Seiko product, which includes stones of a red, orange and blue colour, is achieved by use of the microscope. The most striking characteristic of these new synthetic stones is their remarkable freedom from inclusions, with the sole features being swirled growth lines and clouds of tadpole-shaped gas bubbles (Scarratt, 1984). In the case of the pink and orange varieties, it may be necessary to immerse the stones in methylene iodide before the flow lines become visible. Seiko rubies alone display a characteristic rectilinear (straight) parting which is caused by phase transitions or slight differences in refractive index within the stone (Brown, 1985). Overall, the internal features of these new Seiko stones bear a strong resemblance to those of the Verneuil or Czochralski-grown synthetic stones. However, the swirled appearance of the colour zoning permits distinction from both natural and other types of synthetic corundum.

Apart from the fluorescent reactions (see Table 7.5) of the red and orange varieties, other gemmological characteristics have yet to be published. Perhaps after more samples have been studied additional tests may prove useful for their identification.

The sizes of rough currently being produced allow for only small stones (less than 1.00 carats) to be cut. This is reportedly to hold down the price of the finished jewellery articles in which they are set.

TABLE 7.5

Floating zone synthetic variety	*Ultraviolet fluorescence*	
	Long wave (366 nm)	*Short wave* (253.7 nm)
Seiko synthetic ruby	Strong to extremely strong red	Moderate to extremely strong red to whitish red
Seiko synthetic orange sapphire	Strong to extremely strong red	Moderate to strong red
Seiko synthetic blue sapphire	Unknown	Unknown

Flux

One of the greatest challenges facing the modern gemmologist is the separation of natural and flux-grown synthetic corundum. Unlike the rapidly grown Verneuil stones, flux synthetic stones are grown slowly over a period of several months, developing flat crystal faces in the process. As a result they display internal features such as straight colour zoning which closely mimic the genuine stone. Careful examination of their inclusions will in almost all cases provide the proof necessary to make a positive distinction, but caution is needed if errors are to be avoided. Not only should the gemmologist have experience with the synthetic, but also it is of even greater importance to possess a thorough knowledge of the natural stone, for to neglect this is to commit an unpardonable sin – the condemnation of nature's creation as synthetic. Unfortunately, the author has witnessed this mistake far more frequently than the opposite.

Inclusions provide, with few exceptions, the only reliable means of distinguishing between natural and flux-grown synthetic corundum. As the product of each individual manufacturer commonly displays a distinctive suite of inclusions, the flux-grown synthetic materials will be described individually by producer.

Chatham

Chatham currently produces synthetic flux corundum in red, orange, yellow and blue colours and is the sole manufacturer of orange, yellow and blue sapphires by the flux process. The rubies tend to be of a colour more reminiscent of Burmese than Thai stones, although darker tones are also seen occasionally. Colour zoning in this variety is usually apparent only under magnification, while the orange sapphires are more strongly zoned, displaying areas of orange, yellow and pink, or mixtures of the three. Zoning in the Chatham blue sapphires is so strong that it has prevented their release onto the market, except as non-facet-quality crystal groups. Their colour ranges from a light to an extremely deep (almost black) blue. The yellow variety displays a bright yellow reminiscent of natural Australian stones, with zoning apparent only under magnification.

Of all the inclusions to be found in the Chatham synthetic stones, the most diagnostic is the platinum inclusion, for it serves as a positive means of separation not only from natural corundum, but also from most of the other flux synthetic products. Platinum inclusions result from a partial decomposition of the platinum-lined crucible and are found in a number of different shapes and sizes. Common forms include thin hexagonal or triangular platelets which may be bent or otherwise distorted (Kane, 1982). Large irregular flakes are also seen, as well as spikes, small flat needles and needles of triangular cross-section. In most cases, platinum inclusions can be readily separated from similarly appearing natural inclusions by their random orientation, complete opacity to light and metallic lustre. Platinum is never found as an inclusion in natural corundum.

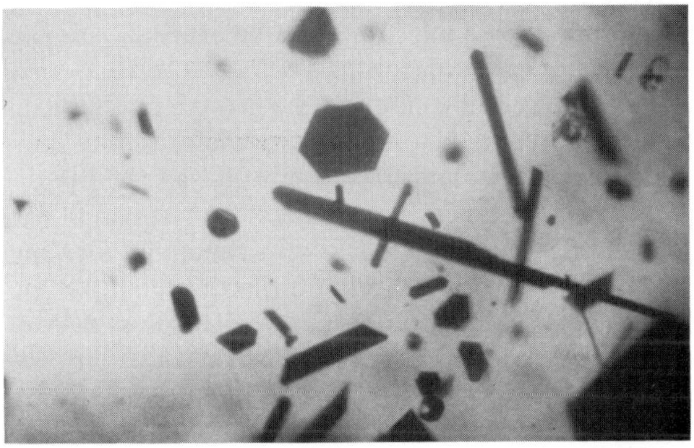

Figure 7.18 Various forms of platinum in a flux-grown Chatham synthetic ruby. Such inclusions display a random orientation and are opaque, with a metallic lustre. They are believed to result from the partial dissolution of a platinum or platinum-lined crucible. (Courtesy of AIGS, Bangkok)

Undoubtedly the most unusual of the various incarnations in which platinum occurs is the 'phantom-like' growth layers sometimes found near the outer edges of Chatham crystals. These platinum-rich films are seen best when looking parallel to their plane and appear as black-edged layers which follow the angular growth lines of the stone. If their presence is suspected, immersion in methylene iodide will quickly confirm their location.

In addition to the above, some Chatham synthetic blue sapphires exhibit thin white needles of platinum that vary in length from short to very long. These can be found as isolated individuals, in small groups, or as extensions of the more typical metallic platinum needles, and occur in both straight and curved forms (Kane, 1982).

The most common inclusions encountered in Chatham synthetic corundum are those of residual unmelted flux and are found in a variety of different patterns. It is these flux inclusions that have caused many a gemmologist to despair because of their uncanny resemblance to the fingerprints, feathers and other liquid inclusions of the natural stone. Generally speaking, the flux in Chatham stones is white or colourless, opaque and of high relief. If spread thin enough it may appear almost transparent, but this thin flux is almost always accompanied by thicker more opaque flux. The patterns in which the flux inclusions are found can be extremely varied, ranging from large globules of a pale whitish-yellow colour to wispy veils and thin mesh-like fingerprints, which are especially common. In fact, the flux motifs are so diverse that it is not advisable to base an identification on the pattern alone. Many who have tried to do so have failed, for natural rubies and sapphires also frequently contain liquid fingerprints which form similar wispy veil-like patterns. Rather than trying to separate flux inclusions by their patterns, they should instead be examined in detail under high magnification (100×–200×), preferably with a fibre optic light source. This will reveal a flux feather to be made up of tiny individual flux grains or droplets, none of which seem in any way connected to neighbouring droplets. In the case of mesh-like flux fingerprints, although the various channels may be interconnected the flux-free areas between the channels show little or no evidence of healing. Natural fingerprints, on the other hand, usually form as the result of a healing process which is more highly developed because of the much slower growth rate. During the growth of a natural crystal, internal stresses often cause fractures to develop. These fractures may then be filled by the growth solution which crystallizes within the confines of the break, healing it up. Inevitably the mother liquid contains impurities not involved in the growth of the crystal, and this residual liquid is trapped in thin channels, forming the familiar fingerprint inclusions of natural crystals. Unlike the flux-type fingerprints, which contain a flux glass (rapidly solidified flux), these inclusions will show evidence of a much slower healing process in the form of delicate patterns preserved on the walls of the healed fracture. Use of the fibre optic light torch reveals details such as fine growth steps and other markings. Minute liquid mosaics, or tiny 'fingerprints within the larger fingerprint' are seen. Vivid interference colours reflect from the thin films coating this once-open wound, testifying to the

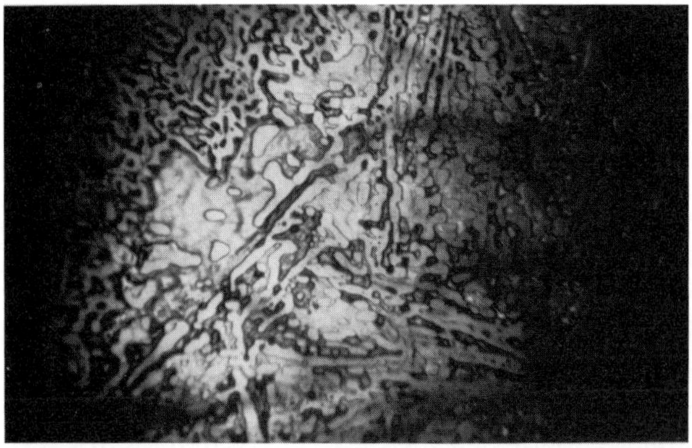

Figure 7.19 A very thick secondary inclusion (fingerprint) in a natural ruby from Thailand/Cambodia. In terms of thickness and pattern the appearance is very similar to the flux inclusions in flux-grown synthetic rubies. Thus a separation is made not just by reference to the pattern or thickness, but instead by determining if the filling is a liquid (natural) or a flux glass (synthetic). The flux inclusion under high magnification shows a crazed surface appearance. (Photograph by the author)

natural repair processes. Although interference colours may also be seen on fingerprints within the flux-grown synthetic stones, they will appear crude by comparison because the thicker flux glass hardened before the healing process could advance very far.

In Chatham stones the flux may sometimes appear slightly yellowish in colour, particularly in the blue variety. Thick opaque white globules of flux, on the other hand, are found most often in the synthetic rubies. The most common flux inclusions in all varieties are the wispy white veils of high relief and variable thickness, while curved or flat veils (which are not wispy) are also seen.

Platinum crystals and residual flux inclusions are by far the most common and diagnostic of the internal features of Chatham synthetic corundum. Other inclusions, however, may be found as well. Dense white clouds composed of tiny dust-like particles are seen in all varieties and are easily confused with the rutile silk or minute particle clouds of many natural rubies and sapphires. These clouds in the synthetic stone are often zoned in straight or angular patterns. Fibre optic illumination may reveal the presence of tiny white needles intersecting at 120° in three directions within these clouds, again similar to natural stones. Transparent near colourless or light-pink crystals of low relief have also been found included in Chatham stones; they may consist of corundum or chrysoberyl. In the synthetic ruby and orange sapphire they are commonly of very low relief, while those found in the blue sapphires tend to be more easily visible. Both the white clouds and the transparent crystals, if considered alone, pose serious problems because of their striking similarity to certain natural inclusions. However, to date, they have always been found in the presence of other more easily identifiable features, such as platinum and residual flux.

Unlike the Verneuil stones, flux-grown synthetic stones will show straight colour zoning which meets at angles. This is similar to the zoning in natural corundum, except that the angles between certain growth planes in flux-grown synthetic stones may differ from those in the natural ones and vice versa. A detailed discussion of the various differences and their usefulness in identification will be presented later in this chapter.

Apart from straight angular colour zoning, Chatham synthetic rubies may also show irregular colour swirls. Known variously as 'treacle', 'heat wave' or 'scotch and water', these colour swirls are reminiscent of those seen in natural rubies from Burma, Sri Lanka and Kenya, and thus cannot be considered as diagnostic for the synthetic stone alone.

Twinning is sometimes encountered in Chatham synthetic corundums along the rhombohedron $(10\bar{1}1)$ planes, but lacks the dislocation (boehmite) needles which usually accompany this inclusion in natural stones. In the synthetic stone, the planes are less numerous and rarely penetrate across the entire stone, as in natural corundums.

In addition to the facetable gem rough and cut stones, Chatham also markets flux-grown crystal groups in red, pink, orange, yellow and blue colours. These crystal groups, which are not suitable for cutting because of their small size, are coated on the backs and sides with a liquid silica-based ceramic glaze to decrease the chances of breakage. Examination of the glaze has revealed the presence of numerous spherical gas bubbles, but these are found only in the glaze and not within the crystals themselves (Kane, 1982).

The visible absorption spectra of Chatham flux-grown synthetic corundum reveals, in the rubies and orange sapphires, features identical to those of the natural stones, while the spectra of the yellow stones have yet to be reported on. Only in the blue sapphires have differences been noted. Chatham blue sapphires show just a single diffuse band at approximately 451.5 nm, and this is absent altogether in some specimens. Thus, if the complete series of iron lines is seen at 451.5, 460 and 470 nm, the gem can be assumed to be natural in origin.

The fluorescent reactions displayed by Chatham flux-grown synthetic corundum overlap completely with those of the natural stones and so are of little or no diagnostic use. However, for the sake of completeness they have been listed in Table 7.6.

TABLE 7.6

Synthetic variety	*Ultraviolet fluorescence*	
	Long wave (366 nm)	*Short wave* (253.7 nm)
Chatham synthetic flux ruby	Moderate to very strong red to orange red; usually very strong except in deeply coloured stones	Moderate to very strong red to orange red; usually very strong except in deeply coloured stones
Chatham synthetic flux orange sapphire	Strong to very strong orangy red, reddish orange or yellowish orange with chalky yellow areas	Very weak to weak orangy red, reddish orange or yellowish orange with chalky yellow areas
Chatham synthetic flux blue sapphire	Inert to very strong chalky greenish yellow, chalky reddish orange, yellowish green, brownish green to sulphur yellow; may have patchy irregular distribution	Inert to strong chalky greenish yellow, dull yellowish green, dull chalky greenish white, chalky reddish orange or strong yellow
Chatham synthetic flux yellow sapphire	Not yet recorded	Not yet recorded

Kashan

The manufacturers of the Kashan synthetic ruby have never quite been able to achieve the vivid red colour of the finest Burmese stones. However, they have succeeded in producing stones of high clarity and, as a result, the Kashan product can at times provide a stiff test of a gemmologist's abilities.

Kashan synthetic corundum is produced in colours ranging from pale pink to deep red. In general, its colour tends to resemble that of red spinel or Thai ruby, as opposed to the Chatham synthetic ruby which simulates better the highly fluorescent red typical of fine Burmese rubies. Contrary to what we might expect, the slightly inferior colour of most Kashan rubies has not been a disadvantage. Instead, due to the relative abundance of Thai rubies (as compared with Burmese) in today's marketplace, a steady demand for a Thai ruby lookalike has developed and Kashan stones fit the bill most closely.

Unlike the Chatham synthetic, Kashan rubies contain no platinum inclusions. Instead, their sole inclusion (other than colour zoning) is residual flux. Analysis of the flux in Kashan rubies has proven it to consist of a sodium aluminium fluoride (Na_3AlF_6) which occurs in nature as the monoclinic mineral cryolite (Gübelin, 1983a). The cryolite flux in Kashan rubies is typically white in colour and stands out in bold relief from the surrounding stone as a result of its low refractive index (n = 1.338). It may occasionally be transparent, but more often is translucent to opaque, with a 'crazed' or irregularly fractured granular surface texture. This crazed surface texture is very useful in separating the flux from liquid inclusions in natural corundum. The size of the flux inclusions may vary considerably. Large 'drippy' flux globules are commonly seen and provide positive evidence of synthetic origin. Coarse flux rods in parallel alignment point outwards towards the crystal faces of the rough material. This parallel alignment of flux is significant and typifies the internal features of many Kashan stones (Gübelin, 1983a,b). Closer examination of many flux droplets reveals an apparent two-phase (solid and gas) nature, while flat crystal faces or striations may be seen on the larger flux forms as a result of the flux filling of negative crystals.

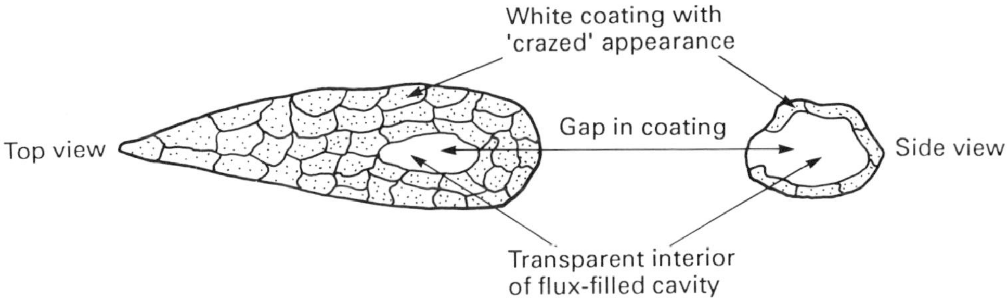

Figure 7.20 The anatomy of primary flux inclusions in Kashan flux-grown synthetic ruby. The inclusions consist of primary negative crystals partially filled with a flux glass and often appear to be two phase in nature. This is apparently because of gaps in the flux lining of the cavity. The flux glass also often exhibits irregular surface cracks or 'crazing' which are distinctive and diagnostic in appearance. (After Burch, 1984)

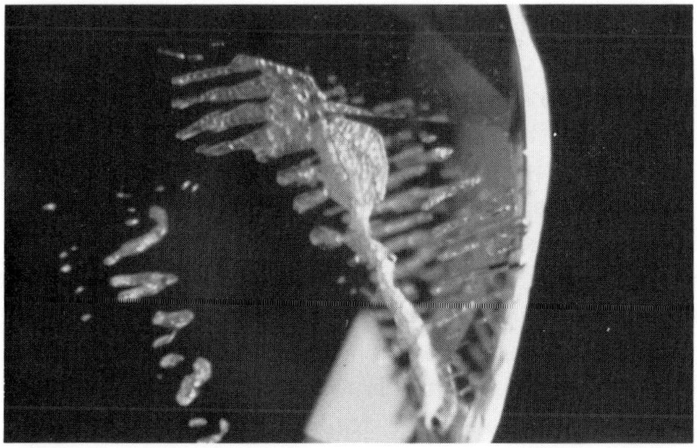

Figure 7.21 Large primary flux-filled negative inclusions in a Kashan flux-grown synthetic ruby. Note the crazed surface appearance. (Photograph by the author)

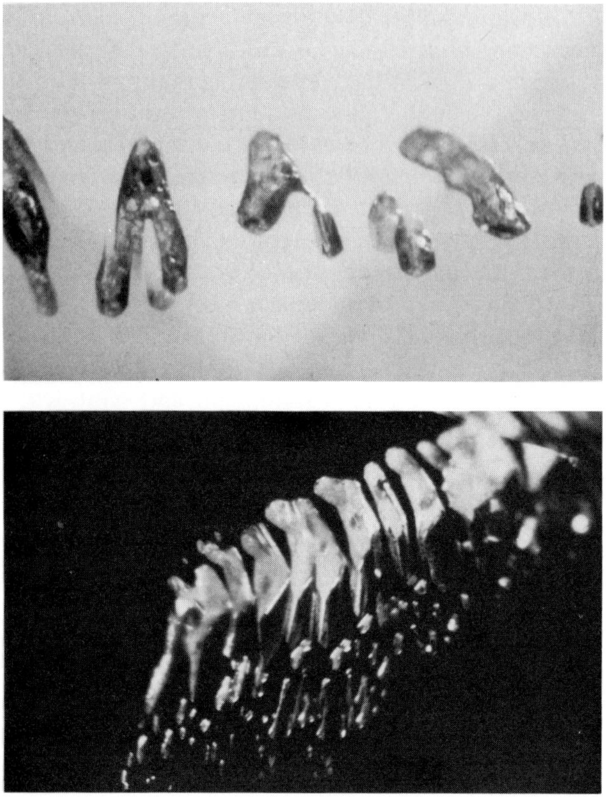

Figure 7.22 Coarse primary flux-filled negative crystals in flux-grown synthetic rubies made by Kashan. These often show a crazed surface texture and may appear to be two phase in nature. (Upper photograph by the author; lower photograph courtesy of the Gemological Institute of America)

In addition to these large flux inclusions, the Kashan product frequently contains distinctive arrangements of smaller dust-like flux particles. Turbid clouds, which are termed 'fog' or 'rain', consist of finely disseminated flux particles that are arranged in roughly parallel zones. Dotted rows of flux particles ('dot and dash' inclusions), 'U'-shaped motifs ('hairpins') and particle streams emanating from a single larger flux particle ('comets') are but a few of the patterns that have been observed regularly. The slender individuals, which resemble but lack the orientation of the rutile needles in natural corundum, require great care in their distinction, as do so many of these particle-type flux inclusions. Fibre optic illumination is, in many cases, essential for the observation of particle flux inclusions, because they may not be visible at all under conventional dark-field light sources.

The variety of possible flux formations is seemingly endless in Kashan stones, with fingerprints, feathers, wispy veils, nets, flags, folds, paint splashes, rags, comets, X's and Y's, etc., to mention just a few. As with the Chatham stones, care must be taken to look beyond the patterns themselves if the synthetic origin is to be

Plate 1 The 'Rosser Reeves Star' ruby now in the collection of the Smithsonian Institution in Washington, DC. Weighing 138.7 carats, it is considered by many to be the largest fine-quality star ruby in existence and is believed to come from Sri Lanka. (Courtesy of the Gemological Institute of America)

Plate 2 A star sapphire of over 20 carats from Sri Lanka. Note the exceptional transparency that the stone has – an absolute essential in top-quality stones. Only star stones with a 'glass body' will allow the colour to reach its maximum potential. (Photograph by Rattana Angkuanpanit, World Jewels Trade Center, Bangkok)

Plate 3 A large and fine yellow/orange sapphire from Chanthaburi Province, Thailand (untreated). This stone displays the preferred 'Mekong Whisky' colour so highly sought after in Thailand. (Photograph by Rattana Angkuanpanit, World Jewels Trade Center, Bangkok)

Plate 4 A fine-quality yellow sapphire from Sri Lanka produced by heat treatment. Such heat-treated yellow (and orange) stones are much more common than the similar colours found in nature. (Photograph by Rattana Angkuanpanit, World Jewels Trade Center, Bangkok)

Plate 5 The very rare and valuable pinkish-orange 'padparadscha' sapphire from Sri Lanka. So far, this colour is available only in nature and not through heat treatment. (Courtesy of the Gemological Institute of America)

Plate 6 A very fine untreated blue sapphire from Burma. (Photograph by Rattana Angkuanpanit, World Jewels Trade Center, Bangkok)

Plate 7 A superfine 10.10 carat untreated Burmese ruby set into a ring. Note the lack of extinction because of the strong fluorescence of the stone. (Photograph by the author)

Plate 8 A fine heat-treated 6.38 carat ruby from Thailand/Cambodia. Note the obvious extinction on some facets because of the stone's weak fluorescence. (Courtesy of the Gemological Institute of America)

Plate 9 Until October 1986 this untreated 4.12 carat Burmese ruby (centre stone) held the per-carat price record for ruby. It sold for an incredible $100 639 per carat in 1979. On October 18 1988 this record was broken at Sotheby's in New York, where a 15.97 carat untreated Burmese ruby sold for $3 630 000, which works out to $227 301 per carat. (Photo courtesy of Valaya Rangsit, Christie's, Geneva)

Plate 10 Although the photograph doesn't show it, this magnificent 11.81 carat Kashmir sapphire once held the per carat record for blue sapphire. It sold in 1979 for an amazing $25 815 per carat. This record was broken in February 1988, at Sotheby's, St Moritz, where a 62.02 carat untreated Burmese blue sapphire sold for $2 828 546 ($45 607 per carat). (Photo courtesy of Valaya Rangsit, Christie's, Geneva)

Plate 11 An 'ottu'-type blue sapphire from Sri Lanka. Such stones are mostly colourless, with just a small patch of blue colour at the culet. (Courtesy of the Gemological Institute of America)

Plate 12 A magnificent rhombohedral crystal of 8.77 carats and a 3.67 carat faceted stone. Both are flux-grown synthetic rubies produced by Ramaura. (Courtesy of the Gemological Institute of America)

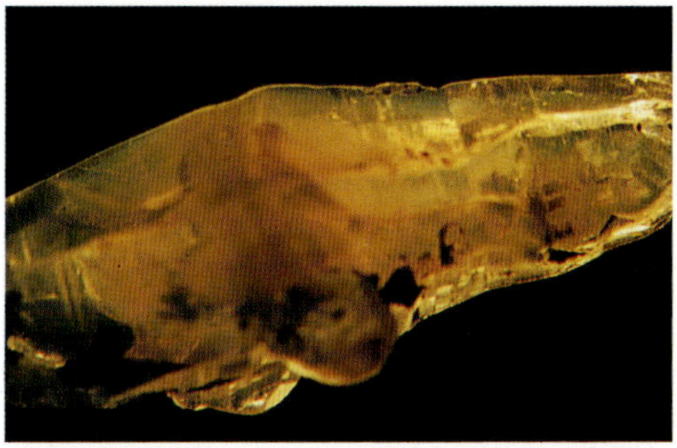

Plate 13 A 'geuda'-type sapphire crystal from Sri Lanka before heat treatment. Note the yellowish 'diesel' effect (believed to be due to rutile) which tells buyers that such a stone will burn to a blue colour. (Photograph by AIGS, Bangkok)

Plate 14 East African rubies are shown before (right) and after (left) heat treatment. The photographs illustrate vividly the dramatic changes that heat treatment can produce. (Photograph by World Jewels Trade Center, Bangkok)

Plate 15 Repeated twinning (rhombohedral) appears as brilliant multi-coloured planes (in a Thai/Cambodian ruby) when the gem is viewed between crossed polars. (Photograph by the author)

Plate 16 When repeated twinning is viewed in polarized light the twinning planes display the opposite pleochroic colour to the rest of this Thai/Cambodian ruby. (Photograph by the author)

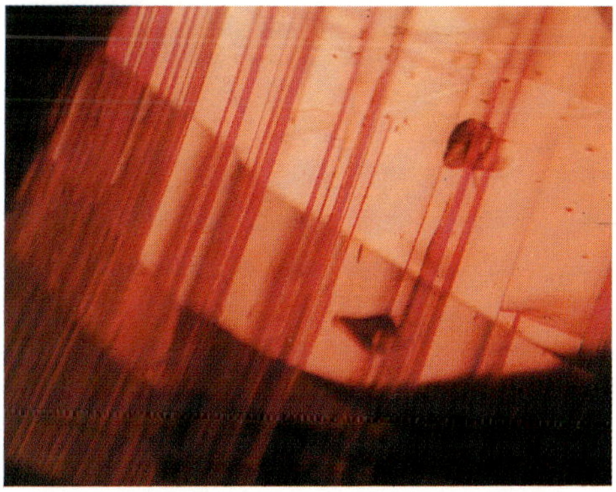

Plate 17 The same stone as in Plate 16 with the polaroid plate rotated by 90°, thus reversing the pleochroic colours. (Photograph by the author)

Plate 18 The separation plane of a doublet is clearly visible here. The crown consists of a thin slice of genuine green sapphire, while the pavilion consists of Verneuil synthetic blue sapphire. (Photograph by AIGS, Bangkok)

Plate 19 A heart-shaped blue sapphire from Kanchanaburi, Thailand. Note the slight cloudiness which is typical of sapphires from this source. (Photograph by Rattana Angkuanpanit, World Jewels Trade Center, Bangkok)

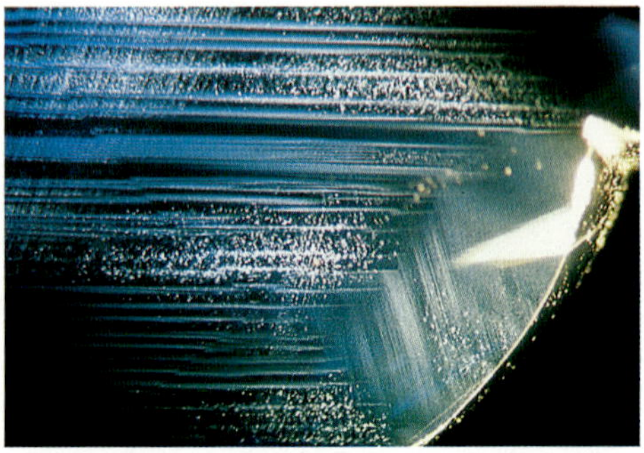

Plate 20 Minute exsolved particles in a hexagonal pattern in a blue sapphire from Kanchanaburi, Thailand. These particles produce the characteristic cloudy appearance in stones from this source. (Photograph by the author)

Plate 21 Rutile silk intersecting at 60°/120° in a Burmese sapphire viewed parallel to the *c* axis. (Photograph by the author)

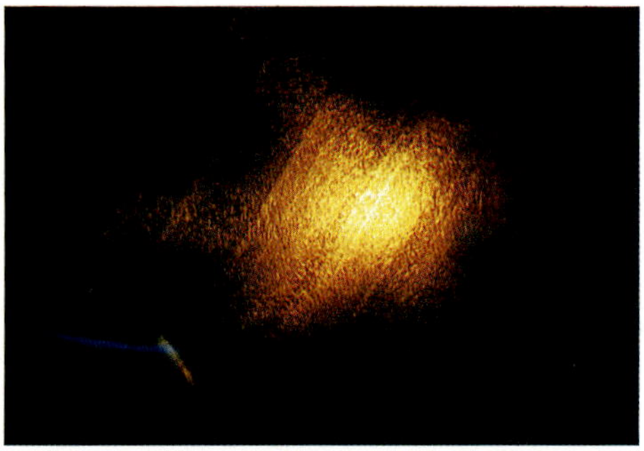

Plate 22 The yellowish star in a 'golden-star' black star sapphire from Chanthaburi, Thailand. This golden-coloured star results from the exsolution of haematite silk in an otherwise yellow sapphire. (Photograph by the author)

Plate 23 Whitish 'texture' clouds of partially dissolved silk form a pattern much like a church window in a heat-treated blue sapphire from Sri Lanka viewed parallel to the *c* axis. (Photograph by the author)

Plate 24 A secondary liquid inclusion (healing fissure) in a Sri Lankan sapphire displays brilliant interference colours when illuminated with a fibre optic lightguide. (Photograph by the author)

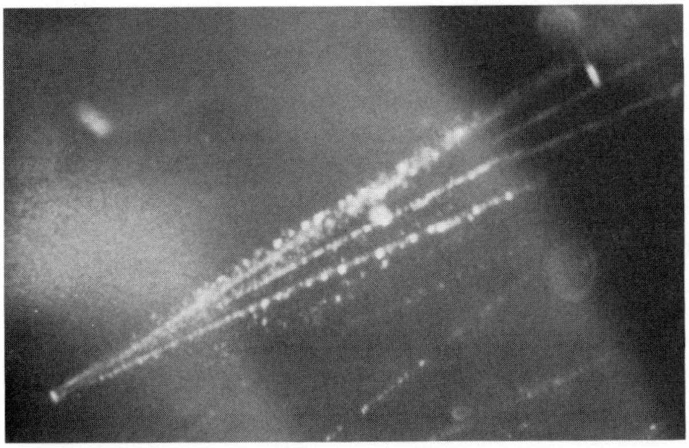

Figure 7.23 'Comet-tail' inclusion in a Kashan flux-grown synthetic ruby. It is composed of a larger flux droplet and smaller flux particles forming the tail along the growth direction. (Courtesy of the Gemological Institute of America)

Figure 7.24 Tiny flux inclusions distributed in somewhat parallel arrangement in Kashan flux-grown synthetic ruby. Such inclusions, typical of Kashan synthetic ruby, are often termed 'rain'. (Photograph by the author using fibre optic lighting)

revealed. The use of high magnification and fibre optic illumination will separate the liquid forms of the natural from the flux inclusions of the synthetic material, which appear at first glance to be so deceptively similar.

Apart from flux, the only other inclusions to be found in the Kashan synthetic rubies are growth features, such as colour zoning. As in the natural stone, the colour zoning in Kashan stones is straight and follows the crystal faces. In the deep-red varieties narrow well-defined lines are seen, while for the lighter toned stones only diffuse angular patches show up.

On rare occasions repeated twinning is also found in the Kashan product. It occurs parallel to the faces of the primitive rhombohedron just as in natural corundum, but the planes rarely penetrate deep into the stone and are generally much less obvious than those of the natural stone. At times, the colour zoning of Kashan synthetic rubies may be so sharp as to resemble twinning, but examination between crossed polars while immersed will separate the two.

One additional factor which may be of use in identifying Kashan stones concerns their pleochroism. The pleochroism of Kashan synthetic rubies and pink sapphires is characterized by an overly yellowish-red extraordinary ray. When examined with a dichroscope at right angles to the *c* axis, natural rubies typically show purplish-red (o-ray) and orangy-red (e-ray) colours. However, with Kashans *of comparable body colour*, the e-ray colour is decidedly more yellowish. (This often gives the deeper red Kashan synthetic rubies an overall colour more akin to that of red spinel than natural ruby.) In order for this test to be meaningful, the synthetic stone must only be compared with the natural stones of similar body colour. Ideally, the pleochroism of an unknown stone should be compared directly with that of known master stones of natural and Kashan synthetic origin. It will be found that the most noticeable difference exists between natural and Kashan rubies of a deep-red colour. As the body colour becomes lighter and more pink, this difference is reduced, being least definite for pale-pink stones (American Gemological Laboratories, 1982). The high titanium content of Kashan rubies is believed to be the cause of this unnatural e-ray colour.

The fluorescence range of Kashan synthetic rubies and pink sapphires overlaps with that of the natural stone and so is of little diagnostic value. When compared with Thai rubies of similar body colour, the short-wave fluorescence of Kashans is generally of greater intensity. As with the dichroism test, known master stones in a range of body colours from various localities will assist the interpretation. However, the differences in the fluorescence of natural and Kashan synthetic corundum are so slight than the gemmologist should not place too much faith in the results.

The typical ultraviolet fluorescence of Kashan synthetic corundum is shown in Table 7.7.

TABLE 7.7

Synthetic variety	*Ultraviolet fluorescence*	
	Long wave (366 nm)	*Short wave* (253.7 nm)
Kashan synthetic flux ruby and pink sapphire	Moderate to very strong red to orangy red; intensity is related to depth of body colour with deeply coloured stones showing weaker reactions	Weak to very strong red to orangy red; intensity is related to depth of body colour with deeply coloured stones showing weaker reactions; short-wave reactions of Kashan synthetic rubies are sometimes stronger than long-wave reactions

Ramaura

Ramaura synthetic flux rubies offer an inclusion suite which is distinctly different from those of both the Chatham and Kashan products. Residual unmelted flux is the most prominent feature, while in addition the growth features such as colour zoning give important diagnostic information.

The flux inclusions in Ramaura stones are distinctive mainly due to their colour, which is often orange yellow. Although they may also appear near colourless or white, especially when thin, the thick orangy flux forms are common and provide a positive means of separation from natural and other flux-grown synthetic rubies. Large flux globules typically occur in 'drippy' forms with a crackled or crazed surface structure. They are generally of high relief and opaque, although transparent patches of lower relief have also been encountered. As with Kashan stones, flux rods are often found in parallel alignment, with some exhibiting a two-phase appearance. However, it is believed that these are probably completely solid in nature. Like the flux of Kashan stones, the larger inclusions often show a distinctive crazed surface texture. Flux-filled negative crystals may sometimes show flat faces or parallel steps aligned according to the colour zoning, with oblique lighting revealing iridescent patches on these steps. In addition, the flux fingerprints are frequently made up of tiny flux particles arranged in a similar way. Other flux forms observed include twisted or folded white fingerprints which radiate outwards from a central point. 'Comets', consisting of large flux grains followed by minute particles trailing out to form a tail, have also been seen, as well as iridescent fractures and healed fractures (Kane, 1983).

Although the flux inclusions in the Ramaura synthetic rubies provide a positive means of identification, many stones are completely free of flux of any kind. Fortunately, however, the growth features of Ramaura stones are distinctive enough to identify reliably those stones in which the flux is altogether absent. All Ramaura stones examined to date display some form of growth features. These include straight colour zoning which runs parallel to the crystal faces and meets at angles. Although natural rubies may also show straight colour zoning, the zoning in the Ramaura product is generally much more regular and narrow, resembling the striae of Verneuil synthetic rubies, except that it is straight instead of curved. Sometimes these straight striae are so narrow as to appear iridescent. Wider bands of colour are also seen when the striae are not lined up perfectly, as well as colour swirls similar to the 'treacle' or 'scotch in water' effect common to many natural Burmese rubies. Broad angular patches of colour are another common feature (Kane, 1983).

The colour zoning of Ramaura stones will often meet at angles, either blurred or sharp, and these junctions are quite distinctive. Single straight growth planes are sometimes encountered which extend deep into the stone and resemble the lamellar twinning planes of the natural stone. However, these run parallel to the colour zoning, unlike the true twinning planes in the natural stone which are parallel to the rhombohedron faces and often feature boehmite needles at their junctions, meeting at 87° and 93°. Different types of illumination are needed to observe the growth features of Ramaura stones, so it is essential that various lighting techniques, as well as immersion, are experimented with. Most importantly, the stone must be examined from every possible position – Hughes' first law of gemmology!

Currently, Ramaura synthetic flux corundum is being produced in various shades of red only. The colour ranges from a pure red to purplish red, orangy red and pinkish red, with intensities from medium to very high. One notable feature of Ramaura stones is their high degree of transparency. This is said to result from the lack of seeds in the growth process, which reduces defects such as impurities and dislocations.

In contrast with other types of synthetic ruby, the ultraviolet fluorescence of Ramaura stones does, in certain cases, provide a positive means of separation from natural rubies. The manufacturer of the Ramaura rubies has taken the unique step of adding a dopant to the flux melt in an attempt to produce a characteristic orange-yellow fluorescence, thus allowing easy identification. Unfortunately, however, in most crystals the dopant is absorbed only in the outer edges, with some pieces showing no trace at all (Kane, 1983). Because the outer areas are generally removed during fashioning, cut stones usually display only the reddish fluorescence typical of natural and other synthetic rubies. Thus, if a stone does show very large zones of orange-yellow fluorescence, this can be considered to be an indication of synthetic origin, while red fluorescence alone would be of no diagnostic use.

The various fluorescent reactions under ultraviolet light for Ramaura synthetic rubies are listed in Table 7.8.

TABLE 7.8

Synthetic variety	*Ultraviolet fluorescence*	
	Long wave (366 nm)	*Short wave* (253.7 nm)
Ramaura synthetic flux ruby	Moderate to very strong red, dull red, chalky red or orangy red; small areas of chalky yellow are also seen in some stones, particularly at the surfaces	Weak to strong red, dull red, chalky red or orangy red; small areas of chalky yellow or bluish white are also seen in some stones

Knischka (ᑫK ruby)

The Knischka synthetic rubies are produced by a variation in the flux growth technique and in many ways resemble the Chatham flux rubies. At present, only relatively small quantities of material have been released onto the market and even though the inclusions provide for ready separation from natural stones, confusion can occur if the gemmologist is unfamiliar with their internal characteristics.

Due to their unique method of growth, the P.K. (ᑫK) rubies provide an interesting and distinctive suite of inclusions. Perhaps the most characteristic of these are the negative crystals which closely mimic the hexagonal bipyramid habit of their host. Found alone or in small groups astride long crystalline tubes, these spindle-shaped voids immediately draw attention to their maker (Gübelin, 1982a,b).

Large gas bubbles are also found within the P.K. (ᑫK) rubies. These are actually the gaseous component of two-phase inclusions with indistinct low-relief outlines. The high refractive index substance in which they are encased is of an unknown nature and is sometimes irregular in outline with a pale-blue milky texture, or in the form of negative crystals.

In addition to negative crystals, Knischka synthetic rubies also contain ghost-like clouds of an unknown nature. These clouds resemble somewhat the dust-like clouds found in natural Burmese rubies and Chatham flux synthetic stones.

Net-like flux fingerprints and feathers are commonly found throughout the Knischka stones and often form wispy veils similar to the flux inclusions of Chatham synthetic rubies. Although usually of high relief, these flux inclusions may sometimes appear less distinct. Their colour is generally somewhat white and they range in transparency from semi-transparent to opaque.

Platinum platelets are a common and highly diagnostic feature of the P.K. (ᖽK) product, but are usually smaller than in Chatham stones. Although the manufacturer claims that these can be eliminated, the few stones examined to date have all showed the small hexagonal plates throughout (Gübelin, 1982a,b).

Growth of the P.K. (ᖽK) rubies takes place on natural, Indian, or Verneuil synthetic ruby seed crystals. It is of course possible that cut stones could include part or all of the seed crystal and so natural inclusions or curved striae would be found within the seed section.

In terms of colour, the Knischka synthetic rubies tend to resemble natural rubies from Burma more than those from Thailand. Straight angular colour zoning is seen, as are swirled colour zones. To date, only intensely red (ruby-red) stones have been reported, with the gems often showing strong purple overtones.

The fluorescent reactions of P.K. (ᖽK) rubies (Table 7.9) may be influenced by the type of seed crystal used. Crystals seeded with natural Indian rubies tend to show slightly weaker fluorescent intensities than those with Verneuil synthetic ruby seeds.

TABLE 7.9

Synthetic variety	*Ultraviolet fluorescence*	
	Long wave (366 nm)	*Short wave* (253.7 nm)
Knischka synthetic flux ruby	Strong to extremely strong red or orangy red	Moderate to extremely strong red or orangy red

Lechleitner

The Lechleitner product consists of a synthetic flux overgrowth on top of either natural or synthetic corundum seeds (Kane, 1985). Those containing Verneuil seeds are termed type I and those with natural seeds type II.

Type I stones are made in a number of colours, including red, pink, orange, blue, yellow, green, colourless and colour-change varieties. In most cases, the colour of the seed crystal matches that of the gem as a whole, but not always. Overgrowths are generally colourless or of a similar colour to the seed, unless the seed itself is colourless.

Under the microscope, type I stones reveal inclusions characteristic of both the flux and Verneuil processes. Throughout the stone are found twisted flux veils and fingerprints, bringing to mind the coarse flux inclusions of the Chatham and Kashan rubies. Curved growth lines and the occasional gas bubble complete the picture. Together with the flux inclusions, they provide the gemmologist with what is quite possibly the most easily recognizable inclusion suite among the flux-grown synthetic stones.

Unfortunately, the same cannot be said of the Lechleitner type II stones. Consisting of thin flux overgrowths on natural corundum seeds, they display, for the most part, inclusions typical of nature's own. Recently the author had the opportunity to examine four type II stones, both before and after cutting. It was, to say the least, an 'educational' experience. The four stones were flux ruby overgrowths on Thai ruby seeds. In the rough, they each displayed tabular habits identical to those of natural rubies, with the exception that their faces were extremely well developed, showing no signs of weathering as would most natural crystals. Repeated twinning on the rhombohedron faces was found in each of the seed sections and carried on through to the surface. This means that the synthetic overgrowth takes on the structural pattern of the seed!

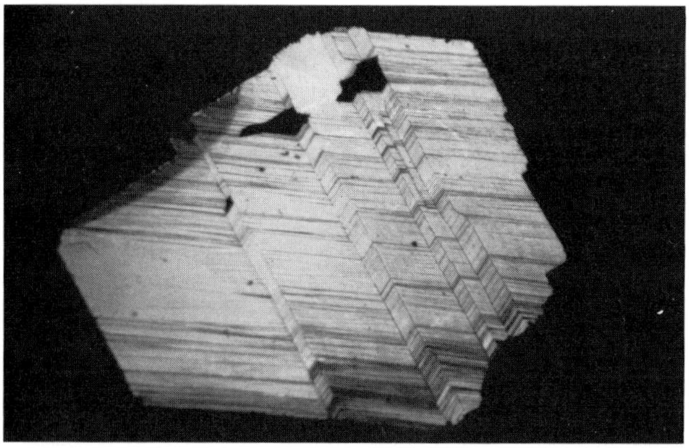

Figure 7.25 Striations and twinning planes visible on the surface of a natural ruby overgrown with a thin layer of flux-grown synthetic ruby by Lechleitner. The synthetic growth has followed exactly the twinned structure of the underlying synthetic ruby. (Photograph by the author)

Within the stones, inclusions varied between those suggesting a natural origin and thick flux fingerprints and feathers. Herein lies the danger of type II gems, for the features of the seed can easily be misleading. Unless the gem is immersed, no junction can be detected between the seed and the synthetic cover. Even with immersion it can be seen only with difficulty as a colourless region separating the thin bright-red overgrowth from the less intensely coloured Thai ruby seed. During the cutting process much of the cover was removed, leaving just a few traces near the girdle and culet. When exposed to long-wave ultraviolet light and examined under magnification, the synthetic overgrowth glowed bright red, contrasting sharply with the weaker red fluorescence of the seed.

Growth planes in natural and flux-grown synthetic corundum

Not so many years ago, the presence of parallel or angular growth zoning in a ruby or sapphire was considered to be positive proof of natural origin. This has changed, though, with the introduction of the flux-grown synthetic stones. Cultivated slowly from a flux bath, the new man-made rubies possess flat crystal faces and so also show parallel or angular colour zoning similar to that of the natural stone. Nevertheless, characteristic differences do exist. Due to differences between the crystallization medium of the natural and flux synthetic stones, the types of faces displayed by the latter often do not match those of the former. Since parallel zoning and other such growth planes are reflections of a crystal's habit, by locating amd measuring the angles between these planes it is possible to determine the types of faces displayed by a crystal, even after it is cut. In certain cases this information will then allow a separation to be made. Measurement of angles between growth planes can be made using a protractor-type eyepiece reticle or the new stone holder designed for this purpose by Dr Karl Schmetzer of Germany.

According to a study by Schmetzer (1985), the faces listed in Tables 7.10, 7.11 and 7.12 are among those possible for corundum.

By comparing the data from these tables it becomes clear that observation of growth zoning parallel to the hexagonal prism faces *a* $(11\bar{2}0)$ and certain hexagonal dipyramid faces is a strong indication of natural corundum because of their absence

TABLE 7.10 Crystal faces and *c*-axis angles of natural and synthetic corundum (After Schmetzer, 1985)

Name of crystal face	*Face symbol*	*Face indices*[a]	*Angle between c-axis and face* (deg)
Basal pinacoid	*c*	(0001)	90
Hexagonal prism (second order)	*a*	$(11\bar{2}0)$	0
Positive rhombohedron	*r*	$(10\bar{1}1)$	32.4
Negative rhombohedron	*d*	$(01\bar{1}2)$	51.8
Negative rhombohedron	*b*	$(01\bar{1}5)$	72.5
Hexagonal dipyramid	*n*	$(22\bar{4}3)$	28.8
Hexagonal dipyramid	*w*	$(11\bar{2}1)$	20.1
Hexagonal dipyramid	*v*	$(44\bar{8}3)$	15.4
Hexagonal dipyramid	*z*	$(22\bar{4}1)$	10.4
Hexagonal dipyramid	*e*	$(44\bar{8}1)$	5.2
Hexagonal dipyramid	*f*	$(14\ 14\ \overline{28}\ 3)$	4.5

[a] On the basis of the morphological unit cell with $a:c = 1:1.365$.

TABLE 7.11 Interfacial angles of natural and synthetic corundum (After Schmetzer, 1985)

Intersecting faces	*Interfacial angles (angles between face normals)* (deg)	*Interfacial angles* (deg)
$r \wedge r'$	93.9	86.1
$n \wedge n'$	52.0	128.0
$a \wedge a'$	60.0	120.0
$r \wedge n$	26.0	154.0
$d \wedge n$	32.0	148.0
$r \wedge d$	47.0	133.0

TABLE 7.12 Common faces of natural and flux-grown synthetic corundum (After Schmetzer, 1985)

Type	*Common faces*	*Less common faces*	*Common plane angle*	*Twin development*
Natural corundum	*c, a, r*	*n, w, v, z, e, f*	(*a, a'*) (*a, c*)	(0001) contact prism –rare ($10\bar{1}1$) contact prism – common ($10\bar{1}1$) repeated prism – very common
Chatham flux-grown synthetic corundum	*c, r, d, n*		(*r, n*) (*n, n'*) (*d, n*)	($10\bar{1}1$) repeated prism – common ($10\bar{1}0$) contact prism – occasional ($11\bar{2}0$) contact prism – occasional
Kashan flux-grown synthetic corundum	*c, r, n*		(*r, n*) (*n, n'*)	($10\bar{1}1$) repeated prism – common
Knischka flux-grown synthetic corundum	*c, r, d* *n, b*	($10\bar{1}9$)	(*r, n*) (*n, n'*)	
Ramaura flux-grown synthetic corundum	*c, r, d*	*n*	(*r, r'*) (*r, d*)	($10\bar{1}0$) penetration prism – occasional

in synthetic corundum. On the other hand, growth zoning parallel to the negative rhombohedrons *d* ($11\bar{2}0$) and *b* ($01\bar{1}5$) can be taken as a strong indication of synthetic origin because of their absence in natural corundum. Other planes offer less help in separating the natural from the synthetic corundum as their occurrence is not exclusively restricted to any particular one.

In addition to the observation of parallel growth planes, measurement of the angles where they intersect can also provide diagnostic information. With natural corundum the most common angle is 120° where the prism faces intersect, 90° for the intersection of the basal pinacoid and hexagonal prism faces, and 32.4° for the intersection of the positive rhombohedron and hexagonal prism faces.

With the exception of faces common to both natural and synthetic corundum, there is one more situation in which problems may arise. This concerns the new synthetic flux overgrowths of Lechleitner. Published reports have indicated that most consist of flux overgrowths on Verneuil seeds. However, the author recently examined four crystals grown by Lechleitner on natural Thai ruby seeds. These displayed faces identical to those of the natural rubies entombed within, including the hexagonal prism face *a* ($11\bar{2}0$). Although each could be identified on the basis of the flux inclusions in the synthetic overgrowth, their habit could confuse the unwary. In addition, Professor Knischka has recently announced that he has succeeded in growing ruby crystals which display hexagonal prism faces. However, Dr Karl Schmetzer has informed the author that a careful check of Professor Knischka's claim found that the so-called 'hexagonal prism faces' actually turned out to be hexagonal bipyramid faces. Similarly, several years ago Kashan proudly announced that needles had been found in their synthetic rubies. These 'needles' were later found to be merely tiny pieces of unoriented stretched flux, not exsolved rutile as implied in the press release. All this serves to underscore the vigilance that gemmologists must possess to separate fact from fiction. Press releases from manufacturers should always be examined sceptically.

Advanced identification techniques

Owing to space limitations, only those tests which are readily accessible to the average jeweller/gemmologist have been described. However, with the ever-advancing technology of modern gem synthesis, it is becoming more and more necessary to apply highly sophisticated instruments to the task of separating natural and synthetic rubies and sapphires. The high costs of these instruments and their complexity of operation place them out of reach of all but the most advanced laboratories. Even so, a short description of some of these techniques seems in order, for it may provide a glimpse of some of the future directions of gem identification.

Ultraviolet spectrophotometry

The spectrophotometer is basically a highly refined spectroscope. However, instead of providing a visual display of a gem's absorption spectrum, it relays this information in the form of a graph. The advantages of the spectrophotometer over a conventional spectroscope are many. Not only does it locate precisely the areas of strong absorption and transmission, but also it measures their extent and strength. In addition, with some models it is possible to measure the absorption and transmission characteristics of a gem in the ultraviolet and infrared regions, as well as in the visible spectrum.

G. Bosshart published the results of his extensive studies of the ultraviolet transmission behaviour of natural and synthetic rubies from a variety of sources in 1982. The ability of the spectrophotometer to quantify the transmission in the ultraviolet region allowed him to establish different population areas for natural and synthetic rubies, not only making their separation possible but also allowing distinction between individual sources. In just one product did a synthetic ruby overlap to any significant extent with the natural stones. This was for Ramaura synthetic rubies, which fell into the natural zones.

With regard to the potential of this instrument for the separation of natural and synthetic sapphires, detailed studies have yet to be performed. However, if the results of the investigations into rubies offer any indication, ultraviolet spectrophotometry may very well prove to be a field worth exploring.

Energy dispersive x-ray fluorescence spectrometry (EDS–XRF)

The analysis of a gemstone by the EDS–XRF technique involves investigation of its fluorescence spectrum while it is being stimulated by x-rays. A beam of primary x-rays is used to bombard the sample material, causing it to emit secondary fluorescence radiation. This produces a spectrum consisting of different fluorescence lines, or energies, corresponding to the various elements contained by the gem. Roughly 80 different elements can be detected by this instrument, ranging from sodium (Na, atomic number 11) to uranium (U, atomic number 92).

If an element is present in sufficient quantities within the gem, it can be detected by EDS–XRF. Other than the usual colouring agents of corundum, the element gallium (Ga) has also been detected. Although it is present in both synthetic and natural rubies and sapphires, the concentrations of gallium are, with one exception, below the detection limits of the instrument in the synthetic corundums. The exception is the Ramaura synthetic ruby, in which gallium has been detected. Unfortunately EDS–XRF is unable to specify the exact quantity present of a particular element, thus necessitating the use of other methods to separate natural rubies from the Ramaura product (Hänni, 1984).

Neutron activation analysis

Instrumental neutron activation analysis (INAA) can be used to determine simultaneously the presence and, more importantly, the quantity of a number of different trace elements within a gemstone. As a result it provides the means not only to separate natural from synthetic corundum, but also to distinguish between gems of different countries of origin (Hänni, 1984).

With INAA, the sample is first irradiated by exposure to a thermal neutron flux from a nuclear reactor. This leads to the formation of unstable isotopes which then release energy through a process of decay. Because the nature of the emissions is distinctive for each element, their measurement (by γ-spectrometry) allows the determination of the types and quantities of trace elements present within the gem.

Unfortunately, INAA does have its drawbacks. The neutron bombardment to which the gem is subjected produces a potentially hazardous radioactivity. In the case of ruby, the chromium atoms take 30 days to decay to the point where the radiation emitted is below the danger level. Thus the gems subjected to INAA must be stored in lead boxes until their radiation no longer poses a health hazard. Another problem with this method of testing is the formation of colour centres, which may lead to a modification in the colour of the gem. So far this has not occurred with rubies, but the possibility exists that other varieties of corundum may suffer a change in colour from the neutron bombardment, particularly those from Sri Lanka.

Chatham flux-grown synthetic rubies have been studied using INAA. The early products showed traces of molybdenum, platinum and iridium, probably contained in inclusions. This indicates that they were grown from a lithium molybdate flux within a platinum–iridium crucible. Gallium was found in both natural and synthetic rubies but, with the exception of the Ramaura product, was present in much higher concentrations in the natural stones (Hänni, 1984).

Note

With each of the above advanced techniques, only a small number of samples have been tested. Thus, as more stones are tested the conclusions drawn will undoubtedly change. Much work remains to be done before any of these techniques can be considered even remotely reliable for the separation of natural and synthetic corundum.

Assembled stones

In addition to synthetic corundums, there also exist two other types of trickery – imitations and assembled stones. Imitations refer to those gems which resemble corundums only in appearance; spinel and garnet represent imitation rubies, for example. Because ordinary gemmological tests (such as refractive index and specific gravity) will allow their separation from corundums, and because they are adequately described in standard gemmology texts, they will not be described here. Assembled stones, however, are another matter. Typically involving the joining together of two or more pieces of natural and/or synthetic or imitation corundum, they can be so deceptive as to fool even the experts. Thus a few words on the subject are in order.

There exist three basic types of assembled corundums: those which are made entirely of natural corundum; those which are made partly of natural corundum; and those which contain no natural corundum whatsoever. Of the three types, the last is rarely a problem because, being made entirely of synthetic or imitation corundum, it can be easily identified as such. (A fourth type of assembled stone can be classified as an imitation. It is a garnet-topped-doublet (GTD) with a thin slice of almandine garnet fused to a base of red or blue glass.)

Assembled stones made of two parts are normally termed ‘doublets’, and those made of three parts are termed ‘triplets’. Doublets made entirely of natural corundum are usually (but not always) rubies. Because the price of rubies (and sapphires) increases dramatically with increasing weight, unscrupulous dealers will sometimes take, say, two 0.75 carat rubies and glue them together to make a single larger piece of over 1.00 carat. One of the most deceptive doublets of this type seen by the author was one in which two rather flat Thai rubies were joined together. The stone was cut shallow and its colour was typical of an average Thai ruby. In fact, everything about the stone suggested an ordinary Thai ruby, so much so that most experienced dealers would not even think to check the stone with a loupe. Furthermore, although most doublets are joined together at the girdle, this piece was joined below the girdle, and the join was well hidden.

A more common type of doublet is made by taking a thin slice of natural corundum and gluing it onto a bigger piece of synthetic corundum. Then the whole is faceted, with the join being placed at the girdle. These stones are deceptive because a quick check with the loupe will reveal the natural inclusions of the crown. Typically, natural green sapphire is used for the crown because of its low cost. If the crown is cut thin enough, the synthetic pavilion need not be the same colour as the crown, but more natural-appearing doublets are made with both pieces being the same colour. The pavilion is normally Verneuil synthetic corundum (due to its low cost), but flux-grown synthetic corundum is also sometimes used.

In most cases the separation plane joining the two pieces is perfectly flat and lies at the girdle, but the author did encounter one doublet consisting of Verneuil synthetic ruby which had an irregular chunk of natural Burmese ruby attached with glass. This was somewhat deceptive because the natural portion contained a large area of rutile silk which was easily visible to the naked eye. Such is the major danger of assembled stones; the natural portion may contain inclusions and/or other characteristics so obviously natural that no one would think to check the stone carefully. These stones often appear when least expected.

Faceted corundums are not the only assembled stones encountered; cabochon-cut star stones are also seen at times. One piece seen by the author was an apparent star ruby which showed strong hexagonal zoning and rutile silk obvious to the naked eye. However, microscopic examination revealed the stone to be a pale (probably light-grey) star sapphire completely coated with red plastic. Although a spot test of refractive index would immediately reveal the deception, the stone looked so obviously natural that most experienced dealers would never even think to check the stone carefully. Under magnification, gas bubbles and red colour swirls were visible, as well as a small hole in the red plastic layer.

Another type of star doublet is where a slice of asterated natural corundum is placed beneath a cabochon of transparent natural or Verneuil synthetic corundum. This gives a star effect to the whole piece. By engraving narrow lines in three directions at 60° angles on the back of a transparent natural (or synthetic) corundum cabochon, a star can also be created. The effect is strengthened by placing a mirror backing on the stone. (It should be mentioned that mirror or foil backings were commonly applied to stones set in jewellery during the Middle Ages. Although deceptive, this was a typical trade practice of the day.)

Probably the most deceptive assembled star corundum seen by the author is the star triplet. A pale star sapphire cabochon of good transparency and good asterism is used as the centre piece. Then a Verneuil synthetic ruby or sapphire cabochon is hollowed out so that it fits perfectly over the pale-coloured star sapphire. A slice of natural corundum is used to seal the back and the stone is then normally mounted in a jewellery piece which hides the join. When well made, this star triplet is very difficult to identify in the mounting. Once again, the inclusions (silk, hexagonal zoning, etc.) of the genuine centre piece are often so obvious that no one would think to check it carefully. If the rear of the mounting is closed, however, hiding the back and girdle of the stone completely, then suspicions should be aroused (usually the back is left open in corundum cabochon mountings to allow more light through the stone and so improve the colour). Microscopic examination from above will also reveal the deception as the double surface of the synthetic top piece (which is completely transparent) is seen; gas bubbles may also be seen in the glue of the separation layer.

The identification of assembled stones of virtually all types usually hinges on one factor – location of the separation plane. A warning is in order here: a mere difference in colour alone between crown and pavilion does not a doublet make. The author will always remember the humiliation he suffered during his student days when in Sri Lanka he wrongly accused a local dealer of selling sapphire doublets because of a difference in colour between crown and pavilion. Unusual reflections, colour zoning or pleochroism may cause the crown to appear a different colour to the pavilion. However, in almost any assembled stone there must be a place where the pieces are joined together. This is what should be sought.

In most cases the separation plane is completely flat and located at the girdle. Glue is normally used to join the pieces together and as the glue (or glass, as previously mentioned) has a greatly different refractive index than the corundum,

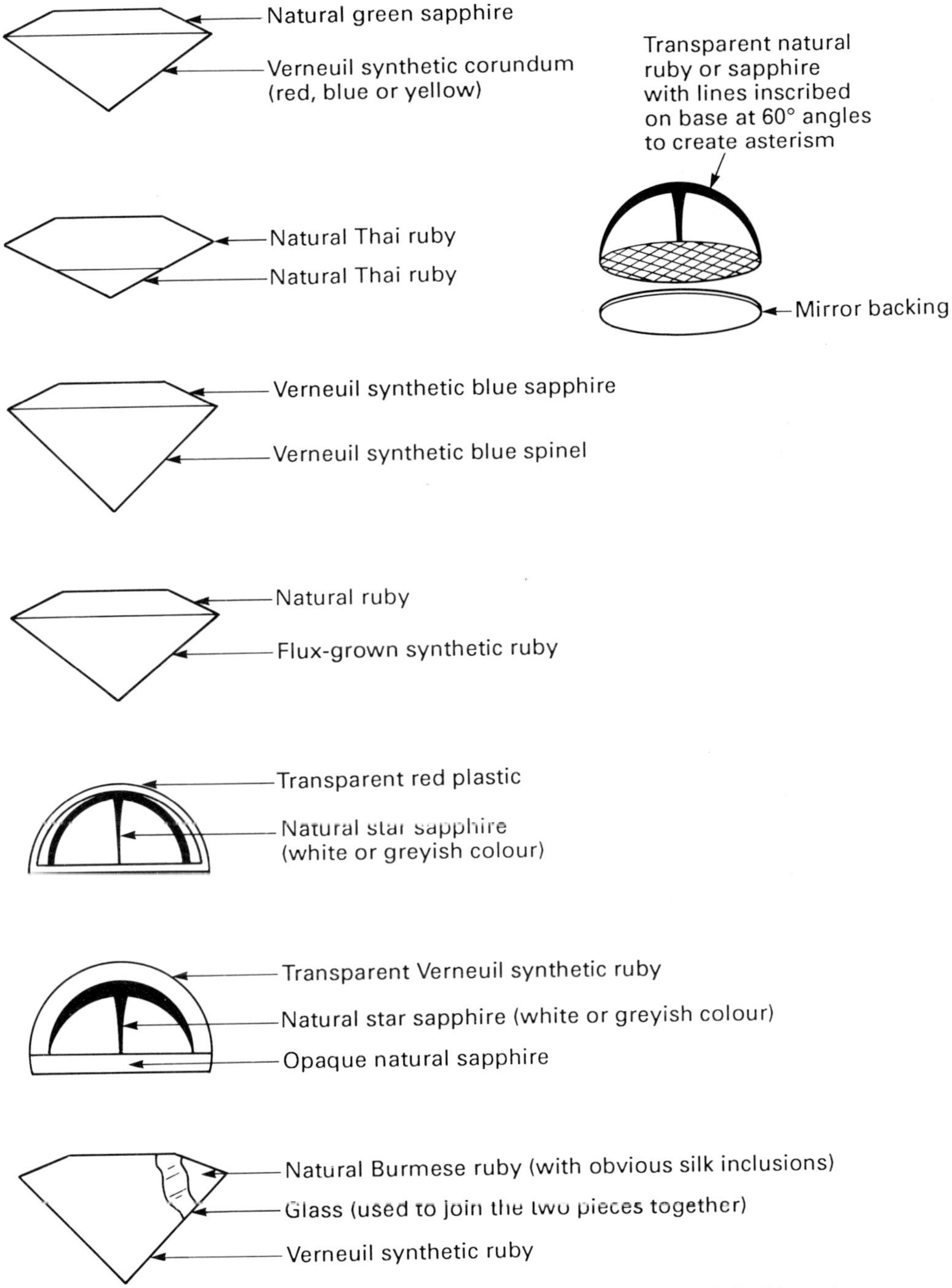

Figure 7.26 Illustrations of the most commonly encountered examples of assembled rubies and sapphires. Of course, other combinations are also possible, but those above represent the most frequently seen

immersion in methylene iodide will quickly reveal the separation plane as a layer of high relief (the corundum will have low relief due to the similarity between its refractive index and that of the methylene iodide). Immersion will also reveal any colour differences between the different layers. (A difference in relief will also be seen in corundum-topped doublets where the pavilion piece is something else, such as spinel or synthetic spinel.)

Further clues are found under magnification. Any inclusions present will be seen to stop at the separation plane. In addition, the layer of glue often displays brush marks, as the glue is usually applied with a brush. Flat gas bubbles are also frequently seen in the layer of glue.

One further remark concerns assembled stones. The author once encountered a most unusual doublet which consisted of a crown of synthetic sapphire and a pavilion of Verneuil synthetic blue spinel. The only possible reason that the author can think of for manufacturing such a doublet (with both pieces being synthetic) is that the cobalt-blue colour of the synthetic spinel was made to appear slightly more natural by the synthetic sapphire crown. In any case, the stone was easily identified by the difference in refractive index between the crown and pavilion pieces.

Bibliography

Alexander, A. E. (1949) Reconstructed rubies in rod and faceted form. *Gems and Gemology,* **6**, 184–185

Alexander, A. E. (1959) The Chatham ruby makes its bow. *The Gemmologist,* **28**, no. 340, 201–204

American Gemological Laboratories, Inc. (1982) Kashan ruby identification study. *Gemline Information Service, Report no.* 100

Anderson, B. W. (1949) Reconstructed rubies. *Gems and Gemology,* **6**, 187–190

Anderson, B. W. (1980a) What the term "synthetic" means. *Retail Jeweller,* **18**, no. 426, 12–13

Anderson, B. W. (1980b) Rapid growth in ruby crystal development. *Retail Jeweller,* **18**, no. 430, 13

Anderson, B. W. (1980c) Reconstructed rubies: some mystery still remains today. *Retail Jeweller,* **18**, no. 436, 11

Anderson, B. W. (1980d) Reconstructed rubies arouse a great deal of interest. *Retail Jeweller,* **18**, no. 438, 12–13

Anderson, B. W. (1980e) How Verneuil pioneered investigations into synthetics. *Retail Jeweller,* **18**, no. 440, 18–19

Anderson, B. W. (1980f) Verneuil publishes details of his revolutionary process. *Retail Jeweller,* **19**, no. 444, 18–19

Anderson, B. W. (1980g) Distinguishing between natural and synthetic gem corundum. *Retail Jeweller,* **19**, no. 452, 6

Anderson, B. W. (1980h) Special methods for distinguishing natural corundums from synthetics. *Retail Jeweller,* **19**, no. 456, 8

Anderson, E. (1981) *Kashan Characteristics and Identification*, Kashan, Inc., Austin, TX

Anderson, B. W. and Jobbins, E. A. (1990) *Gem Testing*, 10th edn, Butterworths, London

Anonymous (1937) Unusual inclusions in synthetic ruby. *Gems and Gemology,* **2**, 91–92

Anonymous (1947) Gemological digests – German synthetic gem production. *Gems and Gemology,* **5**, 435–436

Anonymous (1951) Gemological digests – repeated twinning lines seen in synthetic corundum. *Gems and Gemology,* **7**, 25

Anonymous (1966) Hydrothermal ruby. *Journal of Gemmology,* **10**, 96–98

Arem, J. E. (1973) *Man-Made Crystals*, Smithsonian Institution Press, Washington, DC

Ball, S. H. (1950) *A Roman Book on Precious Stones*, Gemological Institute of America, Los Angeles

Belyaev, L. M. (ed.) (1980) *Ruby and Sapphire*, Amerind, New Delhi

Benson, L. B. (1952) Many reconstructed rubies found to be synthetic corundum. *Gems and Gemology,* **7**, 139–145

Benson, L. B. (1953) "Reconstructed rubies" found to be synthetic corundum. *Journal of Gemmology,* **4**, 1–10

Bosshart, G. (1982) Distinction of natural and synthetic rubies by ultraviolet spectrophotometry. *Journal of Gemmology,* **18**, 145–160

Bosshart, G. (1983) Ramaura – eine neue Rubinsynthese (erste Untersuchungsergebnisse). *Zeitschrift der Deutschen Gemmologischen Gesselshaft,* **32**, 164–171

Breebaart, A. J. (1957) Structure & inclusions of synthetic star-stones. *Journal of Gemmology,* **6**, 72–77

Brown, G. (1981) The Kashan synthetic ruby: an attempt to sift fact from fiction. *Jeweller, Watchmaker & Giftware,* **66**, 45–49

Brown, G. (1984) Inclusions in synthetic corundum by Chatham. *The Australian Gemmologist,* **15**, no. 5, 149–154

Brown, G. (1985) Seiko synthetics. *The Australian Gemmologist,* **15**, no. 11, 418–420

Burch, C. R. (1984) Some observations on a Kashan synthetic ruby. *Journal of Gemmology,* **19**, 54–61

Butcher, J. and White, E. A. D. (1964) A study of the hydrothermal growth of ruby. *Mineralogical Magazine,* **33**, no. 266, 96–98

Chatham, C. F. (1982) Little known facts in the art of growing gem crystals. *International Gemological Symposium Proceedings 1982*, pp. 155–156

Elwell, D. (1979) *Man-made Gemstones*, Ellis Horwood, Chichester

Eppler, W. F. (1958) Notes on asterism in corundum, rose quartz and almandine garnet and chatoyancy in beryl. *Journal of Gemmology,* **6**, 195–212

Eppler, W. F. (1964) Polysynthetic twinning in synthetic corundum. *Gems and Gemology,* **11**, 169–175

Everhart, J. (1985) The strange tale of a troubled company: Kashan files for chapter 11 protection. *National Jeweler,* **29**, no. 3, 28–30

Fitzgerald, E. (1928) *Rubaiyat of Omar Khayyam*, Dent, London

Flanigen, E. M., Breck, D. W., Mumbach, N. R. and Taylor, A. M. (1967) Characteristics of synthetic emeralds. *The American Mineralogist,* **52**, 745–772

Fremy, E. (1891) *Synthese du Rubis*, Libraire des Corps Nationaux, Paris

Gübelin, E. J. (1947) Identification of synthetic gems. *Gems and Gemology,* **5**, 399–402

Gübelin, E. J. (1949) Observations on reconstructed rubies. *Gems and Gemology,* **6**, 186–187

Gübelin, E. J. (1953) *Inclusions as a Means of Gemstone Identification*, Gemological Institute of America, Los Angeles

Gübelin, E. J. (1961a) Hydrothermal rubies and emerald-coated beryl. *Journal of Gemmology,* **8**, 49–63

Gübelin, E. J. (1961b) More light on beryls and rubies with synthetic overgrowth. *Gems and Gemology,* **10**, 105–113

Gübelin, E. J. (1982a) New synthetic rubies made by Professor P. O. Knischka. *Proceedings of the First International Coloured Gemstones Conference and Gem and Jewellery Exhibition* (Colombo, Sri Lanka)

Gübelin, E. J. (1982b) New synthetic rubies made by Professor P. O. Knischka. *Gems and Gemology,* **18**, 165–168

Gübelin, E. J. (1983a) The recognition of the new synthetic rubies. *Journal of Gemmology,* **18**, 477–499

Gübelin, E. J. (1983b) Identification of the new synthetic and treated sapphires. *Journal of Gemmology,* **18**, 677–706

Gunawardene, M. (1983a) Synthetic rubies made by Knischka. *Journal of Gemmology,* **18**, 365–378

Gunawardene, M. (1983b) Uber die synthetischen blauen und orange-farbenen Sapphire von Chatham. *Zeitschrift der Deutschen Gemmologischen Gesellschaft,* **32**, 196–203

Gunawardene, M. (1984a) "Raumara" – A new synthetic ruby made in U.S.A. *Journal of Gemmology,* **19**, 125–138

Gunawardene, M. (1984b) Identification characteristics of synthetic ruby made by Knischka. *Lapidary Journal,* **37**, no. 12, 1702–1706

Gunawardene, M. (1985a) Identification characteristics of flux grown synthetic orange sapphires. *Journal of Gemmology,* **19**, 389–403

Gunawardene, M. (1985b) Gemmological properties of synthetic corundums coated by Lechleitner. *Journal of Gemmology,* **19**, 557–570

Hänni, H. A. (1984) Testing the authenticity of corundum – methods and limitations. *Swiss Watch and Jewelry Journal*, international edn, no. 3, 461–467

Holmes, R. J. (1947) The new Linde stars. *Gems and Gemology,* **5**, 452–456

Hrand Djevahirdjian S.A. (1964) *Hrand Djevahirdjian S.A.: 1914–1964*, Imprimerie Centrale Lausanne, Lausanne

Hrand Djevahirdjian S.A. (1974) *Hrand Djevahirdjian S.A.: Industrie de Pierres Scientifiques*, L'agence Ch. de Felice, Lausanne

Inamori Jewelry Division (1980) *The Inamori Created Ruby*, promotional pamphlet, Kyocera International, Inc., San Diego

Kane, R. E. (1982) The gemological properties of Chatham flux-grown synthetic orange sapphire and synthetic blue sapphire. *Gems and Gemology,* **18**, 140–153

Kane, R. E. (1983) The Ramaura synthetic ruby. *Gems and Gemology,* **19**, 130–148

Kane, R. E. (1985) A preliminary report on the new Lechleitner synthetic ruby and synthetic blue sapphire. *Gems and Gemology,* **21**, 35–39

Koivula, J. I. (1980) Gemological Notes: Brief notes on Chatham flux sapphires. *Gems and Gemology,* **16**, 410–423

Kunz, G. F. (1913) Precious Stones Chapter. *The Mineral Industry During 1912,* **21**, 707–736

Liddicoat, R. T. (1975) *Handbook of Gem Identification*, 10th edn, Gemological Institute of America, Santa Monica, CA

Linde Air Products Company (1943) U.S. develops synthetic sapphire industry. *Gems and*

Gemology, **4**, 88–91

McColl, D. and Oughton, J. H. (1971) More confusion with synthetic yellow sapphire. *The Australian Gemmologist,* **11**, no. 3, 3–5

MacInnes, D. (1973) *Synthetic Gem and Allied Crystal Manufacture*, Noyes Data Corp., Park Ridge, NJ

Mallas, A. A. (1975) *Madrona Gem Club Newsletter*, Spring, 1–4

Mallas, A. A. (1978) Rubies and Kashans – Can you tell the difference? *Kashan Gem Club Pamphlet*, pp. 1–4

Mallas, A. A. (1983) Luxury synthetics – a challenge and opportunity. *Lapidary Journal,* **37**, no. 9, 1298–1302

Mariner, T. H. (1973) Crystal-balling future synthetic gemstones. *Gems and Gemology,* **14**, 241–247

Nassau, K. (1973) The Hoquiam ruby story. *Lapidary Journal,* **27**, no. 1, 26–29

Nassau, K. (1980) *Gems Made by Man*, Chilton, Radnor, PA

Nassau, K. and Crowningshield, R. (1969) The synthesis of ruby. *Lapidary Journal,* **23**, nos 1–4

Nassau, K. and Nassau, J. (1971) Dr. A. V. L. Verneuil and the synthesis of ruby and sapphire. *Lapidary Journal,* **24**, no. 10, 1284–1290; no. 11, 1442–1447; no. 12, 1524–1528

Nicols, T. (1652) *A Lapidary: or, the History of Precious Stones*, Thomas Buck, Cambridge

O'Donoghue, M. (1983a) Orange synthetic corundum. *Journal of Gemmology,* **18**, 736–737

O'Donoghue, M. (1983b) *A Guide to Man-made Gemstones*, Van Nostrand Reinhold, New York

Plato, W. (1952) Oriented lines in synthetic corundum. *Gems and Gemology,* **11**, 169–176

Pough, F. H. (1966) Linde's science re-creates gems in a few hours (that nature requires millions of years to make). *Lapidary Journal,* **20**, no. 1, 56–63

Reinecke, C. (1959) The new rubies of Carroll Chatham (with a discussion of rubies in general). *Lapidary Journal,* **13**, no. 4, 528–530

Root, E. (1980) High technology rubies: purebred gems made for beauty. *Lapidary Journal,* **34**, no. 8, 1728–1730

Le Rubis Synthetique des Alpes (no date) *Sapphire for: Electronic, Optical, Industrial Applications*, Promotion Industrie, Paris

Scarratt, K. (1977) A study of recent Chatham synthetic ruby and synthetic blue sapphire crystals with a view to identification of possible faceted material. *Journal of Gemmology,* **15**, 347–353

Scarratt, K. (1984) How to recognize the new Seiko synthetics. *Retail Jeweller,* **22**, no. 569, 16–17

Schiffman, C. A. (1976) A new type of synthetic ruby. *Journal of Gemmology,* **15**, 105–111

Schmetzer, K. (1985) Ein verbesserter Probenhalter und seine Anwendung auf Probleme der Unterscheidung naturlicher und synthetischer Rubine sowie naturlicher und synthetischer Amethyste. *Zeitschrift der Deutschen Gemmologischen Gesellschaft,* **34**, 30–47

Seemann, A. K. (1944) American-made synthetic crystals. *Gems and Gemology,* **4**, 129–133

Seemann, A. K. (1949) American synthetic crystals. *Gems and Gemology,* **6**, 151–159

Semmes, J. L. (1980) Kashan ruby: a cutter's viewpoint. *Lapidary Journal,* **34**, no. 5, 1194–1197

Smith, G. F. H. (1930) *Gemstones and their Distinctive Characters*, 6th edn, Methuen, London

Smith, G. F. H. and Phillips, F. C. (1972) *Gemstones*, 14th edn, Chapman and Hall, London

Verma, R. K., Sirkar, G. N. and Chatterjee, S. (1956) An automatic Verneuil furnace. *The Gemmologist,* **25**, no. 296, 52–56

Webster, R. (1956) The story of Verneuil. *The Gemmologist,* **25**, no. 295, 31–34

Webster, R. and Anderson, B. W. (1983) *Gems: their Sources, Descriptions, and Identification*, 4th edn, Butterworths, London

Weyl, W. A. (1953) Synthetic minerals. *Economic Geology,* **48**, 288–305

Wodiska, J. (1909) *A Book of Precious Stones*, Putnam, New York

Wood, J. D. C. (1978) An insight into crystal growth techniques. *Journal of Gemmology,* **16**, 11–29

Yaverbaum, L. H. (1980) *Synthetic Gems Production Techniques*, Noyes Data Corp., Park Ridge, NJ

Chapter 8

Methods of fashioning

It is rare that a gemstone coming directly from the mine possesses beauty sufficient for use in jewellery. Instead, most stones require cutting (grinding) and polishing to unlock the colour and lustre for which they are known. This is the job of the lapidary.

Figure 8.1 Ruby cutters polishing cabochon rubies at Mogok, Burma. (From O'Connor, 1905)

The earliest method of fashioning gems was simply to polish the flat crystal faces developed by nature. This later gave way to the cabochon cut, a style of cut imparting a curved upper surface to the gem. Still in use today, the cabochon cut is used for star rubies and sapphires, as well as for those stones too heavily included to be faceted. Only since the Middle Ages have we seen the development of the faceted cut. This is the style that is most popular today for rubies and sapphires, because it allows their colour and brilliance to be seen to best effect.

Faceting

The faceting of rubies and sapphires involves four steps:

1. Sawing (sometimes omitted)
2. Preforming (shaping)
3. Facet grinding
4. Polishing

Commonly, the first two steps are performed by one worker, and then the stone is given to another to finish it. The goal is to produce the most beautiful gem while at the same time retaining as much weight as possible. Sawing and preforming are most important, because they determine to a large degree the final appearance of the stone. Before any saw cuts are made, the rough will be examined carefully to determine how it can best be utilized. Depending on the size, shape, distribution of colour and inclusions, a decision is made as to how it will be cut. With most corundum, especially the smaller sizes, a single stone only is obtained from the rough and so sawing can be omitted. When made, saw cuts are often placed along cracks or flaws so as to eliminate them at the beginning.

The preforming stage is the most important because the essential shape and orientation of the gem are then determined. For this reason preforming is generally the job of the oldest and most experienced worker. The shape of the rough largely determines the shape of the preform and thus that of the finished stone. Most common for corundum are the cushion and oval shapes. Since the cushion provides greater potential brilliance, it is more desirable than the oval. By the same token, the round shape possesses more potential brilliance than the cushion and, since it generally results in greater weight loss, it fetches a premium in the market.

Apart from shape, the main concern of the preformer is colour. Not only must its distribution be taken into account, but also pleochroism must be considered, for rubies and sapphires are strongly dichroic. In shaping the stone, the lapidary is conscious of certain crystal directions along which the colour varies. For both ruby and sapphire, the best colour is seen along the *c* axis, while at right angles to this direction it is slightly less intense. Thus, the lapidary will strive to orient the rough so that the table facet is at right angles to the *c* axis. Fortunately, the shape of both ruby and sapphire in the rough permits the largest weight retention, or 'yield', in just this orientation. Failure to position the *c* axis perpendicular to the table results in a lower colour intensity and, subsequently, a lower value.

Variations from this desired orientation are most often the result of uneven colour distribution. Sri Lankan and Kanchanaburi blue sapphires are often strongly zoned, so much so that many contain only tiny areas of colour, the remainder being colourless. In the 'ottu' blue sapphires from the Island of Gems (Sri Lanka), colour lies just beneath the faces and at the tips of the crystals. The lapidary tries to orient the stones so that the colour is placed at the culet. With most Sri Lankan sapphire rough being spindle shaped, this explains why the gems cut from this rough have such deep pavilions. When properly done with the colour on both sides of the culet, the entire gem faces up blue. However, should the colour lie on one side of the culet only, an unsightly colourless 'window' will be seen in the face-up position.

Should it be impossible to position the colour at the culet, the lapidary may try to lay the crown along one of the faces below where the colour lies. Although the *c* axis is then wrongly positioned to obtain the most intense colour, it is preferable to a partial or complete lack of colour.

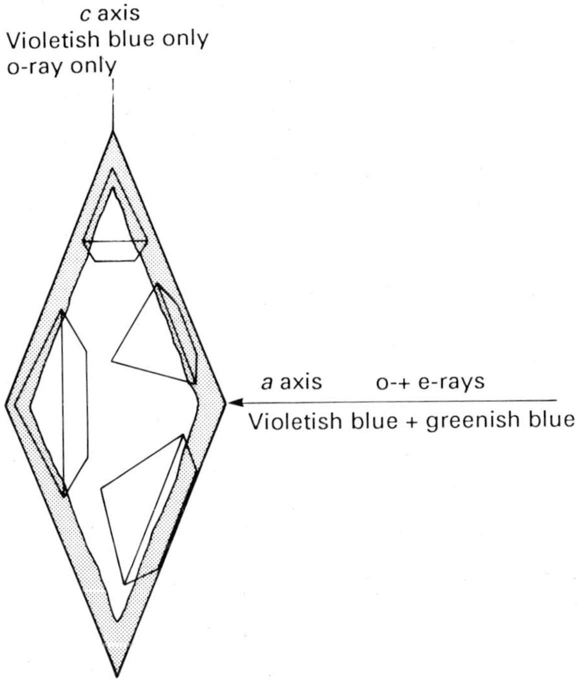

Figure 8.2 In the 'otto' blue sapphires from Sri Lanka (and also Kashmir) the blue colour lies just below the crystal faces, with the core of the crystal being colourless. In order that the faceted stones face up blue, the cutter will orient the rough so that the colour lies at the culet (preferably on both sides of the culet and reaching as far up the pavilion as possible). An alternative orientation is to place the colour band across the crown, preferably extending from girdle to girdle. In many cases, weight retention takes precedence over colour intensity, with the table facet not lying perpendicular to the *c* axis (as it should to achieve the most intense blue colour). Thus pleochroism dilutes the colour

Inclusions may also prompt the lapidary to veer from the desired orientation. Burmese ruby often contains dense clouds of rutile silk. These rutile needles are broad and thin, with the flat side lying in the basal plane. Thus, in stones cut with the *c* axis at right angles to the table, they are very obvious and give rise to bright reflections. However, orienting the table parallel to the *c* axis makes them much less visible. As we can appreciate, the lapidary must constantly strive to balance desirable and undesirable features, all of which are interrelated. The total value of the stone is the key factor in deciding whether to cut a small high-quality stone or a larger, but less costly (per carat), piece. If faced with a decision between cutting a 1 carat stone worth $1000 per carat and a 2 carat stone worth $600 per carat, the choice will be to cut the 2 carat piece, because the total value of the stone would then be greater.

One final factor involved in preforming the gem concerns its depth. The depth of the stone can be expressed as a percentage of the diameter, with approximately 65% being considered most desirable. There are, however, reasons for varying from this ideal. The first concerns the shape of the rough. Thai ruby rough is generally found as flat tabular crystals and, because the high price of ruby makes weight retention vital, the cut stones are often far too shallow for maximum brilliance. Similarly, Sri Lankan sapphire crystals are usually long spindle shapes and so the finished stones end up too deep. The depth of colour also enters into this. Australian sapphires may appear so dark and inky that if cut to proper proportions they would be almost black. Thus, the preforms are made shallow in order to sacrifice brilliance and lighten the colour to acceptable levels. Some Australian sapphires may be purposely oriented with the *c* axis at 45°, or even parallel, to the table, again to lessen the colour intensity in overly dark stones. The latter orientation is often termed 'cross-table' by Australian lapidaries.

Once the rough has been oriented and the preform made, the stone is then given to the cutter for grinding and polishing of the facets. Compared with making the preform, this is a relatively simple task and so is usually carried out by less experienced staff. The cutter will choose a cutting style based upon the type of gem material they are working with, as certain styles lend themselves more readily to some gems than to others.

This choice is based largely on two factors: brilliance and weight retention. Increasing one factor can only be done at the expense of the other, so the cutter strives to achieve a balance, or a happy medium. The facets of a cut stone are so placed as to allow light falling on the gem from above to reflect off the internal pavilion surfaces and then return to the eyes. Ideally, the gem should act as a light trap, catching as many rays as possible; this is achieved by the table facet. Then the pavilion reflects the light rays like a mirror, reversing their direction and throwing them upwards again where they escape back through the table. Some rays will escape through the crown facets, being dispersed, or split, into coloured rays in the process. With ruby and sapphire, their dispersive power, or 'fire', is low (0.018) and their colour masks what dispersion they do show. Thus, the main aim of the cutting style is to produce brilliance and scintillation, the twinkling effect seen when either the stone, light source, or viewer's head moves. Scintillation is achieved by cutting the stone with many small facets, rather than fewer large ones, to break up the light into small angular areas. However, balance is necessary, for if the facets are too small then blurring rather than scintillation results.

The most important factor in the brilliance of a cut stone is the angle of the pavilion facets. If the stone is cut too shallow, light will pass right through it instead of reflecting off the rear facets and creating a 'window'. By the same token, if the pavilion angle is too steep, light reflects off one side and then passes out through the opposite side, creating an effect known as 'extinction'. Windows appear as areas of low colour intensity near the culet while extinction is seen as dark areas, usually near the girdle.

The ideal pavilion angle for a gemstone depends upon its refractive index, which is a measure of its potential brilliance. To understand how this influences a gem's brilliance, we must first understand the concept of 'critical angle'. When light inside a gem strikes a boundary between the gem and air, whether or not it will leave the stone depends on its angle of incidence and the size of the critical angle formed within the stone. The critical angle is the incident angle for the ray which makes a 90° angle of refraction. Light incident with an angle less than the critical angle will leave the stone, but light incident with an angle greater than the critical angle will instead be totally reflected back into the gem, where it travels across to another facet and the same situation prevails regarding its ability to leave the gem. Thus, gems are faceted so that light entering the stone through the table facet will strike the pavilion at an angle greater than the critical angle. This causes it to be totally reflected across the stone to the opposite side of the pavilion. Again, its angle of incidence is greater than the critical angle and so it is totally reflected back towards the table. Now, however, its angle of incidence is less than the critical angle, so it leaves the stone through the table, creating brilliance. Obviously, from the above explanation gems which possess smaller critical angles will permit less light to escape through the back and sides. Thus, they possess a greater potential brilliance than those with large critical angles. The actual brilliance displayed by a gem is a function of its proportions (pavilion angle, etc.) and the size of its critical angle (critical angle = arcsin 1/refractive index of gem).

Critical angles vary in inverse proportion to the refractive index of the stone; the higher a gem's refractive index, the smaller its critical angle and, therefore, the greater its potential brilliance. Since the pavilion angle is largely responsible for a gem's brilliance, the ideal pavilion angle for that gem can be calculated by reference to its refractive index, n. The ideal pavilion angle for diamond (n = 2.417) is 40.75° and that for corundum (n = 1.76) is 41°. The critical angle of corundum is 34.5° and that for diamond is 24.5°.

Should the pavilion facets be cut at an angle less than the critical angle, light will pass right through the stone instead of being reflected, creating a window or an area where one can see through the stone, as in window glass.

The brilliant cut is much used for diamond but rarely for corundum. Although this cutting style allows maximum brilliance, it is only commonly encountered with sapphires from Montana. The reason is that the pavilion facets of the brilliant cut extend all the way from girdle to culet, making it impossible to increase the yield by slightly rounding the pavilion. Instead, the cutting style most often used for rubies and sapphires is the mixed cut, so called because it combines the brilliant-cut crown with a step-cut pavilion. Its chief advantage lies in the fact that the cutter can retain as much weight as possible by rounding off the steps, or facet rows, on the pavilion. A price must be paid, however, for this extra weight retention. If the facet rows nearest the girdle are cut too steep, extinction is created, with light passing out of the side instead of returning to the eyes as brilliance. Owing to the bipyramidal shape of rough Sri Lankan sapphires, they are often cut with deep pavilions and so suffer this malady. On the other hand, if the facet rows astride the culet are cut too shallow (at or under the critical angle of 34.5°) light passes straight through and produces a window. This is the greatest problem today in rubies and sapphires faceted in Asia – a failure to close the culet. Often, a stone which is otherwise well proportioned is spoiled by a window at the final row of facets next to the culet. Recutting the gem to remove this window sometimes results in only a slight loss of

(a)

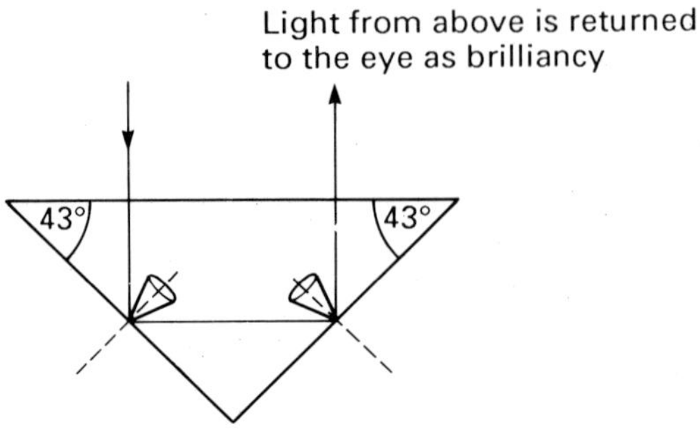
Light from above is returned to the eye as brilliancy
43°
43°

(b)

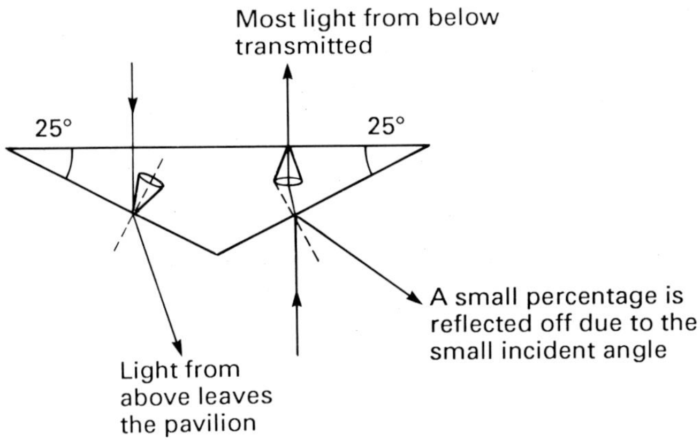
Most light from below transmitted
25°
25°
A small percentage is reflected off due to the small incident angle
Light from above leaves the pavilion

(c)

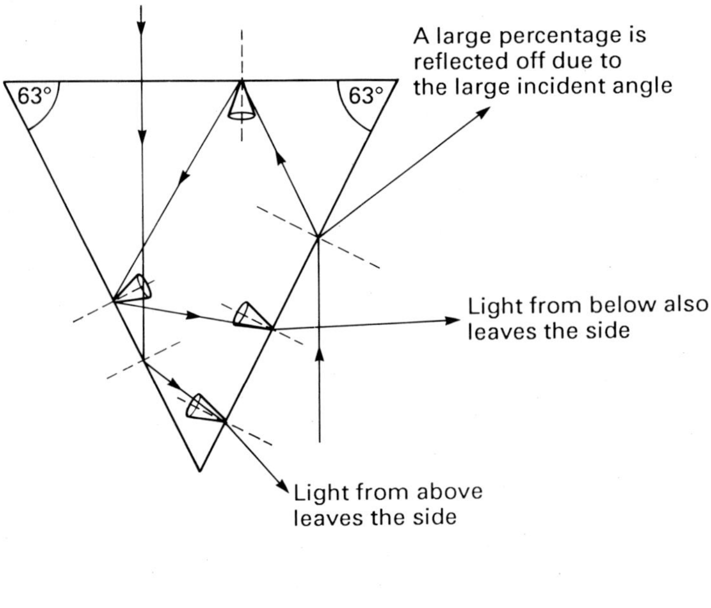
A large percentage is reflected off due to the large incident angle
63°
63°
Light from below also leaves the side
Light from above leaves the side

weight. Thus, it is not weight retention which is the cause of this problem; instead, it indicates merely a lack of proper training.

Again, the shape of the rough plays a vital role in determining the proportions of the finished gem. Thai rubies will often be found to contain large windows because of the thin tabular shape of the rough. In addition, the high value of ruby rough ensures that cutters pay close attention to obtaining the maximum yield. Indeed, amongst the various precious stones large rubies of high quality are cut to worse proportions than any other. This is not indicative of a lack of talent on the part of the cutters – quite the opposite. It is a testament to their prowess at obtaining the maximum value for their clients, as the gems are sold by weight. Often, to recut such a stone to perfect proportions is to lose money, because the increase in quality is not enough to offset the decrease in weight.

The other major facet style currently in use for corundum is the full step cut, also termed the trap or emerald cut. It is used on stones of an emerald shape – rectangular or square in outline with each of the four corners truncated slightly to prevent chipping or fracture. The chief attribute of the step cut is that it allows the

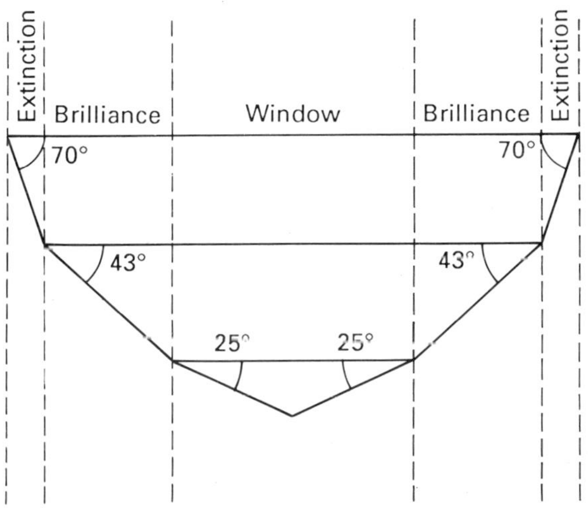

(d)

Figure 8.3 The phenomena of brilliance, windows and extinction. (*a*) shows how facets cut at the proper angles will allow light from above to return to the eye as brilliance. (*b*) shows pavilion facets that are cut too shallow. Light from above passes out of the bottom while light from below is, by and large, able to pass up to the eye. However, because the light path from below is much shorter than on facets showing brilliance, the colour seen will be dull and of low intensity. This effect is termed a 'window'. (*c*) shows pavilion facets which are cut at too steep an angle. Thus light from above reflects across the stone and passes out of the opposite side. Much of the light from below is reflected off the first surface owing to the steep angle of incidence, while that which does enter the gem is unable to leave the table because it is incident outside the critical angle (in some cases it may be able to leave the table but is reduced in intensity because of the loss from reflection as it entered the gem from below). Thus it is reflected back inside and eventually leaves the pavilion. Facets such as these which are cut too steep will appear dark as a result. This is termed 'extinction'. Most corundums are cut with step-cut pavilions in which a combination of the three effects is seen, as in (*d*). Those nearest the girdle are too steep and so show extinction, while those next in line are just right and so show brightly coloured flashes of brilliancy. The row of facets at the culet is too shallow and thus shows a dull low-intensity colour, or window

gem's colour to be seen in all its glory. Long unbroken panels extend along each side, showing off the colour. However, inclusions are also very noticeable with the step cut, particularly in lightly coloured pieces, so the stone should be relatively clean internally for this cut to work best.

In gems with a step-cut pavilion, the cutter can influence the colour, to a certain degree, by varying the number of facets on the pavilion. With lightly coloured pieces, reducing the size of the facets and increasing the number of steps will effectively deepen the colour. Decreasing the number of facets and steps has the reverse effect on dark stones and lightens their colour. An analogy can be drawn by considering candles viewed at a distance. If a single strong flame were held out, it would be spotted more easily than four separate flames, each one-quarter the intensity of the single flame. In both situations the amount of light is equal, but the single flame makes more of an impact on the eye. By creating four facets out of one, the impact of that area can be reduced, causing an apparent reduction in the brightness of the colour, effectively deepening it as a result. This does sacrifice some brilliance overall, but in many cases it is more than made up for by the resulting improvement in colour. Thus, the shrewd cutter will use this principle to good effect, raising the stone's value.

The brilliant, mixed and step cuts are generally the only faceting styles used for rubies and sapphires. Others are occasionally encountered, but this is rare. That known as the 'native cut' is not really a style in itself, but more a lack of style. 'Native cut' is synonymous with 'poorly cut' and comes to us primarily from Burma, India and Sri Lanka. It is a crude attempt at one of the major styles, such as the mixed cut. Often as a result of poor training and a lack of modern equipment, these stones at times bear only a vague resemblance to modern faceted stones. The facets are placed in a haphazard manner and, especially with those from Sri Lanka, are often not even flat. Square tables on round stones and rectangular tables on oval stones are the hallmark of native-cut gems from Sri Lanka. Unfortunately, they still use methods virtually unchanged since ancient times, and the stones certainly show it. This is not to say that all gems cut in these countries are native cut. Today they also do some very fine work, but still many stones must be recut if they are to be sold in the major world consumer markets. The biggest problem with their cutting is an overreliance on weight retention. Proper cutting is just as important as colour or clarity in determining the value of a stone, for the true colour is seen only in the areas where light returns to the eye as brilliance. A ruby may have the most wonderful colour on earth, but what good is it if, as a result of poor cutting, only 10% of the stone's face shows this colour?

Whenever stones which need to be recut are bought, they must be judged or appraised at the weight estimated after recutting to good proportions, just as is done for diamonds. Failure to do so is to risk buying at inflated prices. If a stone weighs 2.50 carats, but after recutting would weigh only 1.50 carats, it must be valued as if it were 1.50 carats, not 2.50. One final tip with regard to purchasing stones for recutting concerns their proportions. It is better to select stones which are bottom heavy, or lumpy, rather than those which are flat. The weight loss in recutting bottom-heavy stones is usually much less than for flat pieces. A good rule when buying stones is, if possible, to avoid buying flat gems.

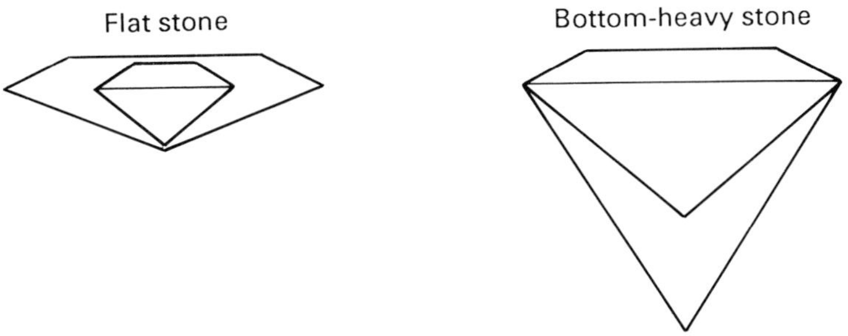

Figure 8.4 In the purchase of 'native-cut' (poorly cut) gems, recutting is often necessary. Generally buyers prefer stones which are originally too deep rather than too shallow because the weight loss is often much less with overly deep stones compared with those that are too shallow. Sri Lankan rubies and sapphires are generally cut too deep, while rubies from Thailand/Cambodia are usually too shallow. This can be traced back to the original shape of the rough material. In recutting Sri Lankan blue sapphires care must be taken not to remove areas containing the blue colour, for this may result in a dramatic loss of colour overall

Some cutting styles are used almost exclusively for synthetic and imitation stones. The one that immediately comes to mind is the 'scissors' cut, a style used for rectangular-shaped stones. The scissors cut, also known as the cross cut, possesses side facets consisting of four triangular facets in an 'x' pattern on the crown. Large quantities of synthetic ruby and sapphire, synthetic spinel and glass are faceted in this style. So rarely is it used for natural corundum that suspicions should be aroused by anything cut in this style that is offered as natural.

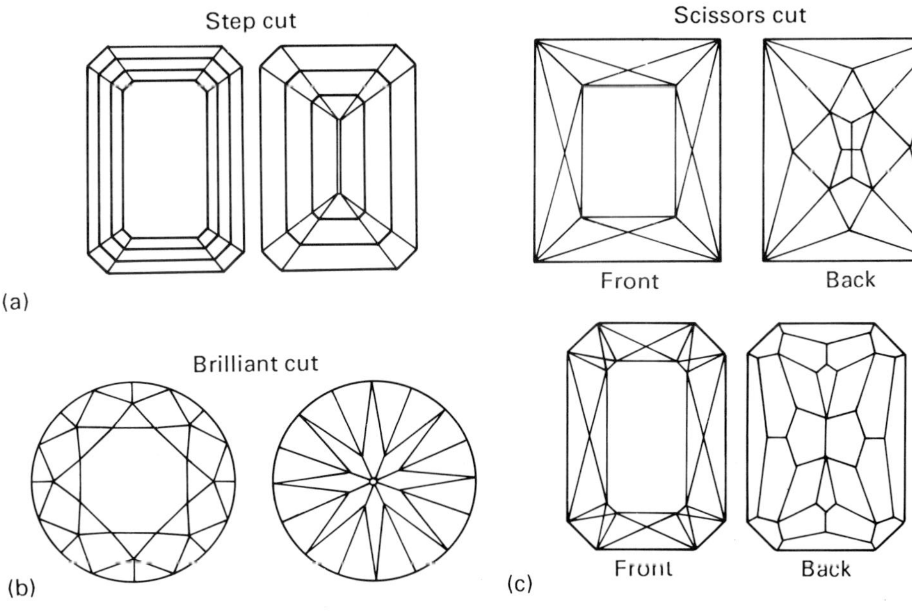

Figure 8.5 Common cutting styles. Most natural rubies and sapphires are cut with the mixed cut (not shown) which features a brilliant-cut crown (*b*) and a step-cut pavilion (*a*). The use of a step-cut pavilion allows greater weight retention. Synthetic corundums are often cut in the scissors style (*c*). This is almost never used for natural corundum

One of the most exciting new developments in the lapidary arts is the creation of the Barion cuts. Developed by South African diamond cutter Basil Watermeyer, the Barion cuts are an adaptation of the round brilliant style of pavilion facets to angular shapes, such as the emerald, square emerald, kite, triangle, pentagon and hexagon. Their key feature is facets at the girdle, termed 'half moons', which allow brilliant style facets to be placed upon non-curved outlines, on the pavilion. The crown is cut in the standard step-cut style. Barion cuts produce as much or more brilliance than even the standard round brilliant cut and allow greater weight retention as well! Rubies and sapphires cut in the Barion style look excellent; it seems that this is truly the cutting style of the future.

Cabochons

One of the oldest methods of cutting gemstones is the rounded shape known as the cabochon. This is a French term which itself was derived from the Latin *cabo,* meaning head. Cabochon-cut rubies and sapphires generally fall into two categories: stones which are too heavily included to be faceted, and star stones.

The cabochon may be cut in a number of ways. Domes vary from very low almost flat surfaces to very high steeply curved surfaces. Usually a medium to high dome is preferable. With transparent gems, the back of the cabochon is generally flat or slightly convex and polished. Convex polished bases are used to provide greater brilliance in stones which possess enough transparency. However, should the stone be near or completely opaque a flat base is considered more desirable, as there is no excess weight below the girdle. In any case, excess weight below the girdle should not exceed about 20% of the whole stone, for it must be remembered that the stone is being paid for by weight. Occasionally stones from Burma, India and Sri Lanka are seen with such heavy bottoms that they look like bowling balls. Just as with faceted gems, these bottom-heavy stones should be valued at the weight they would be after recutting to perfect proportions. Overly dark material is sometimes cut as 'hollow' cabochons, with the base being concave to lighten the

Figure 8.6 Gem cutters in Sri Lanka (Ceylon), 1939

stones. This was a common style of cut for the red pyrope/almandite garnets, termed carbuncles, in days gone by. Today it is rarely used for rubies and sapphires.

The symmetry of the top and bottom surfaces on cabochons is easily checked by spinning the stone on a hard flat table. Any wobbling as the stone spins indicates flat spots or a misshapen dome. The edges should also be examined and should be nicely bevelled to prevent chipping.

In cutting star material orientation is all important, for if it is done improperly the star will be off centre or misshapen. Star corundums must be cut so that the planes of silk lie at right angles to the cabochon base. To do this the rough is examined under a spotlight or other suitable illuminant such as direct sunlight. With the light coming just over one's shoulder, the rough is turned until the silky sheen is exactly on top of the stone. Then the line for the base is marked so that it lies parallel to the floor. An alternative method utilizes a sticky viscous liquid such as honey. When a drop of liquid is placed on the rough gem, a miniature star appears on the drop. If the drop is centred perfectly, the star will appear exactly centred on top of the drop. The base is then cut parallel to the floor in this position.

A very common mistake in cutting star rubies and sapphires is the misalignment of the rays themselves on oval-shaped stones. When cutting ovals, the lapidary must carefully orient the stone so that one ray is parallel to the width of the oval. This creates an 'open star', one in which the rays are well spread. Unfortunately, however, most are oriented with one ray parallel to the length of the oval, not the width. This produces a 'closed star', with the rays pinched together unattractively.

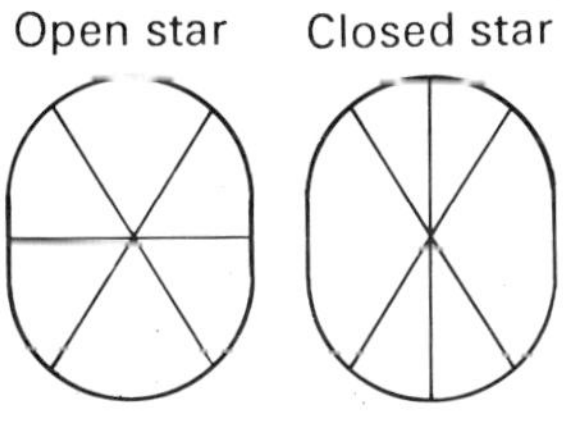

Figure 8.7 In orienting oval-shaped star corundum rough, care should be taken to place one of the rays across the width of the oval rather than along the length. This will produce an 'open star' whose rays are evenly spaced, rather than a 'closed star' whose rays appear pinched together

Many star rubies and sapphires are cut with low-domed cabochons, a defect which greatly affects the appearance of their stars. The cabochon dome acts as a condensing lens to concentrate the light reflections into a star effect. In order for this to be effective, however, the curvature of the dome must be uniform throughout. Low domes generally possess a large flat, or nearly flat, area in the centre. This causes the star to appear diffuse instead of sharp, and it disappears completely unless both the light source and the viewer's eyes are directly above the stone. In contrast, a well-proportioned dome produces a sharp star that is visible from almost all angles.

The black star sapphires from Thailand and Australia owe their asterism to inclusions of haematite, rather than rutile. Included in green, yellow and blue sapphires, they are present in such profusion that the stone is coloured dark brown or black as a result. They are also the cause of the very prominent basal parting unique to this variety. This parting develops so easily that most cutters try to avoid it by cutting low-domed cabochons. Often during cutting, the high-domed stones chip at the edges of these parting planes; hence the reason why the low domes are so common on black star sapphires. The low domes common to black star sapphires are also used because the haematite silk, unlike the rutile silk of Burmese and Sri Lankan stars, is plate-like rather than consisting of slender needles. Haematite plates create a much less intense star than do slender rutile needles. Thus, if a high dome is cut on a black star sapphire, the star appears much weaker and less distinct than when a low dome is used. Because Burmese and Sri Lankan star corundums contain rutile silk, however, they should be cut with higher domes. This illustrates well the fact that poor cutting is not necessarily caused by a poorly trained cutter; instead, it may be the opposite, a problem which is hidden by skilful cutting.

The tendency of black star sapphires to chip at the edges (basal parting) often results in unsightly holes on the back of the stone. Thai cuttters have come up with a novel, if unscrupulous, technique for hiding these holes – filling them with doping shellac. The result is that a fair number of the black star sapphires cut in Thailand possess shellac-filled surface cavities. Most have minor repairs only, but sometimes stones are found with very large shellac-filled pits. Removal of the shellac may reveal a stone which is structurally unsound because of the large pits previously hidden by the filling.

To summarize the cutting of star stones, foremost is orientation. They must be properly oriented so that the star is centred directly on the top of the cabochon. Its dome should be medium to high and without flat spots, allowing the star to remain centred on the top of the stone. The shape should be symmetrical and pleasing to the eye, not too narrow or too wide. Oval stones should display open stars, with one ray lying across the width. The base should be smooth and flat on opaque stones, or slightly convex on more transparent pieces. There should not be any pits or cavities present. Watch out for cavities hidden by filling with doping varnish. Finally, excess weight below the girdle should not exceed 20% of the total.

Double cabochons with a high-domed crown and low-domed convex base are at times set into jewellery with the low dome facing up. This is done with highly transparent stones, with the aim of increasing brilliance. Light enters easily through the low dome and reflects off the deeper bottom to return to the eyes, just as in a faceted stone.

Shape

The shape given to a cut stone does have an effect on its desirability and, thus, on its value. As a guide, the common shapes seen with ruby and sapphire are listed below in order of desirability:

1. Round
2. Square emerald
3. Square cushion
4. Elongated emerald
5. Elongated cushion
6. Oval
7. Pear
8. Heart
9. Triangle
10. Marquise

Other fancy shapes have not been listed because they are rarely used with corundum.

In addition to the shape itself, the proportions of that shape are also important. Round stones should be absolutely round, preferably with an error not exceeding 2%. For guidance on the proper proportions of rectangular emerald, rectangular cushion and oval shapes we can use the laws originally developed by Greek artisans over 2000 years ago. Comparing the width of a stone with its length, a proportion of 1 to 3 yields a stone which most people will find too narrow; a proportion of 2 to 3 gives an attractive rectangle; while for bigger gems, a proportion of 3 to 5 is more satisfying. Comparing the proportions of the crown with those of the pavilion, generally the crown should amount to one-third of the depth and the pavilion to two-thirds. The exception is with overly light stones. Very light stones may be cut with the crown only one-fourth of the total depth, to deepen the colour slightly.

Bibliography

Arem, J. (1977) *Color Encyclopedia of Gemstones*, Van Nostrand Reinhold, New York

Quick, L., Leiper, H. and Kraus, P. (1977) *Gemcraft*, Chilton, Radnor, PA

Sinkankas, J. (1968) *Van Nostrand's Standard Catalog of Gems*, Van Nostrand Reinhold, New York

Sinkankas, J. (1984) *Gem Cutting: A Lapidary's Manual,* Van Nostrand Reinhold, New York

Soukup, E. J. (1962) *Fact Cutters Handbook*, Gems & Minerals, Mentone, CA

Sperisen, F. J. (1961) *The Art of the Lapidary*, Bruce, Milwaukee, WI

Vargas, G. and Vargas, M. (1969) *Faceting For Amateurs*, privately published, Thermal, CA

Vargas, G. and Vargas, M. (1975) *Diagrams For Faceting*, privately published, Thermal, CA

Watermeyer, B. (1982) *Diamond Cutting*, Centaur, Johannesburg

Chapter 9

Quality in corundum and famous rubies and sapphires

Next to diamond, corundum is the most important of all gem species. However, despite this fact there exist very few examples of rubies and sapphires that have been given titles. In the case of sapphires, this is certainly not for want of magnificent specimens of large size, but rubies of large size and fine quality are singularly lacking. While perfect diamonds of many carats abound in history, perfect rubies of even 5 carats are unknown. The simple fact is that diamonds are composed of that abundant element carbon, while ruby is a compound and depends for its quality of colour on the presence of the relatively rare element chromium.

Important rubies

Several examples exist of large rubies with titles. The most famous of these, however, are not rubies at all, but red spinels. Two important examples are found in the British Crown Jewels: the Timor Ruby and, the most famous of all, the Black Prince's Ruby. Despite the fact that these are actually red spinels, their colourful tales are worth telling, so this is what we shall do.

Few other precious stones have such a long and varied history as the Black Prince's Ruby. Its source is unknown; perhaps Burma, Sri Lanka or Afghanistan was its true birthplace, but it first appears in history books in the middle of fourteenth-century Spain. At this time Spain was ruled by a number of petty kings, one of whom was a Moorish prince, Mohammed, of Granada. Don Pedro the Cruel ruled nearby Seville, and it was to him that Mohammed fled after being deposed by his brother-in-law, Abu Said. Don Pedro's army eventually brought Abu Said to heel, and Abu Said and all of his attendants were killed for their jewels when they arrived to negotiate. Among Abu Said's splendid collection was a very large red spinel ocatahedron, the size of an egg. It is known today as the Black Prince's Ruby.

Don Pedro soon found it his turn to flee, his adversary being none other than his own brother, Henry. In 1366, he fled to Bordeaux, where the Black Prince kept court, and beseeched him to help, promising untold treasures in return. Henry was duly defeated and the large red stone passed into the Black Prince's hands, as payment.

The gem reappears next in the hands of the English king, Henry V, at Agincourt on 25 October 1415. Henry V, the gallant king, with his army reduced to 15 000 men, was falling back upon Calais when at Agincourt he encountered the French king and his army of 50 000 men. On the morning of the climactic battle, Henry appeared dressed in splendid attire with gilt armour. Upon his helmet was a rich crown garnished with rubies, sapphires and pearls. Conspicuous amongst the gems was the Black Prince's Ruby.

Henry's helmet served as more than mere decoration, for on that day he was set upon by the French Prince, the Duke of Alencon. The Frenchman struck his helmet a mighty blow with his battle axe and came close to finishing off Henry's career on the spot. Others, excited by the blood-red stone no doubt, also attacked him, one of them even managing to break away a portion of the crown. Miraculously, however, both the stone and Henry survived the battle. After the battle, a French prisoner retrieved the broken fragment and brought it back to England, an act for which he was duly reimprisoned. The helmet worn by Henry at Agincourt was described by Orpen (1890), from which much of the present account was taken, as that residing in Westminster Abbey, short of its jewels. Two deep gashes are readily visible.

From here the precious 'ruby' passed through the hands of numerous English monarchs, including Henry VIII and his daughter Elizabeth I. She did not set the stone into her state crown, but kept it in her private collection. Elizabeth did show it to a Scottish envoy, Sir James Melville, however. One evening the Queen took him into her bedchamber, where

> she shewed me a fair ruby, great like a racket-ball. I desired she would either send it to my queen (Mary, Queen of Scots) or the Earl of Leicester's picture. She replied 'If Queen Mary would follow her counsels she would get them both in time and all she had, but she would send a diamond as a token by me.'

It was for the Black Prince's Ruby that the envoy begged, but Mary was destined to get neither.

King James I had the stone set into his state crown, for the Earl of Dorset describes the stone in an inventory of the crown jewels. His description of the imperial crown concludes: 'and uppon the topp a very greate ballace [ballas ruby, or spinel] perced'. We know this to be our stone for at some point in time it had been drilled ('perced') at the top with a small hole so it could be worn suspended from the neck, a common occurrence with oriental gems. Today this hole is capped with a small ruby.

After the coronation of Charles I, by a fortunate occurrence the great gem was not placed in the jewel house together with the other royal treasures. If it had been it would have been lost for ever, for when Cromwell took power all of the treasures found there were either melted down or sold by order of the Commonwealth.

Among the priceless pieces lost was the gold filigree crown of Edward the Confessor, which was broken up and sold for its weight of bullion. As Orpen (1890) remarked, 'Such vandalism is almost enough to make one a Jacobite.'

In 1660, the Black Prince's Ruby was bought by a jeweller who resold it to Charles II after the restoration of the Stuarts (Prince Michael, 1983).

During the reign of Charles II, the stone, by now set in the Crown of England, had another narrow escape, nearly being stolen by the notorious Colonel Blood who, unbelievably, was later pardoned by the King.

In 1841 the crown was nearly lost once again, this time by fire. Only the quick actions of a police inspector named Pierse saved the day. As the Tower burned, Pierse broke through the iron bars with a crowbar to rescue the treasures kept within, thus preserving these irreplaceable objects. During World War II the royal regalia were once more in danger, this time from Hitler's bombers during the Battle of Britain. However, they survived undamaged and today the giant Black Prince's Ruby can be viewed in the Tower of London, together with the rest of the Crown Jewels, in all its glory.

The Black Prince's Ruby is mounted in the front of the Imperial State Crown, just above the Cullinan Diamond. It is a huge uncut octahedron which appears to have been worn smooth naturally by water. Sitwell (1953) states that the stone is backed by a gold foil, as were many ancient gems, to improve its brilliance. This has not been removed for fear of damaging the gem. The stone measures some 2 inches (5 cm) in length and is of proportionate width (Younghusband, 1919). As mentioned earlier, it is drilled at one end and a small ruby is set atop the opening. According to Younghusband (1919) 'the question is often asked: "What is the value of this stone?" And the answer may safely be given that it is priceless, for no amount of money can buy it.' It is indeed the most famous gem in the world's most famous gem collection.

There exist in gem collections a number of other famous red spinels, of which brief mention seems appropriate.

The Timur Ruby is a large red spinel and rests today in the Crown Jewels in London. It is a large tabular stone of 361 carats which carries Arabic inscriptions of the names of previous owners.

Probably the largest collection of big red spinels is that of the Crown Jewels of Iran, which are housed in Teheran and have been magnificently described by Meen and Tushingham (1968). Included among these is the largest fine red spinel known, a blood-red lump weighing some 500 carats. A companion piece weighing 270 carats is the more important of the two from an historical perspective. Over 350 years ago, Jahangir, the Mogul Emperor of India, had his name carved onto its surface. When his favourite wife chided him for defacing such a magnificent gem, he replied 'This stone will perhaps carry my name farther down through time than the empire of the house of Timur.' This proved to be true as the empire and dynasty of Tamerlane ended within 150 years while Jahangir's name is perpetuated on more than a dozen gems in the Persian collection alone (Meen and Tushingham, 1968).

The larger of the two spinels in Iran is pierced, but the openings are now plugged. Both stones are believed to have originated from the famous mines of antiquity in Badakshan, a district in Afghanistan.

There exist numerous other red spinels in the Crown Jewels of Iran. Many of these stones exceed 100 and some even 200 carats in weight.

The second largest red spinel on record is housed in Moscow, in the Kremlin Museum. It is a giant stone of 414 carats.

As mentioned above, rubies of great size and quality are virtually unknown. Probably the finest collection is that of the Crown Jewels of Iran. Burmese rubies of 10 carats or more are extremely scarce in fine qualities but the Persian collection possesses a number of fine examples. Of particular note is the plaque of 13 magnificent stones described by Meen and Tushingham (1968) and matched by another plaque nearby. Each set is set in a simple gold ring and all are cabochons, ranging in size from 8 to 16 carats. All are believed to be of Burmese origin.

Burma has been and continues to be the source of rubies *par excellence.* Unfortunately, during the many centuries in which the mines at Mogok were under the rule of the Burmese kings, the policy was that all stones of large size (above approximately 5 carats) were the property of the Crown and were to be turned over to the king on pain of death. This resulted in many large stones being broken up into smaller pieces and so these treasures were lost to the world.

Among the most magnificent rubies ever found in Burma was the stone of 42 carats found in 1919. Dubbed the Peace Ruby, it was a superfine piece of rough marred only by a crack estimated to take away a slice of about 8 carats. It was sold on the spot at Mogok for Rs. 3 lakhs (£22 500). Unfortunately it did not live up to its great potential after cutting (Brown, 1933).

Another fine Burmese ruby is that dubbed the 'Chatrapati Manick Ruby'. This stone is an oval cabochon of about 40 carats and of a good deep colour. As of 1934, the stone was reported to be in London and mounted on the front of a diamond tiara.

In 1899, the Burma Ruby Mines Company uncovered a giant ruby of 77 carats rough and valued at £26 666 (Brown, 1933). Another stone weighing 36 carats was sold by King Mindon Min, the father of the last Burmese monarch, for £30 000 (Brown, 1933). Numerous additional examples of large rubies such as the above exist in the literature, but because we do not know where the stones are today it is difficult to assess their quality or to know for certain whether they were in fact rubies. From what we do know about rubies it can be guessed that most or even all of the very large stones (greater than 100 carats) reported in the possession of pre-twentieth-century monarchs were either flawed or actually red spinels.

King Theebaw, the last Burmese monarch, was reported to have a collection of Burmese rubies unsurpassed in all the world. Of this, as a result of the aforementioned policy there can be no doubt that all large stones were the property of the state. John Crawfard, an Englishman who went on an embassy to Ava (the then capital of Burma) in the years 1826 and 1827 had this to say: 'the king lays claim to every Ruby and Sapphire which exceeds the value of 100 ticals; and there is, from all accounts, a large collection of both in the Royal Treasury' (Streeter, 1892). What happened to this magnificent collection after the British annexed Upper Burma? The treasure from the Royal Palace at Mandalay now rests at the Indian Museum in South Kensington, London, but to look at it gives the impression that the Burmese king was a pauper. Although there are a number of rubies in the regalia, they are only of small size or of imperfect quality. Unfortunately for the British and also for gem lovers, on the night that the palace was captured, in 1886, the treasure was smuggled out under the noses of the British soldiers. Guards were posted with orders not to allow anyone to enter or leave the palace. The British soldiers, being a chivalrous lot, did, however, permit female servants to come and go. Throughout the night that is exactly what they did, taking with them the lion's share of the treasure (Stewart, no date). Eventually, no doubt, the stones found their way onto the open market in Lower Burma and India, and then into the private collections of the world's rich. Thus, one of the greatest opportunities was lost for ever – that of the public display of the most fabulous jewels of the Burmese monarchs, a collection put together over centuries. The likes of it will probably never be possible again.

One of the most magnificent Burmese rubies ever found was that known as the 'Nga Mauk' (Gna Monk) Ruby (Streeter, 1892; Keely, 1982). The story of its origin will be related in Chapter 11, but will be briefly told here. During the reign of King Mindon Min, a man found a rough ruby weighing 7 tikals. At this point there is some discrepancy in the accounts. Streeter (1892) claims that the man's wife traded the stone for some fish condiments to a man named Nga Mauk while Keely (1982) does not mention such a trade. It does seem unlikely that anyone living in the Mogok area could fail to recognize such a fine rough ruby, so in this regard Streeter may be wrong. In any case, the owner of the stone then broke it in two, giving one half to the king and sending the other for sale in Calcutta. The king discovered the fraud and, after learning where the other half had been sent, ordered its return, in the meantime also ordering the village and all its inhabitants to be burned as a lesson to others with similar ideas. Eventually, the second half was purchased in Calcutta for an enormous sum and returned to Burma, where it forms a perfect fit with the first piece. The two stones were cut in Burma, one forming a grand stone weighing 98 carats and named 'Nga Mauk', while the other weighed 74 carats and became known as 'Kallahpyan', signifying that it had returned from India (Streeter, 1892). These two pieces were undoubtedly among the treasures smuggled out of the palace when Upper Burma was annexed by the British in 1886.

Also among the great Burmese rubies was a stone found during Mindon Min's rule by a man working on the road to Momeit. It weighed 400 carats in the rough and was cut into three pieces: a stone of 70 carats which was sold to England; a stone of 45 carats sold in Mandalay; and a third portion of unknown weight sold in Calcutta for 70 000 rupees (Streeter, 1892).

'Gnaga Boh', or the 'Dragon Lord', was the name given to a ruby found at Bawbadan weighing 44 carats in the rough and, when cut, 20 carats. This stone was said to be the finest of its size ever reported and was given by the finder to King Tharawadis. It was also amongst the booty smuggled out of King Theebaw's palace in 1886 (Streeter, 1892).

Soon after Theebaw ascended to the throne a very fine stone weighing 100 carats in the rough was found on Pingtoung Hill near Mogok, an area where several 'royal rubies' have originated from. The stone was presented to Theebaw by Oo-dwa-gee, at the time Woon (Governor) of the ruby-mining district (Streeter, 1892).

Jean-Baptiste Tavernier, the famous seventeenth-century gem dealer and traveller, mentions a number of large rubies in his *Travels* (Ball and Crooke, 1925). The illustrations of some of these are reproduced in Tavernier's book. Of course Tavernier was writing at a time when almost any red stone was considered to be a ruby, so it is likely that the larger stones were actually red spinels ('balas ruby'). Nevertheless, a short description is in order.

One piece was described as a ruby that belonged to the King of Persia. It was of the thickness and shape of an egg and drilled in the manner of many gems of the day. Tavernier described it as being 'of very high colour, beautiful and clean, with the exception of a small flaw at the side' (Ball and Crooke, 1925).

Another illustration represented a large stone believed to be a balas ruby which was bought by Ja'far Khan, the Great Mogul's uncle, and presented to the Great Mogul, Aurangzeb. An old Indian who had once been the King's jeweller stated that the stone was not a balas ruby and that it was worth no more than 500 rupees. Aurangzeb eventually did what few buyers are able to do today: he returned the stone for a full refund.

Also included was a ruby which belonged to the King of Bijapur. Its weight was 17.5 carats, it was cut as a hollow cabochon and it was clean and of top quality.

A further stone was a ruby which a Banian merchant showed to Tavernier at Benares (Varanasi). It weighed 50.75 carats and was of the second quality. This stone was also a hollow cabochon, a common cutting style of the day.

Tavernier also described a large topaz (probably a Sri Lankan yellow sapphire) of the Great Mogul. This stone weighed 157.25 carats and was highly prized by Aurangzeb, as Tavernier said he wore no other jewel during his visit.

According to Streeter (1892), the two most important rubies ever known in Europe were both brought there during the year 1875. One, a rich red cushion shape, weighed 37 carats, and the other was a blunt drop-shaped piece of 47 carats. Both stones were recut by J. N. Forster of London; the 37 carat piece yielded 32.3 carats and the 47 carat piece 38.5 carats. The smaller stone eventually fetched £10 000 and the larger £20 000. Streeter, undoubtedly one of the most competent European judges of rubies of the day, apparently did not examine the stones himself, but he states that experts pronounced them to be unrivalled for rubies of such large size. Perhaps even more authoritative proof of their quality was given by the fact that their sale in Burma created intense excitement. The Burmese king was only persuaded to let the stones leave the country because he desperately needed the cash their sale brought. A military guard was considered necessary to escort the persons taking the stones to the ship. Regardless of the king's financial position, we can be quite sure that he would not part with the very best of his collection, for, as with many monarchs, such a collection has value far beyond money. In many cases the royal regalia and associated stones form a most important part of their rule, the foundation of kingly status as it were. Without these trappings he would be truly left as an emperor without any clothes.

Among the star rubies of renown, two large stones immediately come to mind: the 'de Long Star', an oval stone of over 100 carats, and the 'Rosser Reeves Star', an oval of 138.7 carats.

The de Long Star Ruby is displayed in the American Museum of Natural History in New York. Discovered in Burma during the early part of the twentieth century, it was named after the donor, Edith de Long.

Probably the finest large star ruby in existence is the Rosser Reeves Star Ruby. Now displayed in the Smithsonian Institution in Washington, DC, it too was named in honour of its donor, Mr Rosser Reeves. Not only is the stone clearer and more translucent than the de Long Star, but it also possesses a magnificent sharp six-rayed star. The Rosser Reeves Star is also unusual in that it originates not from Burma, as with most fine rubies, but from the gem gravels of Sri Lanka.

Rough ruby specimens of note exist in several museums around the world. Extremely large but impure specimens have been found in a number of localities. The British Museum possesses a rough hexagonal prism 10 inches (25 cm) high and 7 inches (7.5 cm) across which weighs 34 lb (15.5 kg) and comes from North Carolina. Fine specimens, however, come almost entirely from Burma. In 1933, the British Museum acquired a remarkable Burmese ruby specimen measuring 12 × 9 × 4 cm and weighing 1½ lb (690 grams or 3450 carats). Although consisting of a single crystal, it shows the terraced appearance (as a result of oscillation between the rhombohedron and basal pinacoid faces) typical of Burmese ruby crystals (Spencer, 1933). Also displayed at the British Museum is the 167 carat 'Edwardes Ruby Crystal', which was donated in 1887 by John Ruskin (Keller, 1983).

Another fine example of a Burmese ruby crystal is on display in the Los Angeles County Museum of Natural History (Keller, 1983). Weighing 196.1 carats, it too displays the typical etched and terraced appearance and is known as the 'Hixon Ruby'.

What some experts have regarded as the finest cut Burmese ruby of its size is the magnificent 15.97 carat gem owned by New York gem dealer, Alan Caplan (Keller, 1983). The stone is exceptionally clean, is a faceted oval cut and has the rich red colour so well known in Burmese stones.

The largest ruby ever found in Thailand was unearthed in 1985. This giant piece of rough weighed approximately 150 carats and was put on public display during a gem fair held in Chanthaburi in 1986. Thai rubies of large size, however, are generally far too dark in colour to be considered ideal.

Auction records

Next to coloured diamonds, the ruby is the most precious of all gemstones. The 1987 edition of the *Guinness Book of Records* contains quite confusing information as to the exact details. Under the heading of 'Most Expensive Gem' is listed the 42.92 carat Terestchenko blue diamond which sold at auction in 1984 for $4 580 000 ($106 710 per carat). However, the book then goes on to say that the per carat record for a diamond is $113 000 ($4.95 million total) for the 41.3 carat 'Polar Star'. (The actual per carat record for a cut diamond was broken on 28 April 1987 at Christie's in New York, when a 0.95 carat red diamond sold for a fantastic $926 000.) Were this not confusing enough, it is then stated that: 'Since 1955 rubies have been the world's most precious gem attaining a word record carat price of $100 639 at Christie's sale in Geneva in November 1979 in the case of a 4.12-carat caspian-shaped ruby.' The above lapses can be forgiven, since we ruby lovers in Asia do not consider diamond to be a *real* gemstone, either!

Seriously, though, the information given in the *Guinness Book of Records* is correct regarding ruby. According to *The Wall Street Transcript* of 31 December 1979, the highest per carat price for ruby rose steadily throughout the 1970s. In April 1976 a ring containing a ruby of approximately 7.25 carats sold for $230 000 and in October 1975 at Sotheby Park Bernet a suite of nine rubies from the estate of Geraldine Rockefeller Dodge sold for $690 000. The record for ruby was set at $100 639 per carat for an untreated Burmese ruby of 4.12 carats, sold at Christie's Geneva, in 1979. This was later broken in October 1988 at Sotheby's, New York, where a 15.97 carat untreated Burmese ruby sold for $3 630 000 total, a massive $227 301 per carat. During the 1979 Christie's Geneva auction, three separate records were established: the above-mentioned ruby; a Colombian emerald ring of 12.46 carats which went for $48 240 per carat; and a Kashmir blue sapphire of 11.81 carats which sold for $25 815 per carat (this was then a per-carat world record for blue sapphire).

The above sales took place at a time when precious stone prices were nearing their peak and with the worldwide depression in the early 1980s it seemed as though it might be quite some time before prices again reached these levels. However, in 1988 new records were set for both ruby and blue sapphire. The ruby record has already been described above (15.97 carats at $227 301 per carat); the record for blue sapphire was shattered in February 1988 at Sotheby's, St Moritz, when a 62.02 carat untreated Burmese blue sapphire sold for $45 607 per carat ($2 828 548 total for the stone).

As one pundit said: 'Predictions are difficult, especially those on the future.' While the author claims to have no crystal ball into which he can gaze at future gem prices, it would seem that with the increasing supply shortages that have largely been responsible for rising prices in the late 1980s, there is little place to go but up.

Famous sapphires

Although large rubies of quality are extremely rare, fine examples of large sapphires are relatively less so.

Over the centuries Sri Lanka has produced more giant sapphires of quality than perhaps any other source. Several examples are found in the Smithsonian Institution in Washington, DC. The largest of these is the 'Logan Sapphire', a giant of 423 carats set with 20 diamonds. This stone has a rich blue colour, but unfortunately is faceted with a large window. It was donated by Mr John A. Logan and is considered to be one of the finest large sapphires in existence.

Also in the Smithsonian collection is the 'Star of Artaban', a blue six-rayed star sapphire of 316 carats, and the 'Bismark Sapphire', weighing 98.6 carats. Both are of Sri Lankan origin (Punchiappuhamy, 1984).

In the American Museum of Natural History in New York is a fine collection of Sri Lankan sapphires, particularly stars. Largest of these is the 'Star of India', weighing a massive 563 carats. This stone is actually of Sri Lankan origin and shows a fine star, although the colour is a light greyish blue. It was donated by the famous American banker, J. P. Morgan. Another large star in this collection is the 'Midnight Star', a deep-violet stone of 116.75 carats. Also in the same museum is what many consider to be the world's largest true padparadscha of quality – an oval stone of 100 carats (Punchiappuhamy, 1984).

Probably the finest large star sapphire in existence is the 392 carat piece owned by the State Gem Corporation of Sri Lanka. It is an incredibly rich blue colour, much finer than the Star of Asia, and also exhibits a beautiful star.

The British Crown Jewels contain a number of fine small sapphires and two larger stones of note. These are the 'Stuart Sapphire' and the 'Sapphire of St Edward'.

The early history of the Stuart Sapphire is somewhat obscure, although it probably belonged to Charles II and was certainly among the jewels which James II took with him when he fled to France (Younghusband and Davenport, 1919). From him it passed to his son, Charles Edward, the Old Pretender, who gave it to his son Henry Bentinck, later known as Cardinal York. As the Stuart cause was then dead, he left the sapphire with other Stuart relics to George III.

In Queen Victoria's State Crown this sapphire occupied a prominent position just below the Black Prince's Ruby. It was later replaced by the Star of Africa Diamond, and today is set in a similar position on the opposite side of the same crown.

The Stuart Sapphire is more of historical than real value. Although of a fine blue colour, it contains one or two blemishes and is drilled at one end, probably so that it could be worn in earlier times. It is oval in shape, about 1½ inches (3.8 cm) in length by 1 inch (2.5 cm) in width, and is set in a gold brooch (Younghusband and Davenport, 1919).

St Edward's Sapphire is now set in the centre of the cross-patee on top of the same crown as the Stuart Sapphire. It is a stone with a history stretching back farther perhaps than even the Black Prince's Ruby. According to tradition, St Edward's Sapphire was originally set in the coronation ring of Edward the Confessor, who was crowned in 1042. Special powers were ascribed to this sapphire, including those of curing cramps.

St Edward's Sapphire is today a rose-cut gem, but it is thought that this was not the original style of cut, and it is estimated that the recutting was performed during the reign of Charles II. It is a stone not only of exceptional colour and brilliance, but also with a marvellous legend behind it. The legend states that Edward the Confessor had a great respect for St John the Evangelist. One morning Edward happened to meet a beggar near Westminster. Having given away all his money already, he presented his ring to the beggar. Some time later two Englishmen on pilgrimage to the Holy Land ran into a storm in Syria. Suddenly the path ahead lit up and an old man approached, preceded by two youths bearing candles. On hearing that the pilgrims were English and that Edward was their king, the old man guided them to an inn, where he found them food and lodging. The next morning, as they were leaving, he told them he was John the Evangelist and gave them the ring to return to Edward. Bidding them goodbye, John said that he would see

Edward in paradise in six months time. The pilgrims eventually returned to England, where they gave the ring and message to Edward. Edward recognized the ring and then prepared for his death. He died six months later and was buried at Westminster with the ring on his finger. The tomb was later opened in the twelfth century and the ring given to the reigning monarch, according to legend (Sitwell, 1953).

Other fine sapphires in royal collections include the diamond and sapphire parure of Queen Marie Antoinette of France. This is a seven-piece jewellery set containing approximately 29 sapphires, of which 18 are stones of perhaps 20 carats or more. During the French Revolution it vanished, later reappearing to be bought by Napoleon, who gave it to his wife, the future Empress Josephine. Upon her death it passed to her daughter, Queen Hortense, who, after the fall of the Bonapartes, sold it to Louis-Philipe, then Duke of Orleans. Since then it has been worn by the wives of successive heads of the House of France and now belongs to the Count of Paris. Three separate dynasties that have ruled France have owned it: the Bourbons, the Bonapartes and the Orleans (Prince Michael, 1983). A fine illustration of the parure is given in Prince Michael's book the *Crown Jewels of Europe*. Although most of the sapphires possess large windows as a result of overly shallow pavilions, together they are magnificent.

Another fine sapphire necklace is pictured in Prince Michael's book – the 108 sapphire and diamond necklace of Queen Maria Christina of Spain. Originally from Sicily, she was the wife of Ferdinand VII, her mother's brother. After a terrible civil war she fell in love with a bodyguard; the scandal upon their marriage forced her to leave Spain. She had her portrait painted wearing this beautiful necklace which was sold by Christie's in 1982 for $297 000 (Prince Michael, 1983).

E. W. Streeter, in his book *Precious Stones and Gems* (1892), describes a number of fine sapphires. One of these was in the collection of the Jardin des Plantes, in Paris, and weighed $133\frac{1}{16}$ carats. The same stone was also described by S. M. Tagore (1879) who referred to it as the 'Wooden Spoon-Seller's Sapphire', in reference to the poor man who is said to have found it in Bengal, India. Streeter said it was without flaw. This is undoubtedly the same stone as that residing today in the same museum, for it is of a distinctive lozenge shape and possesses only six facets, appearing like a huge sapphire rhomb. It is indeed nearly 'without flaw', containing only one small feather and crystal inclusion, and is probably of Burmese origin as the colour is too even for it to be from Sri Lanka. A colour photograph of this magnificent stone appears in Dr Hermann Bank's book *From the World of Precious Stones* (Bank, 1973). He gives its weight as 135.20 carats and states that it is still exhibited in the Musee au Jardin des Plantes, in Paris.

During the London Exhibition of 1862, two magnificent sapphires were on display (Streeter, 1892). The larger was an oval stone of a somewhat inky colour and free from defects. It weighed about 252 carats and was cut in 1840. The second stone, although smaller at 165 carats, was of much finer colour. According to Streeter, it was by far the finest sapphire of its time to appear in Europe.

Streeter also mentions a fine sapphire in the famous Hope Collection, but no weight is given. It was noted because the gem retained its beauty by candle as well as by daylight. Another sapphire in the Orleans Collection was called in Madame de Genli's tale 'Le Saphir Merveilleux' (Streeter, 1892). It was violet by candle but blue by daylight.

Although many authorities of the nineteenth century, such as Streeter, state that Burmese gems are too dark, such is not the case; next to the gems of Kashmir, they are the finest. One fabulous piece of rough sapphire from Mogok was purchased by the American dealer Albert Ramsay in 1928 (B. O'Leary, 1981, unpublished). This fine stone was found near Gwebin and in the rough weighed 958 carats. The rough was bought for $13 000 and came to be known as the 'Gem of the Jungle'. Nine different stones were cut from it, ranging from 66.53 to 4.39 carats, and including stones of 20.11, 19.19, 13.15, 12.29, 11.39, 11.18 and 5.57 carats. All were personally cut by Ramsay himself and were said to be of exceptional colour. A marvellous account of the purchase and cutting of the Gem of the Jungle was given by Ramsay to the *Saturday Evening Post* in 1934.

S. M. Tagore in his classic work *Mani-Mala* (1879) describes several celebrated sapphires. One of these was a fabulous stone of 951 carats, seen by the British ambassador to the Court of Ava (Burma). Tagore also mentions a curious custom among the Hindus of India. They were said to have a prejudice against sapphires, believing the blue gem to be the bringer of misfortune.

> In consequence of this notion, some of them would invariably keep a stone on trial for several days before they would make final settlement with the sellers. Hence, perhaps, the paucity in the numbers of Sapphires in their possession. (Tagore, 1879)

If only modern-day dealers would be so understanding!

Among the sapphires of colours other than blue, few are described in the literature, although doubtless there exist many fine examples, particularly from Sri Lanka. Australia, being an English-speaking country, has several well-documented examples (Broughton, 1979).

'Anderson's Yellow' was a 21 gram golden-yellow stone found on the Willows field in Queensland in 1949. It produced a cut stone of 70 carats, but was later cut into several smaller gems, the largest weighing 35.75 carats.

In 1951, the Willows field again turned up a large yellow stone. This was named the 'Golden Willow', renamed the 'Golden Queen', and weighed 322 carats in the rough. It produced a 91.35 carat cut stone. The Willows field is noted for producing large yellow sapphires and a number of yellow-green gems in the 50–100 carat range have been found there.

Queensland has also produced very large black star sapphires. The best known of these is the 'Queensland', a 1156 carat cabochon giant. Although the inexperienced may believe such a stone to be extremely valuable, this is not the case as these huge stones are typically heavily included. Their value is chiefly as curiosity items, rather than as true 'gems'. Such stones often surface from the bowels of unscrupulous gem dealers' collections, always with an inflated appraisal claiming the stone to be more valuable than it is. This was the case with the famous 'Life and Pride of America Star Sapphire' which featured in many news reports in 1985 and 1986. The owner, Mr Roy Whetstine, claimed to have bought the stone for $10 at the Tucson gem show. It was later shown to be a hoax when a reporter discovered that the man who appraised the stone had appraised another stone of the exact same weight several years before Whetstine said he found it. All such stories of gigantic rubies and sapphires should normally be treated with great scepticism. One exception to this was at the gem collection of the Jammu and Kashmir State Treasury, where C. S. Middlemiss viewed a number of large fine pieces of Kashmir sapphire rough. He described one piece as being the size of a croquet ball! Even the most jaded observer would drool at the sight of this.

Although a number of large yellow sapphires have come from Australia, this is not the only source, for Thailand, too, produces them. The author has had the pleasure of examining what is perhaps the largest yellow sapphire from Thailand. It was a large emerald-cut stone of approximately 75 or 80 carats and owned by a Thai dealer. The stone was too shallow, possessing a rather large window, but in all other ways was magnificent. It was of the preferred 'Mekong Whisky' yellow-orange colour and was without flaw.

One of the largest pieces of rough sapphire ever unearthed was found about 1967 at Mogok, Burma (Anonymous, 1967). It was bluish grey in colour, measured 27 inches (68 cm) across and weighed a massive 63 000 carats (27.783 lb or 12.6 kg). The Myanma Gems Corporation, a state-owned concern, is the owner of this giant sapphire crystal.

The largest piece of sapphire rough ever heard of was supposedly unearthed in Sri Lanka in 1985 or 1986. During the author's visit to the Island in 1986 he was informed about this giant, but did not have a chance to view it. Said to be part clear, part 'geuda' and part star material, its weight was reported to be a massive 42 kg (93 lb)!

Quality in rubies

Ruby is among the rarest of all the major precious stones, with very few sources that can produce facet qualities in any quantity.

The Mogok Stone Tract in Upper Burma has traditionally been associated with the finest rubies, but even good stones are extremely rare from this fabled area. 'Pigeon's blood' was the term used to describe the finest Burmese stones, but this phrase has little meaning today, for so few people have seen the blood from this bird. Better, perhaps, is to compare the colour of the finest rubies with the red signal of a traffic light. It is a glowing red colour which is diffused throughout the stone as a result of the unique property of fine rubies – a rich red fluorescence to daylight. Burmese rubies possess not just red body colour but, by a freak of nature, fluoresce red as well. It is the body colour plus the red fluorescence which give rubies in general, and particularly those from Burma, their incomparable crimson hue.

In Burmese rubies, the colour often occurs in rich patches and colour zoning can be a problem. The colour swirls found in these stones are termed 'treacle' by gemmologists because of their resemblance to syrup. Not only are faceted rubies obtained from Burma, but also the world's finest star rubies. In these stones a sharp star and, at the same time, good transparency are sought. Stones which contain too much silk will have rather poor colour because the silk diffuses the light. The best stones will have just enough silk to create the star effect, but not so much as to harm the transparency and colour. Such stones are extremely rare and valuable. Good transparency, or a 'glass body', as it is known, is of the utmost importance in creating good colour in star rubies.

Owing to political problems, Burma today produces very little facet-quality ruby, and the major source is now the mines along the Thai/Cambodian border. However, the colour of Thai rubies usually cannot compare with that of the best of those from Burma. Again, the reason is fluorescence. Thai rubies contain traces of iron which quench this red fluorescence. Although heat treatment, which nowadays is performed on virtually all Thai rubies, does improve their appearance considerably, it does not make up for their lack of fluorescence. In Thai rubies, only those facets where light is totally internally reflected will be a rich red; the others appear blackish, as with red garnets. In terms of body colour, the Thai stones are actually a purer red than most Burmese rubies. However, the finest Burmese rubies appear red all over: not only is a rich red seen in the areas where total internal reflection occurs, but, because of the red fluorescence, the other facets are also red. This glowing red colour is what makes fine Burmese rubies so special.

Ulrich Gühler, in his excellent article 'Studies of precious stones in Siam' (1947), was perhaps the first person to describe accurately this difference between Burmese and Thai rubies.

> The colour of the Siamese ruby is inferior to that of the Burmese ruby. The colour tends to be purplish or even brownish and generally it is too dark. A Siamese ruby inspected under natural or artificial light will show its colour only on those facets which are directly hit by the light. Those facets in the shade ('extinction') will appear almost black. A Burmese ruby will appear red throughout, even on those facets which are not directly exposed to the source of light.

A typical complaint with regard to Thai rubies is that they are often too purple in hue. The author, however, has found this to be untrue. In terms of hue position, Thai rubies are actually less purple than rubies from Burma or Sri Lanka. The problem instead is one of a lack of intensity and/or the tone being too dark. Reference to any of the major colour atlases will show this to be the case.

With regard to colour, rubies from most other sources are similar to those from Burma, in that the stones display a strong red fluorescence. The colour of Thai rubies is the exception rather than the rule. Why then, we may ask, are fine rubies not more common? The answer is clarity. Only a handful of sources produce material clean enough to facet.

One of these sources is Sri Lanka. Next to Burma, Sri Lanka produces the finest rubies, although most tend to be too light in colour. As these stones are often of a pink colour the term 'pink sapphire' is usually applied today. However, in the author's opinion this term should be abandoned because it leads to confusion and is historically inaccurate. Research performed by the author has shown that prior to about 1920 all corundums of a red colour, be it light (pink) or dark (red), were termed ruby. Reference to any standard text on precious stones before the 1920s shows this to be true. For example, Max Bauer (1904) stated with respect to ruby:

> The tone of colour differs in different specimens, being sometimes deep and intense ('masculine' ruby), sometimes pale and light ('feminine' ruby). The lighter shades vary from pale rose-red to reddish-white, some specimens being so fairly tinged with red as to appear almost colourless.

Furthermore, since pink refers merely to a light red and since the original sources of ruby, Burma and Sri Lanka, produce stones of which the majority are pink, it seems ludicrous not to consider them as rubies. Probably the best reason for abolishing the term pink sapphire, though, is that there is great confusion among dealers on how to separate ruby from pink sapphire. We do not have any problem with blue sapphire: it is a blue sapphire whether it is dark blue or light blue. So why should light-red corundums not be called ruby? In keeping with this policy, the Asian Institute of Gemological Sciences has abandoned use of the term pink sapphire in its classes and gem identification reports.

Sri Lanka, again, produces fine rubies, but most tend to be too light in tone and low in intensity. As with Sri Lankan sapphires, however, the colour accumulates in large stones and can be quite magnificent in sizes of 5 carats or more. The best pieces will, as a result of the fluorescence, lack the blackish overtones of Thai rubies; instead they appear as glowing embers or as if someone had wiped fluorescent red paint across their face. To help visualize the colour of the finest rubies the ultraviolet fluorescence of a fine highly fluorescent natural or synthetic ruby should be examined. The red glow seen is very close to the colour of a fine ruby. It is a red ever so slightly tinged with orange. That is not to say that it is an orangy red; it is a pure red, but slightly more yellowish (orangish) than the majority of rubies and of much higher intensity, especially intensity. It is intensity, or saturation, that is the most important factor. All fine rubies have one common feature – a very high intensity, or saturation, of the red hue.

A common defect in Sri Lankan rubies (and sometimes in Burmese rubies too) is a slight to moderate bluish tint making the stones appear purplish. However, it must be kept in mind that tone and intensity are of prime importance in judging rubies. A purplish ruby of medium tone and very high intensity is generally worth far more than a stone of more red hue but either too light or dark in tone or lacking in intensity. These comments also hold true for sapphires, or for all precious stones for that matter.

Like Burma, Sri Lanka also produces fine-quality star rubies, some of which are of large size. Again, the deeper red colours in Sri Lankan stones are mainly found in the larger sizes.

Quality in sapphires

Unlike rubies, for which true perfection is virtually impossible, even at the 5 carat level, large sapphires of very fine quality of 100 carats or more do exist.

In the world of blue sapphires one source stands out and is the quality against which all others are measured. That source is Kashmir, India. An approximate ranking of untreated blue sapphires in terms of origin is given below.

1. Kashmir, India
2. Mogok, Burma
3. Pailin, Cambodia
4. Sri Lanka (Ceylon)
5. Thailand
6. Yogo Gulch, Montana, USA
7. New South Wales, Australia

It must be pointed out that this listing is only approximate, with origin alone being of no value. Each source produces all qualities, and of course the lower grades far outnumber the finer stones. The same holds true for ruby.

Kashmir sapphires are noted for their rich blue hue and distinctive 'velvety' lustre, caused by the presence of minute exsolved inclusions. The Kashmir mine, however, produced in quantity only during the years 1881–1890 and has produced very little since. For this reason, Kashmir sapphires are more a state of mind than a reality. To paraphrase J. Coggin Brown, the famous gem expert of the Geological Survey of India, it is likely that most of the fine 'Kashmir' sapphires sold today first saw the light of day at the Mogok Stone Tract in Burma.

Next to Kashmir stones, Burmese sapphires are unsurpassed. Although certain heat-treated Sri Lankan sapphires may rival them in beauty, the Burmese stones are of a deeper richer colour, simply because there is more colour inside those from Mogok. Moreover, the Burmese stones do not require heat treatment for their beauty and in fact are generally damaged by such chicanery. Truly, the Burmese stone comes out of the ground in living colour, a blaze of blue that must be seen to be believed. Star sapphires are very rare from Kashmir, but many fine examples exist from Burma, some of large size. Fine Burmese sapphires tend to be ever so slightly more violet in hue position than those from Sri Lanka.

The Pailin mine in Cambodia has produced a number of fine stones over the past 100 years, although today production is limited because of political problems. Pailin stones, however, tend to be a little on the dark side and stones larger than 5 carats cut are rare and generally too dark in colour. This is in contrast to Kashmir, Sri Lanka and Burma, which have produced a number of fine sapphire giants.

Until the late 1970s, Sri Lankan sapphires generally reached the richer blues only in stones of 10 carats or more. However, the discovery of appropriate heat treatments has changed the situation entirely, so that today deep blues of all sizes are available. In our discussion of fine sapphires we will limit ourselves to untreated gems. Sri Lanka is perhaps the most prolific producer of giant sapphires and has yielded the vast majority of all large sapphires that the world has seen. Blue sapphires from Sri Lanka have a unique beauty, being brighter and more lively than any others. While Burmese stones are of a more intense 'royal blue', the Sri Lankan sapphire is a brighter 'cornflower blue', as there is less colour in the stone. Sri Lanka is also the source of a number of fine star sapphires, some weighing hundreds of carats, and is the greatest producer of star corundums of all colours.

Compared with Kashmir, Burma, Pailin and Sri Lanka, in terms of quality all other sapphire sources are of relatively minor importance. In Thailand, the occasional fine stone is produced, particularly from the mines of Bo Ploi in Kanchanaburi Province. Bo Ploi stones may be of very fine colour and sometimes reach sizes of 30 carats or more, but most are somewhat marred by a certain cloudiness. Many Bo Ploi stones are sold as Sri Lankan, because of their strong colour zoning. Sapphires from Chanthaburi and Phrae tend to be overly dark, although some are quite pleasant.

Yogo Gulch in Montana, USA, produces sapphires of fine colour when found in sizes of greater than 1 carat, but such stones are extremely rare. The lack of larger stones (1 carat or more) has kept Yogo from being a source of major importance.

Australia is one of the biggest producers of faceted sapphire, but the vast majority of stones are dark and inky in colour and require heat treatment. The mines of New South Wales produce the better stones, while the Queensland production consists mostly of darker blues. Nigeria produces sapphires of similar hue to those of Australia.

In terms of colours other than blue, Sri Lanka is without a doubt the best of the lot. Within this small island are found sapphires of virtually every colour, including some which are unique to the island, such as the lovely pinkish-orange padparadscha. Yellow sapphires from Sri Lanka are generally of a light to medium hue, without any brownish overtones. Deeper hues are, like the Sri Lankan blue sapphires, reached only in larger sizes, or through heat treatment. Heat treatment produces deeper yellows, golds and oranges that are virtually unknown or very rare in natural stones from Sri Lanka. The very rare pinkish-orange padparadscha sapphire is found only in Sri Lanka, although brownish-orange stones from the Umba Valley in Tanzania are sometimes sold under this name. Fine padparadschas from Sri Lanka may reach prices that rival even ruby.

Thailand and Australia both produce fine yellow sapphires, with the stones from Chanthaburi in Thailand grading into the highly desirable 'Mekong Whisky' golden yellow to orange colours. These stones fetch very high prices locally in Thailand and are quite beautiful. Australian yellow sapphires tend to be overly greenish, although fine golden-yellow stones are found on occasion in the state of Queensland. Sri Lanka, Thailand and Australia are the only sources which produce deep-yellow sapphires in any quantity, although Burma produces the occasional stone.

The finest green sapphires come from Sri Lanka but are extremely rare. These stones tend to be of a lighter and more lively green hue than those from Thailand and Australia. The latter two countries do produce good green sapphires, but most tend towards an impure blue-green or yellow-green colour which is not very attractive. Green sapphires of good colour over 10 carats in size are relatively scarce.

Violet and purple sapphires are found mostly in places which produce both ruby and blue sapphires. The finest stones come mainly from Burma and Sri Lanka. Purple stones bordering on ruby colour are most valuable and may reach prices approaching those of the lower grades of ruby. Again, colour intensity is of prime consideration here.

Among the most unusual of sapphires are those which display a change of colour. These stones are judged by the quality of their colour change, with the best going from the green side of blue in daylight to a reddish purple in incandescent light. A number of sources may produce these stones, but fine examples are rare. Sources include Sri Lanka, Tanzania (Umba), Burma and Thailand. Many Sri Lankan stones which are sold as colour-change sapphires are of a blue-violet colour in daylight and so, in the author's opinion, should not be considered as colour-change sapphires.

Classification of rubies and sapphires

In the past ten years there have been a number of attempts to classify or grade gems according to their colour. All of these attempts try to develop a useful language for communicating the colour appearance of a gemstone in terms of the three basic components of colour – hue, tone and intensity – and the two minor components – evenness (colour zoning) and pleochroism (doubly-refracting stones only).

Once the colour of a gem is adequately defined, then it is up to the marketplace to determine the relative value of that colour. Attempts to assign relative values or quality grades to each colour will be successful only if the considerations of the real marketplace are taken into account. The major difficulty arises in choosing the correct role model; failure to do this will cause inconsistencies. An example of this failure occurred in the early 1980s when ruby prices reached their peak. The American Gemological Laboratory (AGL) of Cap Beesly in New York City was the first laboratory to grade coloured gemstones for quality. In the beginning, they assessed the quality of a ruby's colour on the basis of purity of hue. Stones receiving the highest grades were those of the purest red hue. Unfortunately this was the wrong role model, because these stones tended to be overly dark non-fluorescent Thai rubies, similar to red garnets in appearance. They may have the purest hue, but because of their dark tones the marketplace did not consider them to be the best. Instead, in the marketplace the top prices were paid for highly fluorescent Burmese rubies, as already described in this chapter. In terms of body colour alone, the Burmese ruby is slightly more purplish than the dark-red Thai ruby, but the fluorescence of the Burma stones more than makes up for this. This factor, though, was not taken into account in the AGL grading, with the result that the dark non-fluorescent Thai rubies were graded higher (better) than the highly fluorescent Burmese stones. In the marketplace, traders were quick to take advantage of this inconsistency. They flew to Bangkok, bought up these dark Thai rubies for low prices, took them back to New York where AGL quality certificates were issued, and then sold them for much higher prices on the basis of these certificates.

To his credit, Cap Beesly of AGL reacted quickly by recalling many of the certificates and readjusting the grades. However, much damage had already been done. This example illustrates clearly the importance of choosing the correct role model for quality grading of corundums (and other coloured gemstones). To make these decisions, gemmologists must work very closely with gem dealers so that the above mistake is not repeated.

With many coloured stone grading systems a numerical grade is given to describe a gem's colour. This, however, still does not allow a person to form an image easily of that colour. In other words, even with the grading report communication is difficult. Mr Henry Ho, gem dealer and founder of the Asian Institute of Gemological Sciences (AIGS) in Bangkok, developed a classification system for rubies and sapphires which can be used with any grading system and which *does* allow effective communication of a gem's appearance. Traditionally, gem dealers have referred to the colour of corundums based upon the typical colours produced at each mine. In other words, a rich but slightly milky-blue sapphire was termed a 'Kashmir-type' sapphire, even if it was mined elsewhere, because of its resemblance to stones mined in Kashmir. Today it is often possible to determine the exact origin by examining the inclusions in the gem, but this has created new problems because good and bad gems come from every source. In other words, what difference does it make if a sapphire is from Kashmir if it is of very poor colour? In the opinion of the author, the answer is none. Only inclusion collectors should pay more for a low-quality stone just because it comes from a famous mine, and how many inclusion collectors are there?

The AIGS system of classifying rubies and sapphires is a logical solution to the above problem. By classifying corundums according to 'type' rather than source of origin, it allows the correct communication of the appearance of the gem. If a gem has an appearance typically associated with a fine Kashmir sapphire, it does not matter where it is from, for it will still be a 'Kashmir' type. Once the type is established, the gem can then be graded according to virtually any of the colour grading systems. By specifying both the type *and* the colour grade, a more accurate mental image of that gem's colour can be formed without even examining the stone. Accurate colour communication then becomes much easier.

The type classifications of Henry Ho for rubies and sapphires are today being taught to students at AIGS in Bangkok and are described in the following sections.

Ruby

Type A

1. Known in the Thai language as 'ploy daeng' (literally 'red stone').
2. Hue position is pure red with slight purplish or brownish overtones in areas where brilliance occurs. The overtones are merely areas of lower colour intensity, or may also be due to pleochroism.
3. Areas showing extinction appear blackish, but not as black as type D.
4. Ultraviolet fluorescence is usually weak.
5. Usually (but not always) of Thai/Cambodian origin.
6. Tone (reflectance) of overall stone not as low as with type D rubies.

Type B

1. Known in Thai as 'daeng chompoo' (literally 'pinkish red').
2. Hue position is pinkish red with either purplish or orangy overtones on facets where brilliance occurs. This pinkish tint is merely a slightly lower intensity than type A rubies. Orangy-red flashes may also be due to pleochroism.

3. Tonal percentage generally does not exceed 80%.
4. Ultraviolet fluorescence is slightly stronger than with types A and D, but not as strong as with type E.
5. Usually of Thai/Cambodian origin.

Type C
1. Known in Thai as 'daeng som' (literally 'orangy red').
2. Hue position is a slightly orangy red.
3. The presence of many inclusions causes a 'scrambling' of the pleochroic rays, giving the stone an orangy-red tint even if the *c* axis is cut perpendicular to the table facet.
4. Often contains wispy liquid feathers ('Lai Thai') or silk (Burmese). May be cut with the *c* axis at an oblique angle to the table facet, making the colour appear more orangy than that of other ruby types.
5. Often from Bo Waen or other Thai/Cambodian mines, Burma, Kenya and Afghanistan.

Type D
1. Known in Thai as 'daeng dum' (literally 'blackish red').
2. Hue position is pure red on the brilliant areas, while almost or completely black on the areas suffering extinction. The window colour is also slightly darker than in other ruby types.
3. Weakest ultraviolet fluorescence of all the different types, causing the extinct areas to appear pitch black.
4. Usually of Thai/Cambodian origin.

Type E
1. Hue position is a pink to purplish red, more purplish than type B.
2. Ultraviolet fluorescence is extremely strong, the strongest of all types.
3. As a result of the strong fluorescence, the areas showing extinction are masked, appearing reddish instead of black.
4. Usually lighter in tone than the other types, again partly as a result of the fluorescence.
5. Usually of Burmese or Sri Lankan origin. This colour is generally *not* found in Thailand/Cambodia.

Blue sapphire

Type A
1. Hue position is a rich 'royal blue', almost verging on the violet and deeper than type C.
2. Often appears very dark or black where extinction occurs.
3. Origin is mainly Burma, Sri Lanka or Pailin (Cambodia).
4. This colour completely lacks the 'inkiness' of type D.

Type B

1. Hue position is a deep blue similar to types A or D but slightly milky or silky.
2. May display strong colour zoning.
3. Usually filled with minute particle or silk clouds, making the stone appear slightly sleepy or foggy.
4. Usual origin is Bo Ploi (Kanchanaburi, Thailand) or Kashmir.

Type C

1. Hue position is a pure 'cornflower-blue' colour of lighter tones than type A.
2. Usual origin is Sri Lanka, but may also come from Yogo Gulch (Montana, USA) or Burma.
3. Strong colour zoning often present.

Type D

1. Hue position is an 'inky blue' with very dark tones (70–100%).
2. Often shows slight greenish overtones.
3. Hexagonal colour banding often visible.
4. Shows strong iron lines (451.5, 460 and 470 nm) in the spectroscope. This produces less transmission in the blue, creating the inkiness in the colour.
5. Usual origin is Australia, Thailand (Chanthaburi), Pailin (Cambodia) or Nigeria.

Yellow sapphire

Type H

1. Sri-Lankan-type yellow sapphires. Hue position is a pure yellow without greenish or brownish overtones.
2. Usually shows an orangy ultraviolet fluorescence.
3. Iron lines in spectrum usually absent.
4. Typical origin is Sri Lanka (unheated).
5. Tones arc typically lighter than with types K and L.

Type K

1. Hue position is deep-yellow orange with brownish overtones.
2. Darker tones than type H.
3. Strong iron lines usually present. Usually inert to ultraviolet light.
4. In Thailand, the best colour is compared with that of the local 'Mekong Whisky'.
5. Body colour is much deeper and areas of extinction are darker than with type H.
6. Typically free of inclusions, often cut with shallow pavilions (Thai or Australian). Heat-treated Sri Lankan stones usually contain minute exsolved haematite particles which mechanically colour the stone.
7. Usual source is Chanthaburi (Thailand), Australia or Sri Lanka (heat treated).

Type L

1. Hue position is greenish yellow and lacks the brownish overtones of type K.
2. Tones are lighter than with type K but not as light as those of type H.

3. Strong iron lines usually present. Inert under ultraviolet light.
4. Usually of very high clarity.
5. Colour banding visible most clearly at 90° to the *c* axis (Thailand and Australia) and often displays green bands near the table and culet.
6. Usual source is Australia or Chanthaburi (Thailand).

Note
The differences between the various types are usually clear cut. However, some stones lie on the border of two different types. In this case, a split type grade (A/C, for example) should be given.

Consumer markets

One of the major advantages of the type classification of rubies and sapphires is in understanding the various preferences of the major consumer markets around the world. Contrary to what might be expected, individual consumer markets often have specific preferences in terms of ruby and sapphire colours. Bangkok gemmologist Dr William Sersen described it as 'individual preferences on a national scale'. In other words, Japanese buyers generally prefer rubies of a slightly pinkish-red colour, while the Italians prefer slightly darker red stones. Type classifications are invaluable for understanding these preferences and so are used below in this manner.

Japan
Japanese buyers generally prefer stones of a slightly lighter colour overall.
Rubies: Prefer type B stones (those of a pinkish-red colour). Eye-clean and well-proportioned pieces are desired.
Sapphire: Good quality type C stones. Bright and lively appearance. Eye clean and well proportioned.

Germany
The German taste is quite similar to the Japanese, with emphasis being placed on lighter (but not too light) stones of good clarity and of very good proportions and finish.
Rubies: Type B stones are preferred.
Sapphires: Type C stones are preferred.

France
Colour is of prime importance while clarity and proportions are secondary.
Rubies: Stones with a slightly orangy-red (type C) colour are preferred. Moderate inclusions (visible to the eye) are acceptable. Stones with a 'big face' (shallow pavilion) are also acceptable.
Sapphires: Stones of a fine pure blue are desired, consistent with types A, B and C.

England
Due to the lower buying power of British buyers, stones of lower grades may also be acceptable, in addition to the finer grades.

Rubies: Darker type D rubies are often purchased, as well as the better quality (types A, B and C) stones. Some buyers (those catering for customers with less money) purchase the darker stones, while others buy in accordance with Swiss and French tastes.
Sapphires: As with rubies, some buyers purchase darker (type D) stones while others follow the Swiss and French, purchasing the more expensive type A and C stones.

Italy
Slightly darker colours than French tastes are preferred. Proportions and clarity are of secondary importance.
Rubies: Types A and C are desired. Stones with a big face are often sought after.
Sapphires: Any sapphires with distinct colour flashes and of reasonable clarity and proportions. This is consistent with types A, B, C and better quality type D.

Switzerland
The Swiss market is extremely discriminating in taste, preferring stones of excellent colour and proportions and, to a lesser extent, clarity.
Rubies: Two types are sought after: orangy-red type C stones and the type E Burmese stones.
Sapphires: Types A, B and C are preferred.

USA
This is the largest market for rubies and sapphires. Good proportions and eye-clean clarity are of prime importance; colour is secondary.
Rubies and Sapphires: Virtually any type is saleable as long as it is eye clean and well cut. Brilliance is of great importance.

Australia
The Australian taste closely follows that of the British, with darker stones often selling well.
Rubies: Type D are preferred.
Sapphires: Dark type D stones of 'Australian' colour are preferred.

Middle East
In the past, stones of the lower grades sold best, with price being of great importance. A small number of dealers also sell the superfine qualities. Today tastes are moving towards slightly better qualities. Most Middle East buyers in Bangkok are from the Lebanon, Dubai and Iran.

Summary
In summary, the major consumer markets each display specific preferences in the purchase of rubies and sapphires. Use of the type classification system described above is very useful in delineating these national preferences.

It should be noted that the above preferences are only general. Buyers tend to purchase stones in the price range which is most easily saleable in their country. In other words, Swiss buyers tend to prefer the more expensive types (C and E for ruby) because their customers have enough money to buy the best. With other markets, where most customers are less well off, buyers will purchase the less expensive types (type D rubies, for example), as these are more easily saleable.

With regard to the preferences of ruby and sapphire connoisseurs, there is general agreement as to which types are best, *no matter what country that person comes from.* For rubies, connoisseurs prefer a type C stone (a slightly yellowish red), but of good clarity (eye clean) and strong fluorescence. W. K. Ho, a well-known Bangkok dealer from Burma, once said to the author that the best ruby will be 'a Burmese ruby that looks like a Thai stone, or a Thai ruby that looks like a Burmese stone'. What he meant by this was that the best ruby would combine the best features of both Burma and Thailand. It would have the good clarity and purer red hue (slightly more yellowish red than the majority of Burmese rubies) typical of Thai rubies, but with the strong red fluorescence and high colour intensity (saturation) typical of better grade Burmese stones. In summary, it would be a type C ruby of high clarity and strong fluorescence. This coincides with the stones that fetch the highest price on the Bangkok market, and indeed around the world for that matter.

As for sapphires, connoisseurs in Bangkok and around the world prefer those stones with the highest degree of colour saturation, while still being bright (not too dark) and of good clarity. This coincides with the best features of types A and C, and, to a certain degree, type B as well. In other words, the best sapphire will have a highly saturated blue colour (almost verging on the violet, as in type A), but with the brightness typical of type C. Some also prefer a hint of milkiness (typical of type B), just enough to add a slight velvety texture but not enough to affect the clarity overall. These stones come mainly from Burma today, but at one time also came from Kashmir.

As with any coloured gemstone, the colour depends to a certain extent on the light source used to illuminate it. Most gem dealers do their buying during the day and so use daylight, but even this changes slightly throughout the day. If artificial light is used, rubies look best under incandescent light while sapphires look best under fluorescent tube lights. Thus, rubies in particular should never be bought under incandescent lights, as this lighting makes them appear much more red. Instead, rubies should be bought under daylight or fluorescent tube lights. This is done so that the buyer will not get a surprise when the stone is examined under another lighting condition. The opposite is true for buying sapphires.

To a certain extent the latitude of the buying location may also affect a stone's colour. Countries like Burma, Thailand or Sri Lanka lie closer to the equator; the sunlight is thus more intense than in places like Europe, Japan or the United States, which are farther from the equator. As a result, gems bought in the tropics will appear slightly darker when taken to these more northerly countries. It is a slight but nevertheless a noticeable difference. To neutralize these factors many dealers today do their buying under special 'daylight' lamps designed to simulate true 'northern daylight', with a colour temperature of approximately 5800K.

Bibliography

Anonymous (1932) A 15 carat sapphire. *The Gemmologist,* **2**, no. 13, 28

Anonymous (1933) A rare ruby from Burma – gem valued at nearly £10,000. *The Gemmologist,* **2**, no. 19, 220

Anonymous (1934) Princess Marina's engagement ring – a Kashmir sapphire in a modern setting. *The Gemmologist,* **4**, no. 38, 59

Anonymous (1935a) Sapphire for Lady Alice. *The Gemmologist,* **5**, no. 50, 44

Anonymous (1935b) Star sapphires in America. *The Gemmologist,* **5**, no. 52, 110

Anonymous (1936) Romance of royal gems – The Black Prince's "Ruby". *The Gemmologist,* **6**, no. 67, 167–169

Anonymous (1938) The world's largest star ruby exhibited. *The Gemmologist,* **7**, no. 80, 654

Anonymous (1949) Half-pound star sapphire. *The Gemmologist,* **18**, no. 219, 254

Anonymous (1950) 12-ray ruby. *The Gemmologist,* **19**, no. 222, 2

Anonymous (1951a) Huge sapphire attracts thousands. *The Gemmologist,* **20**, no. 239, 137

Anonymous (1951b) Famous sapphires. *The Gemmologist,* **20**, no. 240, 165

Anonymous (1952a) Carved sapphire bust of Lincoln. *The Gemmologist,* **21**, no. 255, 190

Anonymous (1952b) Engraved sapphires. *The Gemmologist,* **21**, no. 255, 190

Anonymous (1959) "Ecce Homo" (Behold the Man) – a 4,060 carat carved ruby. *The Gemmologist,* **21**, no. 255, 190

Anonymous (1966) The Rosser Reeves ruby given to the Smithsonian Institution. *Lapidary Journal*, April, 136

Anonymous (1967) World's largest star sapphire. *Gems and Gemology,* **12**, 158

Anonymous (1981) Coloured gems find new favour. *Jeweller Watchmaker & Giftware*, January

Anonymous (1986) Sotheby's sets records with ruby and diamond. *Jewelers' Circular–Keystone*, December, 75

Ball, V. and Crooke, W. (1925) *Travels in India by Jean-Baptiste Tavernier*, Oxford University Press, London

Bank, H. (1973) *From the World of Gemstones*, Pinguin, Innsbruck

Bassett, A. M. (1981) The magnificent mineral museum at the Univ. of Paris. *Lapidary Journal*, August, 1010–1014

Bauer, M. (1904) *Precious Stones*, Charles Griffin, London

Brown, J. C. (1933) Ruby mining in Upper Burma. *Mining Magazine*, June, 329–340

Church, A. H. (1883) *Precious Stones – Considered in their Scientific and Artistic Relations*, Chapman and Hall, London

Clarke, V. W. (1934) The Chhatrapati Manick ruby still exists. *The Gemmologist,* **3**, no. 30, 178–179

Gühler, U. (1947) Studies of precious stones in Siam. *Siam Science Bulletin,* **4**, 1–39

Ho, H. (1981) *Gem Investor's Seminar On Rubies & Sapphires*, Asian Institute of Gemological Sciences, Bangkok

Holmes, D. (1976) Fit for a king . . . or a Burton. *Jewelers Circular–Keystone*, October, 89–94

Jones, W. (1902) *Crowns & Coronations*, Chatto & Windus, London

Keely, H. H. (1982) The ruby mines of Burma. *Gems,* **14**, no. 3, 6–11; no. 4, 10–14; no. 5, 10–13; no. 6, 12–14

Keller, P. C. (1983) The rubies of Burma: a review of the Mogok Stone Tract. *Gems and Gemology,* **19**, 209–219

Meen, V. B. (1966) Exotic Oriental treasures in the Topkapi Palace, Istanbul, Turkey. *Lapidary Journal,* **20**, no. 2, 297–302; no. 3, 416–417

Meen, V. B. and Tushingham, A. D. (1968) *Crown Jewels of Iran*, University of Toronto Press, Toronto

Mosey, I. (1971) A visit to the Topkapi Museum and Treasury, Istanbul, Turkey 1970. *Journal of Gemmology,* **12**, 214–218

O'Donoghue, M. J. (1970) The Townshend Collection of precious stones in the Victoria and Albert Museum. *Journal of Gemmology,* **12**, 1–5

Orpen, G. (1890) *Stories About Famous Precious Stones*, Lothrop, Boston

Prince Michael (1983) *Crown Jewels of Europe*, Harper & Row, London

Punchiappuhamy, T. G. (1984) Well known gems of Sri Lanka. *Journal of the Gemmologists Association of Sri Lanka*, no. 1, 17–20

Ramsay, A. (1925) *In Search of the Precious Stones*, London

Rush, J. and Rush, R. H. (1979) Connoisseur's Corner – notes on the art and antique market. *The Wall Street Transcript*, 31 December, 56

Russell, A. (ed.) (1987) *1987 Guinness Book of World Records*, Bantam, New York

Sitwell, H. D. W. (1953) *The Crown Jewels*, Dropmore Press, London

Spencer, L. J. (1933) Nation acquires large ruby. *The Gemmologist,* **2**, no. 18, 176–178

Stewart, A. T. Q. (no date) *The Pagoda War*, Faber & Faber, London

Streeter, E. W. (1892) *Precious Stones and Gems*, 5th edn, Bell, London

Tagore, S. M. (1879) *Mani-Mala, or a Treatise on Gems*, Bose, Calcutta

Chapter 10

Formation of corundum deposits

Major and minor deposits

Corundum deposits are relatively abundant around the world, but localities producing gem-quality rubies and sapphires are much less common.

Gem-quality corundum is recovered almost entirely from secondary or alluvial deposits. Of the major gem deposits, only at Yogo Gulch, Montana, in the United States and at Kashmir in India is corundum mined directly from the primary source rock.

According to Pratt (1906), many years ago corundum was regarded as a comparatively rare mineral, whereas now it is known to be widely distributed naturally. It was once thought to occur only in basic magnesian rocks such as peridotites and their serpentine derivatives, but it has been found abundantly in syenites, basalts and other igneous rocks, as well as in various gneisses, schists and crystalline limestones.

Corundum has been found in both igneous and metamorphic rocks, as well as in alluvial deposits. Pratt (1906) states that where it occurs in igneous rocks it is either as a direct constituent of the more acidic intrusive rocks, or as segregated masses near the periphery of the basic intrusive rock, or associated with inclusions picked up by the intrusive rocks. Within metamorphosed rocks, the corundum is either a major component of the igneous rocks subjected to metamorphism or the result of regional metamorphism by which shales, bauxite or other aluminium-rich minerals have been converted into corundum. It may also be a result of a contact metamorphism where the corundum simply occurred in the narrow zone adjoining the intrusive rock. In regard to alluvial deposits, the corundum is found either in the recent alluvium composed of soil, or in sands, or in gravels.

Table 10.1 describes the most important occurrences of gem-quality corundum around the world, while minor occurrences are described in Table 10.2.

TABLE 10.1 Major world occurrences of gem corundum

Locality	*Mode of occurrence*	*Varieties*
Mogok, Burma	Ruby: found in alluvial gravels derived from metamorphosed crystalline limestones Sapphire: found in alluvial gravels derived from granitic pegmatites, or rarely from corundum syenites (Iyer, 1953)	Ruby (most common), blue sapphire; also yellow, violet, green sapphire, stars, colour-change sapphires, etc.
Chanthaburi, Kanchanaburi, Trat and Phrae Provinces, Thailand, and Battambang Province (Pailin), Cambodia	Ruby and sapphire: found in eluvial and alluvial gravels derived from nepheline normative and highly alkaline basalts, especially basanitoid basalts (Vichit, 1978)	Ruby, blue, yellow, green and black star sapphires
Sri Lanka	Ruby and sapphire: found in alluvial stream gravels derived from granulites, pegmatites, garnetiferous gneisses, skarns, cordierite gneisses and associated rocks (Heilmann and Henn, 1986)	All colours, except green (rare); also stars
Kashmir, India	Blue sapphire: found in feldspar pegmatites surrounded by modified actinolite–tremolite bodies and in alluvial gravels derived from these rocks (Atkinson and Kothavala, 1983)	Blue sapphire, rarely pink sapphire; also star sapphires
Queensland and New South Wales, Australia	Sapphires: found in eluvial and alluvial gravels derived from alkali basalts (Coldham, 1985). Recent work suggests that 'basaltic' tuffs may in fact be the source of the sapphire in New South Wales (Pecover, 1987)	Blue, green, yellow and black star sapphires
Jagdalek, Afghanistan	Ruby: occurs *in situ* within a metamorphosed dolomitic limestone cut by granitic intrusions of Oligocene age (Bowersox, 1985)	Ruby
Umba Valley, Tanzania	Ruby and sapphires: found in elluvial and alluvial gravels derived from limestones or pegmatite veins within or bordering a serpentinite pipe (Bridges, 1982)	All varieties
Mangari District, Kenya	Ruby: found *in situ* or in eluvial gravels derived from desilicified veins cutting through altered and serpentinized ultramafic rocks, or in or near the contact zone of a serpentinized pipe associated with desilicified pegmatoidal segregations (Bridges, 1982)	Ruby
Yogo Gulch, Montana, USA	Blue sapphire: commonly found in eluvial or alluvial gravels derived from a lamprophyric igneous dike intruded into a fault within the limestone country rock. The dike rock is also mined itself (Brown, 1982)	Blue and lilac-coloured sapphire
Kaduna, Nigeria	Secondary deposits associated with alkali basalts (Kiefert and Schmetzer, 1987)	Mainly blue sapphires, also greens, yellows and bicolours (green/yellow or blue/yellow)

TABLE 10.2 Minor world occurrences of gem corundum

Locality	*Mode of occurrence*	*Varieties*
North Carolina, USA	*In situ* and alluvial deposits mostly associated with basic magnesian rocks such as peridotite, pyroxenite and amphibolite (Pratt, 1933)	Mainly ruby, but also sapphires of various colours
Hunza Valley, Pakistan	*In situ* deposit of a ruby-bearing dolomitic marble (Gübelin, 1982)	Mainly ruby, some purple to violet sapphires
Mingxi, Fujian Province, China	Alluvial deposits associated with alkali basalts (Keller and Keller, 1986)	Yellow-green, green, blue-green and blue sapphires
Penglai and Wenchang, Hainan Island, China	Alluvial deposits associated with alkali basalts (Keller and Fuquan, 1986)	Blue and green sapphires
Sichuan Province, China	Undescribed (Keller and Fuquan, 1986)	Translucent to opaque ruby
Mercaderes – Rio Mayo area, Cauca, Colombia	Alluvial deposits associated with alkali basalts (Keller *et al.*, 1985)	Mainly blue and green sapphires, also yellows, pinks, reds and colour-change stones
Rio Coxim, Mato Grosso, Brazil	Unknown (Eppler, 1964)	Blue sapphires
Chimwadzulu Hill, Malawi	*In situ* and alluvial deposits derived from an epidotized amphibolite (Rutland, 1969)	Virtually all colours, from pinks and reds to blues, greens, yellows and oranges
Barauta, Zimbabwe	*In situ* and alluvial deposits derived from a pegmatite (Sweeney, 1971)	Blue sapphire

Other deposits

Corundum itself is a relatively common mineral, with deposits being found all over the world. However, most deposits do not produce material clean enough for faceting, or even for cutting cabochons. To the best of the author's knowledge, only those localities described in Table 10.1 are producing gem material in significant quantities. Those localities producing lesser quantities are described in Table 10.2. It should also be mentioned that a number of other localities exist where the occasional corundum gem is produced, and of course the production situation is constantly changing, with new finds being made and older mines being depleted. However, space limitations preclude the description of every single corundum deposit around the world. Some of the other localities which produce the occasional corundum gem include Idaho (USA), Madagascar, Mozambique, South Africa, Namibia and the Soviet Union.

The geology of four major corundum deposits

It is most unfortunate that few of the major gem-producing corundum deposits have been studied and reported on in depth, because they are often located in

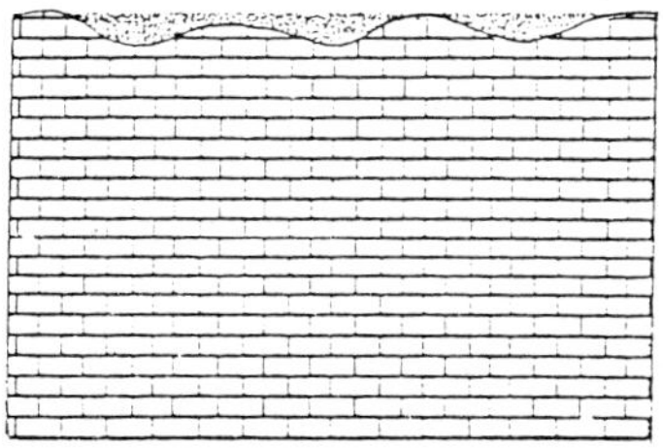

(1) Following the deposition of the limestone and withdrawal of the sea, erosion ensued, with the development of a localized karst topography.

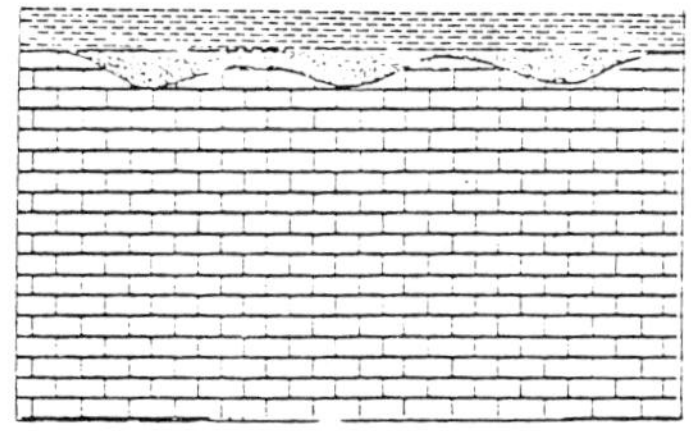

(2) Sediments of the Big Snowy Group, consisting of shales, sandstones, and silt-stones, were deposited on top of the karst topography.

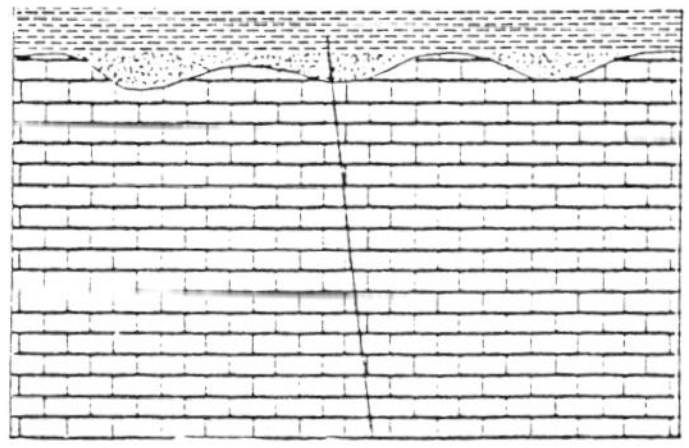

(3) A longitudinal fault, trending northeast-southwest transected the limestone.

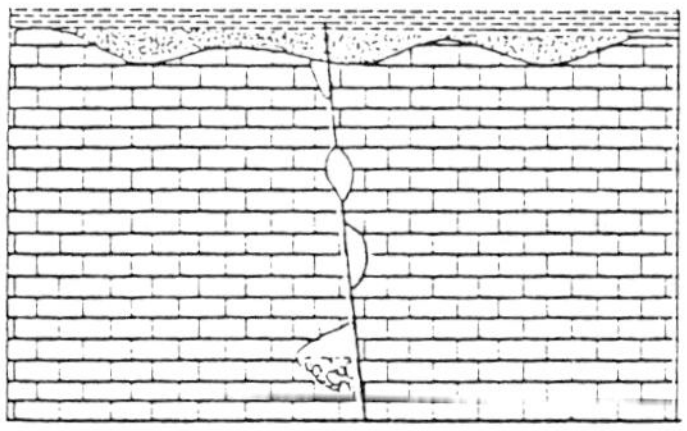

(4) Minor solution of the limestone by circulating ground water occurred along the fault plane. Solution breccia developed in several areas.

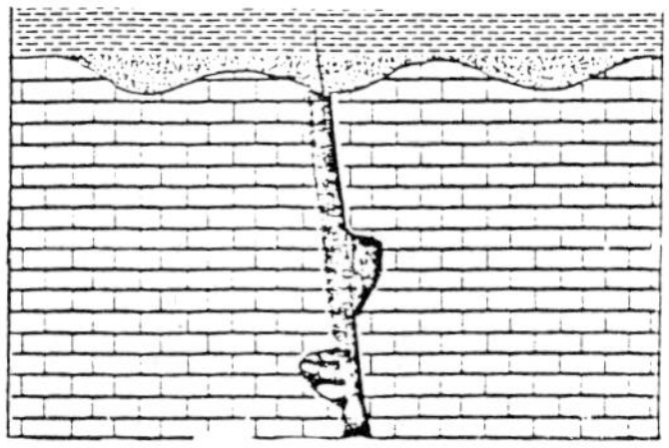

(5) The Yogo sapphire-bearing dike was emplaced along the fault plane. In areas of existing solution breccia, the dike material flowed around the breccia fragments.

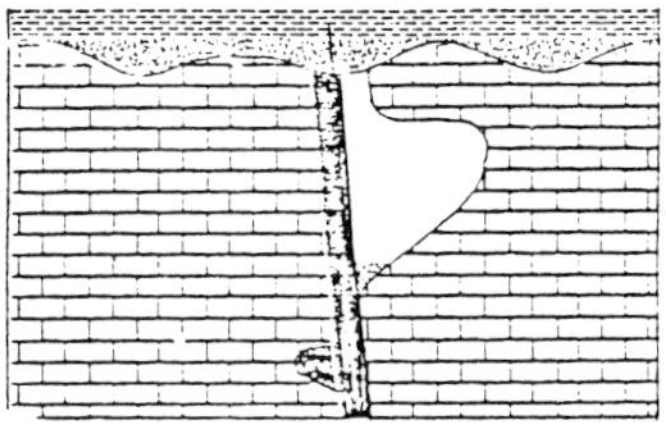

(6) Extensive solution by ground water occurred along the emplaced dike. Solution channels and large caverns developed adjacent to the dike.

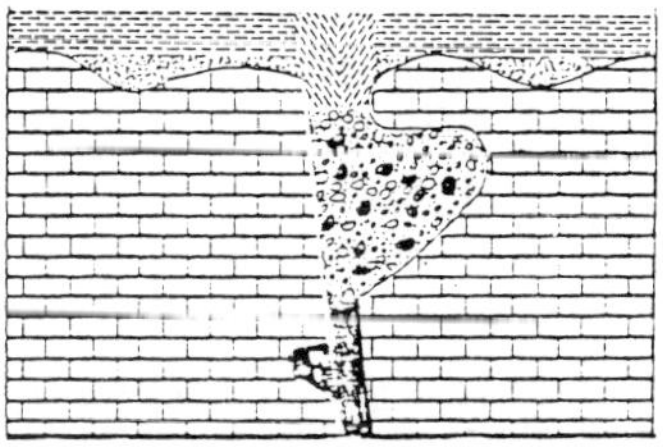

(7) Transverse faulting occurred, offsetting the dike in several areas. Intense ground movement caused the collapse of most of the large caverns. Dike rock, limestone, and sediments from the overlying Big Snowy Group collapsed into the former caverns.

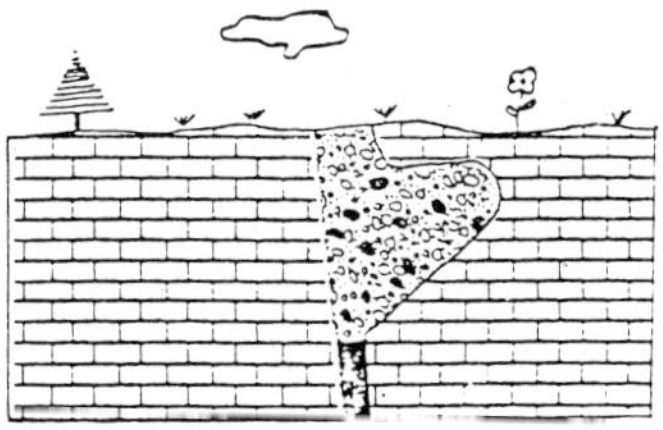

(8) Erosion removed all the overlying easily weathered material, leaving today the Madison limestone exposed at the surface.

Figure 10.1 The development of the sapphire-bearing dike at Yogo Gulch, Montana. (After Brown, 1982)

politically unstable or inaccessible areas. From the information in existence we will describe the geology of four of the most important occurrences in greater detail. These four deposits are: Yogo Gulch, Montana, USA; Mogok Stone Tract, Burma; Chanthaburi/Trat Provinces, Thailand; and New South Wales, Australia.

Yogo Gulch, Montana, USA

The Yogo Gulch sapphire deposit in Montana has been described in detail by Delmer Brown (1982), to whom much of the following information is owed. At Yogo, the sapphires occur in a steeply dipping mafic igneous dike of lamprophyric composition. It can be traced in the field along a roughly linear course for approximately 5.6 km. Both the dike itself and its associated alluvial gravels have been mined for sapphires. In terms of width, the dike varies between 1 and 2 m on average and is believed to extend down into the earth for a substantial distance, making the Yogo Gulch deposit the single largest proven *in situ* occurrence of sapphire in the world.

Brown (1982) believes that the field outcrops of the Yogo dike represent the extreme uppermost portion of the original dike. Evidence of this is the relative lack of alluvial deposits of sapphires except at two isolated localities. This is in contrast to earlier theories that most of the dike has been eroded away, leaving only the 'roots' exposed at the surface. It tends to confirm the belief that the dike stretches far down into the earth.

The Yogo dike magma, and thus the sapphires too, are believed to have originated at considerable depths (within the Precambrian basement complex) and cooled gradually during the long vertical ascent into the overlying sediments. Due to its ever increasing viscosity resulting from cooling, it was not able to penetrate any further upwards than the contact of the limestone with the overlying shales.

Intergem, the Denver-based company which used to work the Yogo sapphire deposit, was mining the dike through surface diggings, but plans were afoot to extend this eventually to underground workings if they could obtain steady financing. They did not manage to do this and in 1987 the company went bankrupt. Many others have failed in their attempts to work the Yogo deposit profitably and only time will tell whether or not another company will make a success of the most important sapphire mines in North America.

Mogok Stone Tract, Upper Burma

The Mogok Stone Tract, which is situated in Upper Burma, has been known for centuries as the source of the world's finest rubies. Indeed, so prolific a producer is Mogok that it has become virtually synonymous with the term 'pigeon's blood', a description of the rich crimson colour of the world's best rubies.

The earliest detailed study of the geological origins of the gems of Mogok was carried out by the expedition of Mr C. Barrington Brown in 1887. J. W. Judd studied the collections of Barrington Brown and wrote a joint paper with him in 1896 on the origins of the rubies. Later, other geologists were to study the area, the most important report of which was that produced by L. A. N. Iyer (1953) of the Geological Survey of India.

Iyer identified 13 mappable rock units in the Mogok area. These 13 units, however, can be grouped into five basic divisions:

1. Intrusive igneous rocks
2. Mogok gneiss, consisting of metamorphic schists and gneisses
3. Pleistocene and recent (Quaternary) alluvium
4. Ultramafic intrusives
5. Marbles

It is believed that the origin of the Mogok rubies is related to the origin of the Mogok limestone, as specimens of ruby are commonly found embedded within a metamorphosed limestone matrix. Judd's theory originally attracted a great deal of attention. He believed that the limestone was formed from basic igneous rocks by the breaking up of the basic feldspars, leading to the formation of an unstable scapolite. Complicated chemical actions and changes eventually allowed for the formation of calcite and dolomite, as well as other minerals such as corundum. Thus, with Judd's theory the source of the alumina in the corundum and spinel is the limestone itself. He also stated that, unlike most crystalline limestones which have yielded corundum and spinel, the Mogok rock does not appear to contain silicates combined with fluorine and silicon, and that tourmaline and chondrodite, which are so common in North American limestones, appear to be almost entirely absent in the Mogok limestone.

Iyer (1953), however, found this not to be the case, since fluorine- and silica-bearing minerals are found in the Mogok limestones and calciphyres. Instead, he attributed the formation of the Mogok limestone to a sedimentary origin of Precambrian age.

Within the Mogok region the rubies generally occur in coarsely crystalline limestones (marbles) and the sapphires in pegmatites, but rubies have also been found in a white feldspathic rock. Iyer pointed out that in all the ruby-bearing areas the limestone has been thoroughly soaked by the pegmatite material related to acid intrusions and that in all areas the effects of the thermal metamorphism are accompanied by pneumatolytic phenomena.

In summary, the rubies of Burma are derived from the crystalline limestone on the contact of, or interbedded with, a complex series of gneissic rocks. This limestone is intruded by veins of granite. The alluvial gravels in which the rubies are actually mined are derived from these rocks and consist of disintegrated granite and marble accompanied by typical contact minerals. Sapphire occurs mainly in gravels derived from pegmatites, but has been found within the limestones as well. Occurring together with the rubies are large quantities of spinel, which actually outnumber the corundums in places. Although in certain areas, such as at Kathe, the sapphire is more abundant than ruby, generally sapphires are outnumbered by as much as 100 to 1.

Within the valleys and on the hillsides is found the gem-bearing gravel layer, known locally as 'byon'. This rests upon a soft decomposed rock of distinctive appearance and is made up of a brown or yellow clayey or sandy material. It is this layer of alluvium left by solution of the marble during weathering which contains the rubies, sapphires, spinels and so many other gems for which Mogok is justly famous. Iyer (1953) states that when large pieces of quartz are found within the byon, rubies are close by.

These quartz lumps indicate the last stage of pneumatolysis, as does the presence of tourmaline with the ruby. Indeed, the author recently examined a stunning Burmese ruby which contained very large green prisms of tourmaline, similar to those which are frequently enclosed within the sapphires from Kashmir.

Byon is also found as a rotten rock between the limestone and the tongues of granitic intrusion, i.e. the granite–limestone contact, which has been found to be very favourable for the byon.

Chanthaburi/Trat ruby mines, Thailand

The corundum of eastern Thailand occurs in both eluvial and alluvial gravels derived from Tertiary to Pleistocene basalts intruded into older sedimentary or metamorphic rocks. In many ways they resemble the corundum-bearing basalts of Cambodia and Australia.

The corundum deposits of Thailand and Cambodia differ from those of nearby Burma in that the matrix of the latter locale is a crystalline limestone (ruby) or syenite (sapphire) rather than basalt. Study of the xenocrysts from Thailand and Cambodia by Jobbins and Berrange (1981) suggest that the corundums, clinopyroxenes, garnets, spinels and other associated minerals originally formed at great depths by metamorphism and/or metasomatism on the edges of a basic pluton. They were then intruded rapidly with the magmas at a much later date.

The basalts of Indo-China may be divided into two broad groups – small and large. The smaller bodies carry the corundums, zircon and garnet, whereas the larger are barren of corundum, containing only zircon. It has been hypothesized that the larger bodies, which are also older, may have contained corundum initially, but that later eruptions were barren and the gems lie buried under later lavas.

Evidence of the origin of corundum in basalt is given by the distribution of the gems. Ruby and sapphire are found in the gem gravels weathered from the lavas, with rich concentrations of gem fragments in the *in situ* weathered basalt (eluvial deposits) and the soil profiles overlying the basaltic bodies. However, rubies and sapphires actually found in the basalt matrix are very rare. This suggests a sparse distribution of corundum in the lavas themselves, with a marked increase in concentration in the gravels as a result of weathering. Euhedral crystals are relatively rare, with most exhibiting a slight to large degree of rounding of the crystal faces, due to partial corrosion by the lava. A high surface polish is common in grains recovered near the vents.

The exact origin of corundum in the basalts has yet to be fully explored. Barr and Macdonald (1977) have speculated that the corundum-bearing basalts could have been generated in the mantle, or possibly at a depth of about 65–95 km. However, the alumina content of the basaltic magma is problematic in the former scenario. Instead, Vichit (1978) has suggested that corundum could crystallize in the latter stages from a basaltic magma generated at depths of 65–95 km by partial melting of mantle materials. This magma could then move upwards to a depth of 25–35 km, with fractionation processes increasing the alumina content substantially, as well as increasing the percentages of nepheline normative minerals. Wall-rock contamination is one possible factor involved. At this point, a decrease in temperature could produce crystallization of the corundum, with an eruption carrying the magma to the surface where it would be quenched. This scenario would explain the origin of corundum samples found embedded in a basalt matrix, as well as the lack of corundum in accompanying samples of schist, gneiss and syenite.

The corundum varieties found in the Chanthaburi/Trat area are mainly ruby and blue, green, yellow and black star sapphires. Other minerals found together with these include pyrope garnet, zircon, quartz, spinel (mostly black) and a number of opaque black stones collectively termed 'Nin', which are sometimes faceted for use in mourning jewellery.

In 1979, Dr P. Aranyakanon noted that certain corundum varieties are distributed along three distinct zones in Chanthaburi and Trat Provinces. The first zone is located less than 5 km due west from Chanthaburi town and includes mines at Khao Ploi Waen, Khao Wao, Bang Kha Cha and Khao Sra Khaeo. Mostly six- and 12-rayed black star sapphires are found here, including the distinctive 'golden-star' black star sapphires, as well as smaller amounts of deep-blue, green and rarely 'Mekong Whisky' yellow sapphires. Ruby is absent entirely.

The second or middle zone is situated astride the border separating Chanthaburi and Trat Provinces, behind the mountain of 'Khao Sa Bap' which dominates the view south-east of Chanthaburi town. Mines are at Ban Tok Prom, Ban Bo I-Ram, Ban Welu Klang, Ban Welu Lang, Ban Si Siat, Ban Sato Wang Cham, Sak Lao, Ban Sato Noi, Bo Waen, Ban Ang Et and Ban Na Ta Maei. This is the region called 'Muang Klung' in Bauer (1904). Ruby and sapphire are found in the second zone, with the exception of Ban Bo I-Ram, where mainly deep-blue sapphires are mined in some quantity.

The third zone is situated in Trat Province and the deposits actually stretch across the border into Cambodia. The mines are situated in particular around the towns of Ban Nong Bon, Bo Rai and Ban Bo Na Wong, including Ban Ta Bat, Nab Na Yai, Noen Fai Mai, Ban Tak Waeng Lang, Ban As Yai and Ban Thung Chuck. Only ruby is found in the third zone, with sapphire being almost entirely absent.

In Chanthaburi and Trat Provinces, the thickness of the gem-bearing gravels varies from place to place, depending upon the nature of the bedrock and slope of the deposits. At Khao Ploi Waen sapphires can be found at the surface or at depths of between 3 and 8 metres.

The gem-bearing soil at Bo Rai is from 4 to 10 metres deep or more, while the author has witnessed mining at Ban Nong Bon reaching depths of 20 metres or more. The thickness of the gravel layer ranges from 30 cm to 1 m.

New South Wales, Australia

Much of the world's ruby and sapphire is recovered from deposits associated with lava flows. At sites such as the Chanthaburi/Trat fields of Thailand, Pailin in Cambodia, New South Wales and Queensland in Australia, Nigeria and south-eastern China, the corundum has long been thought to have formed at depth and then been carried to the surface by basaltic lavas. Very few examples of corundum actually *in situ* in the basalts have been found, however.

In Australia recent discoveries have suggested that the basalt may not actually be the source rock of the corundum. S. R. Pecover (1987) of the New South Wales Geological Survey has reported that sapphire has now been found in 'basaltic' tuffs (a hard volcanic rock consisting of consolidated fragments of lava) in the Inverell–Glen Innes region. 'Basaltic' is used in their description because the tuffs are found associated with basalts.

The following reconstruction of events has been suggested: apparently the corundum formed deep below the earth's surface and was carried up by tuffaceous rocks. These tuffs were later overlain by basaltic lavas. As the tuffs are more easily weathered than the basalts, *in situ* specimens of corundum in the tuffs were not found. Instead, the corundums were found together with the weathered remains of the basalts; thus, it was assumed until recently that the basalt was the source rock of the corundum.

Although deeply weathered, the tuffs are widespread and occur at or near the base of the volcanic pile. According to Pecover, they generally consist of fine to very coarse red, brown, grey and white clastic rock types which overlie basement rocks. Extensive deposits of laterites and bauxites have also been found at the base of the volcanic pile. In some places the tuffs may contain corundum, zircon, pleonaste and ilmenite.

The ironstone formations also represent tuffs in various stages of deep weathering and ferricrete formation. Sapphire is often 'specked' by fossickers from these ironstones after rain, while ironstone nodules are regularly found in the concentrates at sapphire washing plants. Recent bulk sampling of red ashflow tuffs has yielded sapphire, zircon, pleonaste and ilmenite.

Because of the similarity of the Australian sapphire deposits to those in Thailand, Cambodia, Nigeria and China this discovery is of great importance. Recognition of tuffaceous rocks as the main host of corundum could lead to the discovery and expansion of corundum mines worldwide.

Bibliography

Aranyakanon, P. and Vichit, P. (1979) *Gemstones in Thailand*, unpublished report for the Economic Geology Division of The Ministry of Industry, Thailand

Atkinson, D. and Kothavala, R. Z. (1983) Kashmir sapphire. *Gems and Gemology,* **19**, 64–76

Barr, S. M. and Macdonald, A. S. (1977) Geochemistry and petrogenesis of late Cenozoic alkaline basalts of Thailand. *Bulletin of the Geological Society of Malaysia*, preprint

Barrington Brown, C. and Judd, J. W. (1896) The rubies of Burma and associated minerals. *Philosophical Transactions of the Royal Society of London,* **187**, 151–228

Bauer, M. (1904) *Precious Stones*, Dover, New York

Bowersox, G. W. (1985) A status report on gemstones from Afghanistan. *Gems and Gemology,* **21**, 192–204

Bridges, C. R. (1982) Gems of East Africa. *Proceedings of the International Gemological Symposium, 1982*, pp. 266–275

Brown, D. L. (1982) *Geology of the Yogo Sapphire Deposit, Judith Basin County, Montana*, Delmer L. Brown, Lakewood, CO

Coldham, T. (1985) Sapphires from Australia. *Gems and Gemology,* **21**, 130–146

Eppler, W. F. (1964) Sapphire from Rio Coxim, Mato Grosso, Brazil. *Journal of Gemmology,* **9**, 199–204

Gübelin, E. J. (1982) Gemstones of Pakistan: emerald, ruby and spinel. *Gems and Gemology,* **18**, 123–139

Heilmann, G. and Henn, U. (1986) On the origin of blue sapphire from Elahera, Sri Lanka. *The Australian Gemmologist,* **16**, no. 1, 2–4

Iyer, L. A. N. (1953) The geology and gem-stones of the Mogok Stone Tract, Burma. *Memoirs of the Geological Survey of India,* **82**, 1–100

Keller, A. S. and Keller, P. C. (1986) The sapphires of Mingxi, Fujian Province, China. *Gems and Gemology,* **22**, 41–45

Keller, P. C. and Fuquan, W. (1986) A survey of the gemstone deposits of China. *Gems and Gemology,* **22**, 3–13

Keller, P. C., Koivula, J. I. and Jara, G. (1985) Sapphire from the Mercaderes–Rio Mayo area, Cauca, Colombia. *Gems and Gemology,* **21**, 20–25

Kiefert, L. and Schmetzer, K. (1987) Blue and yellow sapphires from Kaduna Province, Nigeria. *Journal of Gemmology,* **20**, 427–442

Pecover, S. R. (1987) New concepts on the origin of sapphires in Northeastern New South Wales. *The Australian Gemmologist,* **16**, no. 6, 221

Pratt, J. H. (1906) Corundum. *USGS Bulletin*, no. 269, 1–175

Pratt, J. H. (1933) Gems and gem minerals of North Carolina. *The American Mineralogist,* **18**, 148–159

Rutland, E. H. (1969) Corundum from Malawi. *Journal of Gemmology,* **11**, 320–323

Sweeney, J. W. (1971) Rhodesian sapphire deposits. *Lapidary Journal*, November, 1076–1077

Chapter 11
World sources of corundum

The Mogok Stone Tract, Burma

When one speaks of ruby, the Mogok Stone Tract in Upper Burma immediately springs to mind. Lying approximately 640 km north of Rangoon, Mogok has for the past 800 years been the premier source, and for many connoisseurs the *only* source, of fine rubies. It is an area steeped in legend and history and its story embraces not only gems, but also the early exploration and expansion of the European colonial empires into Asia.

The town of Mogok is located in the Katha district of Upper Burma and is the centre of the ruby trade in the area. Consisting of thickly jungled hills rising to a height of some 2300 m above sea level, the Ruby Mines District covers about 1000 square kilometres, although only a portion (175 square kilometres) is gem bearing. Considered to be one of the most scenic areas in Burma, it is home to a number of colourful ethnic groups, as well as to a variety of wildlife, including elephants, tiger, bear and leopard.

Figure 11.1 The spoils of the jungle. Burmese miners with a slain leopard from the jungles at Mogok. A variety of wildlife is found in the thick jungle surrounding the Mogok ruby and sapphire mines. (From O'Connor, 1905)

History

The exact time when rubies were first discovered in Mogok is unknown. Vague references (Ehrmann, 1957) exist, suggesting on the basis of stone relics unearthed that the area was first settled by Mongolians about 3000 BC. However, this has not been documented. Ehrmann (1957) describes a local legend stating that modern Mogok was founded in AD 579 by headhunting tribesmen from nearby Momeik. After losing their way they discovered a 'mountain break full of beautiful rubies' when investigating a commotion made by many birds. This story is very similar to that told of many gem deposits and is believed to derive from the 'valley of precious stones' in Sri Lanka of Sinbad the Sailor, or perhaps Al Kazwini's relation of Alexander's valley of serpents and diamonds in India (Kunz, 1913). In the Burmese version, a fever-stricken and serpent-ridden valley was found teeming with rubies. As the valley was far too dangerous for mere mortals to enter, the stones were obtained by casting lumps of fresh meat into the abyss. This attracted large birds of prey who snatched up the meat and brought it out, together with the rubies adhering to it. They were then retrieved from the birds' nests and droppings.

Figure 11.2 Native gem diggers at Mogok about the turn of the century. (From O'Connor, 1905)

From the earliest times of European contact with East Asia, the country of Burma has been associated with rubies (George, 1962). Nicolo di Conti and Ludovico di Varthema both visited Burma in the fifteenth century and referred to the rubies found there. The first Englishman on the scene was Ralph Fitch in 1586, whose journey led to the founding of the British East India Company. He said

> Caplan is the place where they finde the rubies, saphires, and spinelles: it standeth sixe dayes journey from Ava in the Kingdome of Pegu. There are many great high hilles out of which they digge them. None may go to the pits but onely those which digge them.

Not only did Fitch comment upon the rubies, but also he told of a curious local custom:

> In Pegu, and in all the countreys of Ava, Langeiannes, Siam, and the Bramas, the men weare bunches or little round balles in their privy members: some of them weare two and some three. They cut the skin and so put them in, one into one side and another into the other side; which they do when they be 25 or 30 years old, and at their pleasure they take one or more of them out as they thinke good . . . The bunches aforesayd be of divers sorts: the least be as big as a litle walnut, and very round: the greatest are as big as a litle hennes egge: some are of brasse and some of silver: but those of silver be for the king and his noble men. They were invented because they should not abuse the male sexe [Evidently the noise of the balls striking one another would alert others in the vicinity as to the activity taking place.] for in times past all those countries were so given to that villany, that they were very scarse of people.

A. Hamilton (1744), who travelled to India and Burma in the eighteenth century, also had some interesting remarks about the Burmese. In reference to the sarongs worn by ladies he said:

'Under the Frock they have a Scarf or 'Lungee' doubled fourfold, made fast about their Middle, which reaches almost to the Ancle, so contrived, that at every Step they make, as they walk, it opens before, and shews the right Leg and Part of the Thigh.

This Fashion of Petticoats, they say, is very ancient, and was first contrived by a certain Queen of that Country, who was grieved to see the Men so much addicted to 'Sodomy,' that they neglected the pretty Ladies. She thought that by the Sight of a pretty Leg and plump Thigh, the Men might be allured from that abominable Custom, and place their Affections on proper Objects, and according to the ingeiuous Queen's Conjecture, that Dress of the 'Lungee' had its desired End, and now the Name of Sodomy is hardly known in that Country.

Hamilton also mentioned the products of Burma:

The Product of the Country is Timber for building, Elephants, Elephants Teeth, Bees-wax, Stick-lack, Iron, Tin, Oyl of the Earth, Wood-oyl, Rubies the best in the World, Diamonds, but they are small, and are only found in the Craws of Poultry and Pheasants, and one Family has only the Indulgence to sell them, and none dare open the Ground to dig for them . . . About twenty Sail of Ships find their Account in Trade for the limited Commodities, but the Armenians have got the Monopoly of the rubies, which turns to a good Account in their Trade; and I have seen some blue Sapphires there, that I was told were found on some Mountains of this Country.

After reading such tales, it is little wonder that the East was and still is today looked upon by outsiders as a place of wonder and exotic mystery. But the above tale was nothing compared to that related by the famous French traveller and diamond merchant, Jean Baptiste Tavernier, about the King of Bhutan. In essence (no pun intended), the story he related was that when the king went to 'do the deeds of nature', the result would be carefully collected, dried and then sold to Bhutanese nobles who used 'the odeur' as a type of snuff. Tavernier knew this to be fact because a Bhutanese merchant showed him the contents of his snuff box! So shocking was this to the Victorian British that the passage was removed entirely from later editions of his book edited by Valentine Ball (1925).

Of the many accounts of the gems of Pegu, as Burma was then known, perhaps the most interesting was that of Caesar Fredericke of Venice who journeyed to Asia in 1563 (Hakluyt, 1904). The following is his description of the gem trade in Pegu.

. . . it is a thing to bee noted in the buying of jewels in Pegu, that he that hath no knowledge shall have as good jewels, and as good cheap, as he that hath practized there a long time. There are in Pegu foure men of good reputation, which are called Tareghe, or brokers of Jewels . . . through the hands of these foure men passe all the Rubies: for they have such quantitie, that they knowe not what to doe with them, but sell them at most vile and base prices. When the Marchant hath broken his mind to one of these brokers or Tareghe, they cary him home to one of their Shops, although he hath no knowledge in Jewels: and when the Jewellers perceive that hee will employ a good round summe, they will make a bargaine, and if not, they let him alone . . . when any Marchant hath bought any great quantitie of Rubies, and hath agreed for them, hee carieth them home to his house, let them be of what value they will, he shall have space to looke on them and peruse them two or three dayes: and if he hath no knowledge in them, he shall alwayes have many Marchants in that Citie that have very good knowledge in Jewels; with whom he may alwayes conferre and take counsell, and may shew them unto whom he will; and if he finde that hee hath not employed his money well, hee may returne his Jewels backe to them who hee had them of, without any losse at all. Which thing is such a shame to the Tareghe to have his Jewels returne, that he had rather beare a blow on the face than that it should be thought that he solde them so deere to have them returned.

Thus 'spake' Caesar Fredericke, and after reading his tale we can only wish that modern-day gem merchants would be so understanding. Perhaps businessmen have not really changed all that much. Fredericke was probably just an early example of the gullible tourist taken for a ride. Had he made his journey today, he might have returned to Europe with wooden elephants, in addition to his gem purchases.

Fredericke also gave one of the earliest accounts of the fascinating Asian technique of negotiating prices in secret by covering the hands of the buyer and seller with a cloth.

> There are many Marchants that stand by at the making of the bargaine, and because they shall not understand howe the Jewels be solde, the Broker and the Marchants have their hands under a cloth, and by touching of fingers and nipping the joynts they know what is done, what is bidden, and what is asked. So that the standers by knowe not what is demaunded for them, although it be for a thousand or 10. thousand duckets. For every joynt and every finger hath its signification. For if the Marchants that stande by should understand the bargaine, it would breede great controversie amongst them.

In the year AD 1597, the Burmese King, Nuha-Thura Maha Dhama-Yaza, ratified a royal edict exchanging small parts of Burma under his control for the Mogok Stone Tract, previously under the control of a Shan Sawbwa, or prince. Both the Burmese text of this order and an English translation are reproduced in Figure 11.3 (George, 1962).

From that point on Mogok was part of Burma proper. The first European to visit the actual mines in the Mogok area sometime before 1833 was a Portuguese priest, Père Guiseppe d'Amato. His brief account of their workings was published in the *Journal of the Asiatic Society of Bengal* (d'Amato, 1833). He stated:

> The territory of Kyat-pyen is situated to the east, and a little to the south of the town of Mon-Lha, distant, say 70 miles. It is surrounded by nine mountains. The soil is uneven and full of marshes, which form seventeen small lakes, each having a particular name. It should be noticed, however, that the ground which remains dry is that alone which is mined . . .

သက္ကရာဇ် ၉၅၉ ခု၊ တော်သလင်းလပြည့်ကျော် ၁၂ ရက်နေ့၊ ရတနာပူရတွင်၊ ရွှေဝမြို့ကြီး တည်တော်မူသည်။ မိုးကုပ်ကျပ်ပြင်သည်၊ ရတနာတမည်ဖြစ်သည်၊ ရွှေဝမြို့ကြီးအဝင်အပါ ဖြစ်စေရမည်။ မိုးမိတ်စော်ဘွားစီရင်စုဖြစ်သည်။ မိုးကုပ်ကျပ်ပြင်ကို နှုတ်ယူရမည်။ မိုးကုပ်ကျပ် ပြင်အစား တကောင်းမြို့ကို အရံအကျေးနှင့်တကွ အပ်တော်မူသည်။ မိုးမိတ်မြို့က သိမ်းစေ။ မိုးကုပ်ကျပ်ပြင်မှာ ရွှေဝမြို့ကြီးတည်ရာ ရတနာတမည်ဝင်စေ။ ကျောက်ထူးအကြီး ငယ်သား အဝင်အပါတို့ကို ဝန်တို့က၊ စာရင်းအင်းချယူ၍ ရွှေတိုက်တော်သွင်းရမည်။ မိုးမိတ်မြို့က မခန့် မထားနှင့်၊ နှုတ်တော်မူသည်။ တကောင်းမြို့ကို အစားလဲတော်မူ၍၊ အပ်တော်မူလေပြီးသည်။ ရွှေဝမြို့ကြီးတွင် ထမ်းရွက်စေ။

သက္ကရာဇ် ၉၅၉ ခု၊ သတင်းကျွတ်လပြည့်ကျော် ၅ ရက်နေ့။
နားခံတော်ပြော့ကြီးမ။

Translation

Shwe-Wa-myo (Ava) was established on 12th *Tawthalin Labyigyaw* of 959 B.E. It is the Ratna Pura (Ratna = Gem, and Pura = City).

Mogok and Kyatpyin are names for Gem. They should be included in the Shwewamyo. These two were part of Momeik *Sawbwa's* State but should be excluded and Tagaung *Myo* with its surrounding villages be included in the State instead.

It is ordered that the Momeik *Sawbwa* take possession of Tagaungmyo and that Mogok and Kyatpyin be given over to Shwewamyo. The *Wuns* concerned must take over the rubies with a list of all descriptions (big and small) and pay into the Government Treasury.

No appointments whatever are therefore to be made by the *Sawbwa* to Mogok and Kyatpyin which have been given to Shwewamyo in exchange for Tagaungmyo.

NAKHANDAWPYAWGYIMA.

5th Thadingyut Labyigyaw, 959 B.E.

Figure 11.3 Royal edict of AD 1597 transferring ownership of the Mogok Stone Tract from a Shan Sawbwa to the Burmese King

> The miners, who work at the spot, dig square wells, to the depth of 15 or 20 cubits [an ancient measure of length based on the forearm], and to prevent the wells from falling in, they prop them with perpendicular piles . . .
>
> When the whole is secure, the miner descends, and with his hands extracts the loose soil, digging in a horizontal direction. The gravelly ore is brought to the surface in a ratan basket raised by a cord, as water from a well. From this mass all the precious stones and any other minerals possessing value are picked out, and washed in the brooks descending from the neighbouring hills.
>
> Besides the regular duty which the miners pay to the Prince, in kind, they are obliged to give up to him gratuitously all

Figure 11.4 A native miner at a 'twin-lon' mine near Mogok, Burma. (From O'Connor, 1905)

jewels of more than a certain size or of extraordinary value . . .

The precious stones found in the mines of Kyat-pyen, generally speaking, are rubies, sapphires, topazes, and other crystals of the same family [the precious corundum.] Emeralds are very rare, and of an inferior sort and value. They sometimes find, I am told, a species of diamond, but of bad quality.

The Chinese and Tartar merchants come yearly to Kyat-pyen, to purchase precious stones and other minerals. They generally barter for them carpets, coloured cloths, cloves, nutmegs and other drugs. The natives of the country also pay yearly visits to the royal stones. I have avoided repeating any of the fabulous stories told by the Burmans of the origin of the jewels of Kyat-pyen.

There is another locality, a little to the north of this place, called Mookop, in which also abundant mines of the same precious gems occur.

Note – While I am writing this brief notice, an anecdote is related to me by a person of the highest credit, regarding the discovery of two stones, or, to express myself better, of two masses [amas] of rubies of an extraordinary size, at Kyat-pyen. One weighed 80 biches, Burmese weight, equivalent to more than 80 lbs.! the second was of the same size as that given to Colonel Symes [an early English visitor to Ava]. When the people were about to convey them to the capital to present them to the king, a party of bandits attacked Kyat-pyen for the second time, and set the whole town on fire. Of the two jewels, the brigands only succeeded in carrying off the smaller one; but the larger one was injured by the flames: the centre of the stone, still in good order, was brought to the king. I learned this from a Christian soldier of my village of Mon-Lhá, who was on guard at the palace when the bearer of the gem arrived there.

Burmese monarchs worked the Stone Tract as a royal monopoly in an often despotic manner. All rubies above a particular size were considered to be property of the king and failure to surrender them was punishable by torture and death. The story of the Nga Mauk Ruby provides an example. Nga Mauk, a poor miner, uncovered a large fine ruby which was later divided into two excellent pieces along an incipient flaw. One piece was duly given to the king, but the other half was secretly sold. Unfortunately (according to Keely (1981)) the king learned of the deception when he proudly showed his half to the dealer who had bought the other part. Enraged, he sent his minions to exact punishment. All the villagers in the area were rounded up into a makeshift stable and burned alive. Even today, some 150 years later, the remains of this horrible cremation can be seen at a spot called Laung Zin, which means 'firey platform'. As for the famous Nga Mauk Ruby, it disappeared when the British annexed Upper Burma in 1885 (Keely, 1981; Streeter, 1892).

During the latter part of the nineteenth century, production from Mogok declined

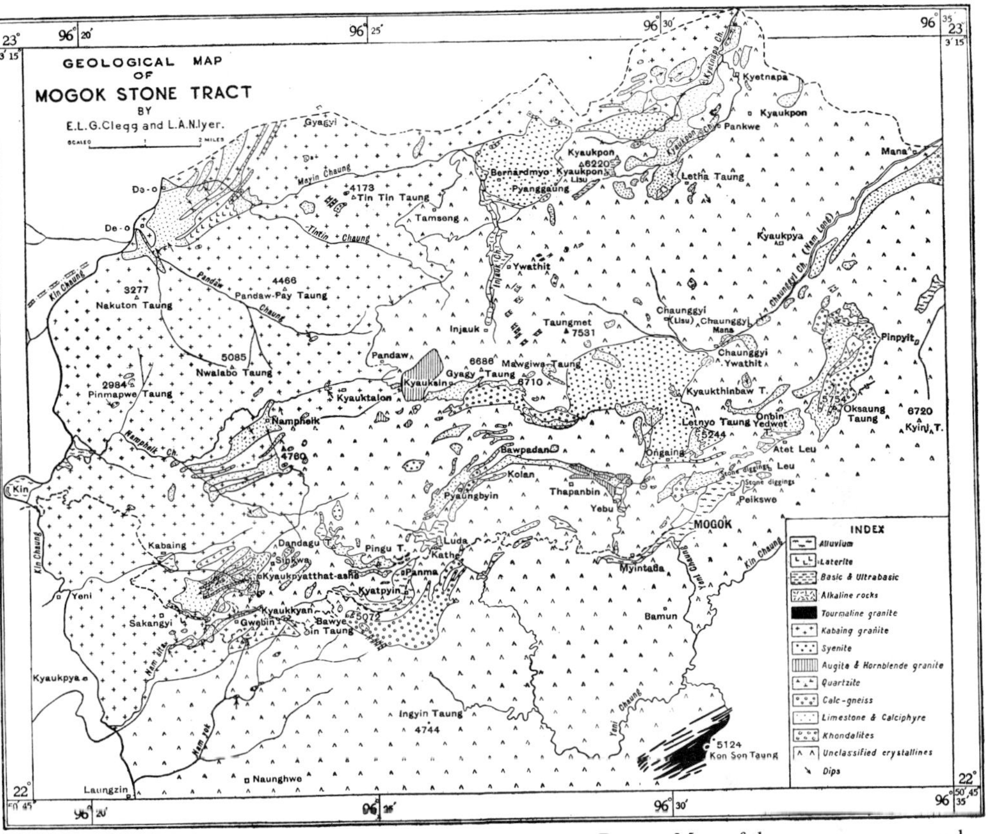

Figure 11.5 Geological map of the Mogok Stone Tract in Upper Burma. Most of the gems are recovered from alluvial deposits situated near the towns of Mogok and Kathe. (From Iyer, 1953)

drastically due to the despotic rule and heavy handed policies of the Burmese monarch, Theebaw, and his dictatorial wife, Supayalat. Commercial disputes and reported corruption at the Court eventually caused the British to annex all of Upper Burma, including the Mogok Stone Tract, in 1885. When British troops arrived in Mogok they found the area nearly deserted, most residents having already fled the repression of the king. Unfortunately, the king's fabulous jewels were never recovered. Owing to the oversight of British officers, the jewels were all smuggled out of the palace by female servants under the noses of British troops on the night the palace was captured (Stewart, no date).

The Ruby Mines Company

After the annexation of Upper Burma in 1885, it was eventually decided to lease out the ruby mines to private interests while still protecting the rights of native miners.

Mogok was occupied by British troops in December 1886, but it was not until 1889 that the lease began. The reasons for the delay were entirely political. Edward Moylan, a disbarred barrister, managed to convince many in London that the lease holder, E. W. Streeter, had acquired the lease through dishonest means. Moylan, who was writing for the famous *Times* of London, succeeded in destroying the reputation of many a man with his poison pen. In the end, his true motive was revealed: he was working for another group who hoped to win the lease themselves, and so the lease was awarded to Streeter & Co. (Stewart, no date). Streeter later sold the lease to the Burma Ruby Mines, Limited for £55 000.

Extraordinary scenes were witnessed when the company was first floated. Huge crowds of people rushed to buy stock in London and the issue was sold out within hours. The general public and the company directors were under the mistaken impression that fabulous riches were just waiting to be unearthed in Mogok. No one gave a thought to the difficulties which might be encountered in mining gems in such an inhospitable and remote location. Instead, only the potential profits were considered.

The company started its career with a paid-up capital of £150 000. All kinds of unforeseen difficulties were encountered during the first few years and an unusual amount of time was spent in preliminary operations. The annual rental fee to be paid to the Government of India was originally fixed at the high sum of £30 000 per year plus 30% of any profits made, in return for the sole rights to mine with machinery. Native mining was also allowed by native methods in areas not utilized by the company, for which a tax of 30% on all finds was collected. This tax soon proved unworkable and so a fee per workman was charged instead. Revenue from native workings later became an important source of funds, but in the beginning little was collected.

Tremendous difficulties were encountered from the outset. Roads had to be constructed, machinery imported and set up, diseases took their toll of men and livestock; these were but a few of the problems. With losses mounting, the company renegotiated its lease with the government several times, but in the end it was not enough. Slumps in the market caused by the introduction of synthetic rubies and the onset of World War I made sales difficult. Finally, in 1925, faced with mounting losses, the company went into voluntary liquidation. No buyers were forthcoming and so in 1931 the lease was surrendered. Thus ended the only major attempt at mechanized mining of the world's richest source of rubies.

In retrospect, a number of reasons can be found for the failure of the company. Those mentioned above and many others have been discussed at length in the literature (Halford-Watkins, 1932b). One valuable opinion, however, has not been previously discussed. In a confidential report (Brown, 1927) written to the Government of India on the future of mining at Mogok, the head of the Geological Survey of India, Mr J. Coggin Brown, stated his opinion on the failure of the company

At this juncture I cannot refrain from writing an opinion which I have already

Figure 11.6 The Burma Ruby Mines Company mine at Mogok, with both British and native workers. (From Claremont, 1906)

expressed verbally, that the influence of the De Beers diamond concern has had more to do with the present (1927) position of mining for coloured gems in Burma than appears on the surface. The reasons for this are obvious, and it is significant that there has always been a powerful representative of the Great South African concern on the Board of the Burma Ruby Mines, Limited.

In the opinion of the present author, Brown was referring to the competition for a share of the gem market between De Beers and the Ruby Mines Company. Apparently, Brown believed that the Burma Ruby Mines Company was sabotaged by De Beers. This idea is not as far-fetched as it may seem upon first glance. We must bear in mind that in the early part of the twentieth century De Beers was not the huge and powerful company that it is today. The present domination of the gem market by diamond has been created largely over the past 70 years by the monopolistic practices and skilful marketing of De Beers. Early in the twentieth century the diamond market faced the very real problem of over-supply because of the discovery of vast new deposits in South Africa. Thus it is not difficult to see that, from the viewpoint of De Beers, the success of the Burma Ruby Mines Company represented a very substantial threat. Of course, rubies cannot be sold if they are not being mined. According to Brown, poor decisions taken by the Board of Directors of the Burma Ruby Mines Company greatly contributed to the eventual failure of the company. (It is important to note that this opinion was probably only expressed by Brown because his report was confidential. In his public writings on the Burma Ruby Mines Company, Brown did not express this opinion.)

In his summary, Brown discussed the future potential of ruby mining in Burma:

> The operations of the Company, apart from an abortive attempt to mine gems from the limestone, and one or two half-hearted efforts to prove the hill deposits, have consisted entirely in working the valley alluvials, confining their attentions to the Mogok, Kyatpyin and Kathe Valleys. There are, however, other valleys in the stone tract and the question arises whether these have been sufficiently explored. It is exceedingly doubtful if they have, in particular the Kin and Khabine deposits. It is notoriously difficult to prospect this type of mineral deposit and no two geologists of experience would agree as to the reliability of the results so obtained. But it was surely the duty of the Company to put these questions beyond doubt. This has not been done, not through any fault of the local technical command but owing to the inanition of the Board of Control . . .
>
> It has been stated to me repeatedly, and I see no reason to doubt the fact from my own view as a geologist, that there are great possibilities in the hillside deposits. They will certainly be more difficult to evaluate and occasional failures might result, but, in a speculative business like gem mining perhaps this does not matter much . . . The history of the Company proves that time after time the selection of a field for a new enterprise instead of being a matter of scientific certainty, has been a pure speculation, based for the most part on reports and rumours of the success of native miners.

In the many years that have passed since the failure of the company, the working of the mines has reverted back to the age-old methods of the native miners. Machinery was left unattended and eventually became useless. A fantastic drainage tunnel which was constructed on the orders of the company's brilliant engineer, A. H. Morgan, was damaged by heavy flooding in 1925 and never repaired, resulting in the formation of two large lakes which today mark the landscape of Mogok. In 1962, a change of government in Burma resulted in the nationalization of virtually all industries, including that of gem mining. Since then, the government has worked the mines in a rather desultory fashion, with the little output that falls into government hands being sold once a year in February at auction in Rangoon. Illegal mining also takes place and accounts for the lion's share of current production. These stones eventually find their way onto the world market through Thailand and India. Today, production remains very small, a fact that has pushed up prices for Burmese rubies to record levels.

The prospects for the future appear dim. While it does seem that plenty of material remains in the ground waiting to be mined, apparently only a change of government could bring about a change in mining policy and methods. In the meantime, production from other sources such as Thailand has filled, to a degree, the world's appetite for ruby. The ruby from Thailand may assuage the pangs of hunger, but it does not satisfy the heart. And so the world waits, as it has since the early 1960s, for the day when once again the glowing red stones of Burma will take their rightful place as the world's premier ruby.

Mining methods

The rubies of Mogok occur in a crystalline limestone matrix, while the sapphires are derived from pegmatites. Weathering has transported the rubies down from the hills to the valley floors where they have settled in the bottom of the streams and rivers to form part of the alluvium. It is from these ancient river gravels that the majority of the stones have been recovered.

Three major types of mines exist in Mogok: the twin-lon, or pit method, for excavating the valley alluvials; the hmyawdwin, or open trench method, for excavating hillside deposits; and the ludwin system for the extraction of gem-bearing materials that fill limestone caves.

Twin-lons

The twin-lon or 'twin' mining involves the sinking of a small round shaft or hole down to the gem-bearing gravel, which is locally termed 'byon'. Each mine is generally worked by two to three men and is similar to the mining in Thailand and Sri Lanka. Two of the men take turns digging the shaft while the third stands above ground, lowering a small basket to haul up the earth. This basket is attached to the end of a long bamboo pole with a counterweight to assist in lifting the earth. The depths of these twins vary with the depth of the gravel and range from 3 to 25 m. Light is provided by the ingenious manipulation of a looking glass at the mouth of the shaft so that a beam is thrown down to the bottom. Candles may also be used.

Once a layer of byon is reached, horizontal tunnels are driven for distances of up to 9 m to remove as much earth as possible. Shallow shafts need no shoring up, but the deeper ones are firmed with posts. Even with these precautions cave-ins do occur on occasion and may be fatal. Flooding from groundwater is a constant problem, and the first job each day is to remove the previous night's accumulation of water. Ingenious pumps made from bamboo have been devised for this purpose. Generally, twin-lon operations can only be carried on in the dry season (November–May).

Larger excavations are termed 'lebin' and 'kobin', and the biggest 'inbye'. The inbye is

Figure 11.7 'Twin-lon' mines and native miners at Mogok, Burma. (From O'Connor, 1905)

Figure 11.8 A Burmese miner at Mogok dewatering an excavation using an ingenious bamboo pump. (From O'Connor, 1905)

rarely seen due to the expense of the timber needed.

Hmyawdwins

These are open cuttings on the sides of hills. J. Coggin Brown was quoted earlier as feeling that great potential still remained for this type of mining. A stream of water, sometimes brought from great distances by means of bamboo channels, is directed under pressure to the upper end of the working. This carries the earth as mud into the tail race of the excavation, with the lighter material being swept away. The heavier concentrate is then carried to a suitable site for washing. As this method requires plenty of water, it is used mainly in the rainy season (June–October).

Ludwins

These are excavations into the sides of the hills and follow the gem-bearing material through the crevices and caves in the limestone. European methods were once applied to these deposits by the Ruby Mines Company, but failed. It is within these caverns and crevices that some of the richest finds have been made. One cavern proved so vast in size and the depth of byon so great that hmyawdwins and twins were actually set up inside the cavern itself! Unfortunately the roof caved in, putting an early halt to the proceedings. The danger attending this method has often been overstated by writers, however (George, 1962). Ludwins are the least common of the three different methods of mining in Mogok.

Washing

Once sufficient byon has been obtained, it is transported to the washing area. A shallow circular enclosure is formed with big rocks, the floor of which slopes slightly at one end. Into this is placed the byon and a stream of water is directed onto it while the whole mass is stirred. Water and lighter debris flow out of a small opening at the lower end, leaving behind the gems and heavier material. Eventually the opening becomes clogged with heavy gravel. This gravel is then removed for further washing on circular bamboo trays, similar to the method used in panning for gold.

Almost inevitably, some of the gems escape from the opening and are carried away to the tailings. According to local custom, these tailings may be searched through by anyone, with any stones found being the property of the finder. Under the company, however, this was later restricted to women only; any man who raised a stone from the ground, unless a worker or licence holder, was subject to imprisonment. Thus, it is the women who search in this way; they are termed 'Kanes' women. According to J. F. Halford-Watkins (1932b), this custom resulted in the wholesale theft of large numbers of stones from the company mines, as well as providing a convenient method for the disposal of stolen goods. The way it worked was as follows. A dishonest workman might catch sight of a stone. He would then pass it secretly to a nearby Kanes woman or tell her where to search for it. Moments later, there would be a shout of joy from the woman as she 'uncovered' a stone which, by custom, was hers to keep. The company went to great lengths to prevent the theft of stones by enclosing the sorting areas, requiring workers to wear steel masks so that stones could not be swallowed, etc. However, all this was to no avail because of the Kanes custom.

Final sorting of the gravel was and is normally done by the mine owner, his relatives, or other trusted staff, with another person present to ensure honesty. All others are kept away. As stones are found they are placed into a bamboo container or, today, a plastic bag. At the end of the day the contents are put into a packet which is sealed on the spot if the owner has partners who are

Figure 11.9 Native 'Kanes' women washing ruby gravel at Mogok, Burma. (From O'Connor, 1905)

Figure 11.10 British sorters at the company mines at Mogok, Burma. (From O'Connor, 1905)

not present. This seal will only be broken in the presence of all the partners, again to maintain honesty.

At this point a few words should be mentioned in regard to the theft of gems. In a gem-mining operation, it is absolutely vital that theft be kept to a minimum (it can never be eliminated entirely). If a significant portion of the production is stolen, these stones will appear in the marketplace, always undercutting the prices asked for legitimate production. The author was once told of a unique method of dealing with this problem. Two brothers purchased a gem mine in Africa. While one brother handled the daily affairs at the mine, the other set up incognito at a nearby town as a gem buyer, purchasing all of the stones stolen by his workers! Although they were buying stones that were rightfully theirs, the brothers were happy with the arrangement, for it did allow them to control the entire production of the mine and, thus, to influence the price better.

Grading and trading

In the morning, stones found during the previous day's work are placed onto a polished brass plate for grading in direct sunlight (Halford-Watkins, 1932b). First the inferior material, termed 'sonzi', is removed and separated into three types: ruby, sapphire and spinel. This is later crushed for use as abrasive.

Those left on the plate are now graded roughly by hand or sieve according to size. The owner removes the very best stones, to be graded personally at a later time, and the remainder are graded by sorters. Lower qualities are arranged in little piles first, and then the better stones. In the end, all are passed to the owner for final classification, assisted by the ubiquitous brokers who play an important role in the valuation and sale. Valuations and bids are all done with the secret hand language so quaintly described earlier by Caesar Fredericke.

The role of brokers is very important, both at Mogok and further north at the jade mines near Mogaung. Each dealer employs brokers to act as their 'eyes and ears'. Their job is not only to assist in valuations and sales, but also to obtain information about what valuable stones have been recently

Figure 11.11 Native workers under British supervision at the company mines at Mogok, Burma. (From O'Connor, 1905)

Figure 11.12 Main street in the town of Mogok, Burma, around the turn of the century. (From O'Connor, 1905)

mined, who the owners of these stones are and, just as importantly, who the stones are being offered to and the prices bid. According to Mr W. K. Ho, a major Bangkok dealer from Burma, the owners of valuable stones would do their best to keep all details secret, for if a piece were to be offered to one man who bid, say, 50 000 Kyats, when the spies of other dealers learned of the bid, no one would offer more. In other words, if Ming (a fictitious name) believed it was worth only 50 000, then why should anyone else offer more? This situation resulted in purchases taking place in a cloak and dagger atmosphere as both buyer and seller sought to conceal their activities. Stolen stones in particular were generally offered at remote jungle rendezvous, sometimes in the dead of the night with only a hand torch as illumination. Sometimes legitimate goods were also offered in this way, being represented as stolen in the hope that this would increase the buyer's feeling that he was getting a real 'steal' of a deal.

Perhaps the best summary of the gem business in Burma is that of the British officer, Major F. L. Roberts, former Deputy Commissioner of Myitkyina (Chhibber, 1934). Although speaking in reference to the jade business, he could just as easily have been discussing the ruby trade:

> From the time jade is won in the Jade Mines area until it leaves Mogaung in the rough for cutting there is much that is underhand, tortuous and complicated, and much unprofitable antagonism. In my opinion the whole business requires cleansing, straightening and the light of day thrown on it.

From the tone of his statement, it sounds like Major Roberts would be just the man for the job, too.

Gem classification

Through the many generations of trading in Mogok, a local classification system has evolved as follows (George, 1962; Halford-Watkins, 1932b):

1. Individual stones are termed 'lon-bauks'. Parcels of rubies are graded by size and colour.
   (a) Stones of the first water (deep rich crimson). Burmese names:
      1. *Anyun*, 2 carats weight and over.
      2. *Lethi*, average 1.75 carats.
      3. *The-bauk* (*haibauk*), average 0.75 carats.
      4. *Saga-the*, average 0.50 carats.
      5. *Ame-the*, average five stones per carat (0.20 carats each).
   (b) Stones of the second water, i.e. bright crimson.
      1. *Ani-gyi*, 2–6 carats weight.
   (c) Stones of the third water known as *Ani-te* or *Bombaing* (because they

were fancied in Bombay, India), a bright light crimson.

1. *Ani-te*, 2–6 carats weight.

(d) Fourth class, *Ahtet-Kya*.

1. *Ahte-Kya* (literally 'fallen from the top'), mixed stones of the above grades but slightly defective in shape or water.
2. *Kyauk-me*, very dark stones which were sold mainly in Madras, India.

(e) Parcels of inferior stones are classified as follows:

1. *Gaungsa* or *Yawya*, pale inferior stones of mixed sizes (up to 6 carats).
2. *Asa-yo*, dark inferior stones of mixed sizes (up to 6 carats).
3. *Asa-yo Kya*, inferior to *Asa-yo*.
4. *Akyan-the*, similar to *Asa-yo* but smaller.
5. *Apya*, flat stones of fine quality.
6. *Apya-Kya* or *Apya-sa*, flat stones of second quality.
7. *Apyazone*, third-quality flats.
8. *Awa*, large defective stones.
9. *Gair*, large, impure, almost opaque stones.
10. *Ani-the*, small stones of second water and good quality.
11. *Akyaw-the*, small, pale, good.
12. *Apyu-the*, small, pale, inferior and rough.
13. *Atwe*, very rough and impure.
14. *Zon-si*, spinels and rejections from other classifications.
15. *Mat-sa*, opaque sapphire.
16. *Thai*, tiny stones (literally 'sand').
17. *Pingoo-cho*, first-quality star rubies (literally 'spider's thread').
18. *Pingoo* or *Pingoo-sa*, silky rubies with or without a star.
19. *Gaw-done* or *Gaw-cho*, star sapphires.
20. *Anyan*, ordinary mixed water-worn spinels.
21. *Anyan-nat-thwe*, rose spinel octahedra of perfect lustre and crystallization (literally 'spinels polished by the spirits').
22. *Anyan-seinche*, tiny spinels of the same type as number 21.
23. *Nila*, large sapphires.
24. *Nila-sa*, mixed inferior sapphires.

The Burmese word for ruby is *padamya*, which means 'plenty of mercury'. Other terms for ruby are derived from the word for the seeds of the pomegranate fruit, as is the Thai word for ruby, *taubptim*.

Figure 11.13 U Hmat, the 'Ruby King' at the town of Mogok in Burma. (From O'Connor, 1905)

Traditionally, the Burmese have referred to the finest hue of ruby as 'pigeon's-blood' red (*Kothway*). Various writers have compared this colour with that of the centre of a live pigeon's eye (Brown and Dey, 1955), or a rich crimson with slight yellowish overtones. The best analogy this author can make is by comparison with the colour of a red traffic signal or stoplight. It is a glowing red colour, because of the strong red fluorescence of Burmese rubies, and is unequalled in the world of gemstones. Thai rubies may possess a more pure-red body colour, but lack this red fluorescence, usually leaving them dull by comparison. It must be stressed that the true pigeon's-blood red is extremely rare, more a colour of the mind than physical. Perhaps a Burmese trader expressed it best when he said 'asking to see the pigeon's blood is like asking to see the face of God' (Nordland, 1982).

The second-best colour in Burma is termed 'rabbit's blood', or *Yeong-twe*. It is slightly darker and more bluish red. Third best is a very deep hot pink termed *Bho-Kyeit*. This was the favourite colour of the famous Mogok gem dealer, Mr A. C. D. Paine, after whom the rare gem painite was named. Fourth best is a light-pink colour termed *Ley-Kow-Seet* (literally 'bracelet-quality' ruby). At the bottom of the ruby scale is the very dark-red colour termed *Ka-la-ngoh*. This word has a very interesting derivation for it means literally either 'crying Indian quality' or 'even an Indian would cry', so termed because these rubies were even darker than an Indian's skin. Most dark rubies were sold in Bombay or Madras, India. The *Ka-la-ngoh* stones were said to be so dark that even Indians would cry out in despair when confronted with this quality.

Burmese rubies are considered to be the finest in the world, bar none. This is mainly in reference to their rich crimson colour, which is actually a mixture of the slightly bluish-red body colour and the purer red fluorescent emission. It is this red fluorescence that is the key, for it tends to cover up the dark areas of the stone caused by extinction from cutting. Thai rubies possess a purer red body colour, but lack the strong fluorescence. In Thai rubies, where light is properly reflected off the pavilion facets (internal brilliance), the colour is good. However, where the facets are cut too steep, light leaves through the side instead of returning to the eye, creating darker areas (extinction). All stones possess this extinction to a certain degree, but in the Burmese rubies the strong red fluorescence masks it. The best Burmese stones actually glow red and appear as though mother nature brushed a broad swath of fluorescent red paint across the face of the stone. This is the 'carbuncle' of the ancients, a term derived from the glowing embers of a fire.

In fact, rubies from most sources possess a strong red fluorescence similar to those from Burma, with the Thai rubies being the exception in this regard. However, those from Sri Lanka are generally too pale in colour while with the other sources, such as Kenya, Pakistan and Afghanistan, material clean enough for faceting is very rare. Thus the combination of fine colour (body colour plus fluorescence) and facetable (i.e. clean internally) material has put the Burmese ruby squarely atop the ruby mountain. Some connoisseurs consider Burma to be not just the best source, but the *only* source of stones fit to be called ruby. When we consider that today probably 90% or more of the rubies in the market owe a good measure of their clarity and colour to heat treatment, this statement does not seem so outlandish (unfortunately, even many Burmese rubies are heat treated nowadays).

Mining areas

Information on the current mining situation in Mogok is extremely difficult to obtain. The biggest problem is that foreigners are not permitted to travel to Mogok because, owing to the ongoing rebellion, the government cannot guarantee their safety. E. J. Gübelin (1966) was one of the last outsiders to visit the area, but that was over 20 years ago. Fortunately the present author recently had a chance to discuss the current mining situation with a long-time resident of Mogok (for obvious reasons this person must remain unnamed). Recent photographs of the area were also obtained and, although not generally of good quality, they did reveal that little has changed in Mogok over the past 20 years.

Private mining has long been banned throughout Burma, but in a town of several thousand people whose sole means of income is mining gemstones, the government is forced to turn a blind eye. Just as in former times, a gem market is held once a week in Mogok at the parade grounds and traders come from far and near to attend. Cheaper goods are displayed openly, while more expensive stones are sold behind closed doors in small sheds which line the edges of the grounds.

The government operates several large mines in the Mogok area. The mine at Myintada near the town of Mogok is famous for fine-quality star rubies and sapphires, with facetable ruby and fancy spinels also being found in quantity. Near the town of

Kathe (famous for sapphires) is the government mine at Pingu Taung, where fine sapphires are found. To the west of Pingu Taung is another government mine at Kyaukpyatthat-ashe. In addition to cut sapphire and ruby, a number of other gems, and even uranium, are mined.

Illegal mining (and cutting) generally takes place in more inaccessible regions and is sometimes supported by armed rebel groups. Hot-pink spinels are mined at Kabaing, while just to the south, at Sakangyi, hot-pink rubies are obtained. Also near Kabaing, at Kin, is located a mine famous for star sapphires. The world's finest red spinels are obtained near the town of Kyatpyin, while many miles north, at Pandaw, the best pink spinels are found. These are but a few of the localities where mining is proceeding today. Rubies are found in virtually all of these localities, along with spinels and zircons.

Rubies have been found at localities in Burma other than Mogok, but these other mines did not prove profitable and were abandoned. In the Sagyin Hills, just 16 miles north of Mandalay, rubies were once obtained from clay-filled hollows and fissures in the crystalline limestones. This locality is famed for fine marble as well. Another ruby-bearing tract existed near Namseka, 15 miles south-west of Mainglon. Mining has not taken place at these localities for many years, though.

Burmese sapphires

Although it is the rubies for which Burma is justly famous, some of the world's finest blue sapphires are also mined in the Mogok area. Today the world gem trade recognizes the quality of Burmese sapphires, but this was not always the case. In the nineteenth and early twentieth centuries, Burmese sapphires were considered to be too dark for general approval. G. Herbert Smith wrote that: 'It is a typical case of giving a dog a bad name. The Burmese sapphire has been ignominiously but unjustly dismissed as of poor quality whereas, in actual fact, nowhere in the world are such superb sapphires produced as in Burma.' Although this statement must be qualified by adding that the finest Kashmir sapphires are in a class of their own, those from Burma are also magnificent. J. Coggin Brown said:

> It has been stated that Burmese sapphires as a whole are usually too dark for general approval, but this is quite incorrect; next to the Kashmir sapphires they are unsurpassed. Speaking generally, Ceylon sapphires are too light and Siamese sapphires too dark, and it is more than probable that many of the best "Ceylon" stones first saw the light of day from the mountainsides of the Mogok Stone Tract.

Not all Burmese sapphires are deep in colour; the best display a very rich, intense, slightly violetish blue, but some are quite light, similar to those from Sri Lanka. The major difference from Sri Lankan sapphires is in the saturation of colour, with those from Burma possessing much more colour in the stone. Often, colour banding, so prominent in Sri Lankan stones, is entirely absent in sapphires from Burma.

Although rubies are found with much greater frequency at Mogok (rubies form about 80–90% of the total output), the sapphires may reach larger sizes. Cut gems of over 100 carats are not unknown. Large fine star sapphires are also found at Mogok, in addition to star rubies.

The sapphires of Burma occur in intimate association with rubies in the alluvial deposits, but are found in quantity at only a few localities, particularly west of the town of Kathe. One magnificent gem mined in 1929 weighed an incredible 959 carats in the rough! It was named the 'Gem of the Jungle' and produced nine fine stones, ranging in size from 66 to 4 carats.

Other than blue, sapphires also occur in violet, purple, colourless and yellow forms at Mogok. The violet and purple stones may be very fine, but the yellows tend to be on the light side and are not common. Green sapphires are almost unknown.

In addition to corundum, the Mogok area produces a number of fine gems of other species. In this regard, Mogok is, next to Sri Lanka, probably the most prolific source of gems in the world. The following are among the species mined in the Mogok Stone Tract:

Spinel	including the world's finest red, pink and orange spinels, plus fancy colours

Zircon	yellow, green, orange and red
Garnet	pyrope, almandite, spessartite and hessonite (grossularite)
Iolite	good violetish-blue colours
Tourmaline	yellow, red, brown and orange
Chrysoberyl	alexandrite, colourless, yellow
Topaz	colourless
Danburite	fine yellow and, rarely, pink
Scapolite	colourless, pink and violet cat's eyes and faceted gems
Quartz	colourless and violet (amethyst)
Fluorite	violet
Moonstone	with distinctive rainbow or blue schiller
Apatite	green and light blue, including cat's eyes
Diopside	green, including cat's eyes
Enstatite	green
Kornerupine	green
Peridot	green, world's finest and largest, from Bernardmyo
Andalusite	orange
Fibrolite	fine blue gems, including cat's eyes
Kyanite	blue
Lazurite (lapis lazuli)	blue
Sodalite	blue
Taaffeite	colourless
Painite	dark red
Beryl	aquamarine

Bibliography

Adams, F. D. (1927) A visit to the gem districts of Ceylon & Burma. *Annual Report of Smithsonian Institution – 1926,* pp. 297–318

d'Amato, P. G. (1833) A short description of the mines of precious stones in the district of Kyatpyin, in the Kingdom of Ava. *Journal of the Asiatic Society of Bengal,* **2**, 75–76

Ball, V. (1925) *Travels in India by Jean Baptiste Tavernier*, Oxford University Press, London

Bauer, M. (1968) *Precious Stones*, Dover, New York

Brown, C. B. and Judd, J. W. (1896) The rubies of Burma and associated minerals; their mode of occurrence, origin and metamorphoses: a contribution to the history of corundum. *Philosophical Transactions of the Royal Society of London*, Series A, **187**, 151–228

Brown, J. C. (1927) *Gem mining in the Mogok Stone Tract of Upper Burma (Confidential report)*, Office of the Superintendent, Rangoon, Burma,

Brown, J. C. (1933) Ruby mining in Upper Burma. *Mining Magazine*, June, 329–340

Brown, J. C. (1958) Sapphires of Burma. *The Gemmologist,* **27**, no. 318, 1–6

Brown, J. C. and Dey, A. K. (1955) *India's Mineral Wealth*, Oxford University Press, Bombay

Chhibber, H. L. (1934) *The Mineral Resources of Burma*, Macmillan, London

Claremont, L. (1906) *The Gem-Cutter's Craft*, Bell, London

Ehrmann, M. (1957) Gem mining in Burma. *Gems and Gemology,* **19**, 3–30

Enriquez, C. M. (1923) *A Burmese Arcady*, Seeley, Service, London

Enriquez, C. M. (1934) Burma rubies. *The Gemmologist,* **4**, no. 41, 135–137

Father Sangermano (1893) *The Burmese Empire a Hundred Years Ago*, Archibald Constable, London

George, E. C. S. (1962) *Burma Gazetteer – Ruby Mines District*, vol. A, Government Printing and Stationery, Rangoon, Burma (reprinted from the original 1915 edition)

Gübelin, E. J. (1966) The ruby mines of Mogok, Burma. *Lapidary Journal*, June, 418–422; July, 522–525

Hakluyt, R. (1904) *The Principal Navigations Voyages Traffiques & Discoveries of the English Nation*, reprint, vol. V, pp. 365–449 (Caesar Fredericke); pp. 463–504 (Ralph Fitch)

Halford-Watkins, J. F. (1932a) World production of rubies. *The Gemmologist,* **1**, no. 7, 209–211

Halford-Watkins, J. F. (1932b) The ruby mines of Upper Burma. *The Gemmologist,* **1**, no. 9, 263–272; no. 11, 335–342; no. 12, 367–373

Halford-Watkins, J. F. (1935a) Burma sapphires – in defence of a much-abused name. *The Gemmologist,* **5**, no. 50, 39–43

Halford-Watkins, J. F. (1935b) Rubies and ground nuts. *The Gemmologist,* **5**, no. 51, 63–69

Halford-Watkins, J. F. (1935c) Burma sapphires – locations and characteristics. *The Gemmologist,* **5**, no. 52, 89–98

Hall, D. G. E. (1968) *Early English Intercourse with Burma*, Frank Cass, London

Hamilton, A. (1744) *A New Account of the East Indies*, London

Iyer, L. A. N. (1953) The geology and gem-stones of the Mogok Stone Tract, Burma. *Memoirs of the Geological Survey of India,* **82**, 1–100

Keely, H. H. (1981) The ruby mines of Burma. *Gems,* **14**, nos 3–6, 6–8, 8–11, 10–13, 12–14

Keller, P. C. (1983) The rubies of Burma: a review of the Mogok Stone Tract. *Gems and Gemology,* **19**, 209–219

Kessel, J. (1960) *Mogok – The Valley of Rubies*, Macgibbon & Kee, London

Kunz, G. F. (1913) *The Curious Lore of Precious Stones*, Dover, New York

Noetling, F. (1891) Note on the reported Namseka ruby-mine in the Mainglon State. *Records, Geological Survey of India,* **24**, part 2, 119–125
Nordland, R. (1982) On the treacherous trail to the rare ruby red. *Asia*, October, 35–42
O'Connor, V. C. (1905) *The Silken East*, Dodd, Mead, New York
O'Connor, V. C. (1907) *Mandalay and Other Cities of the Past in Burma*, Hutchinson, London
Simpson, R. R. (1922) Notes on a visit to the Burma ruby mines. *Transactions of the Mining and Geological Institute of India,* **17**, part 1, 42–58
Spaulding, D. L. (1956) The ruby mines of Mogok, Burma. *Gems and Gemology,* **8**, 335–342
Stewart, A. T. Q. (no date) *The Pagoda War*, Faber & Faber, London
Streeter, E. W. (1892) *Precious Stones and Gems*, 5th edn, Bell, London
Streeter, G. S. (1887) Burma's Ruby mines. *Murray's Magazine,* **1**, 669–678
Symes, M. (1800) *Account of an Embassy to the Kingdom of Ava*, Debrett, London
Temple, R. C. (1928) *The Itinerary of Ludovico di Varthema of Bologna from 1502 to 1508*, Hakluyt Society, London
Wynne, T. T. (1897) The ruby mines of Burma. *Transactions of the Institute of Mining & Metallurgy,* **5**, 161–175
Yule, H. (1968) *Narrative of the Mission to the Court of Ava in 1855*, Oxford University Press, London (reprint)

Thailand and Cambodia

The rubies and sapphires of this region are principally derived from an area along the Thai/Cambodian border, with mines lying on both sides of the frontier. For this reason, the deposits of both countries will be discussed together. References to the mines in eastern Thailand should also be understood to include those across the border in Cambodia.

History

Although early references to the corundum deposits of Thailand (or Siam, as it was previously known) do exist, most travellers stated that the stones found (ruby and blue sapphire) were of low quality when compared with their Burmese counterparts. One of the first to make mention of gems from Thailand was the Chinese traveller, Ma Huan, in AD 1408 (Guhler, 1947):

> A hundred li (20 miles) to the southwest of this Kingdom (Siam) there is a trading place Shang-Shwei, which is on the road to Yun-hou-men, Wen hau-mun (possibly a canal between Chanthaburi and Trat Provinces in eastern Thailand). In this place there are 500–600 foreign families who sell all kinds of foreign goods; many "Hung-masz-ken-ti" stones are sold there. This stone is an inferior kind of ruby, bright and clear like the seeds of the pomegranate. (The Thai word for ruby, *tubtim*, means pomegranate.)

Nicolas Gervaise, writing in 1688 (Gervaise, 1928), mentions what may have been an occurrence of sapphire in Siam: 'There are blue stones found in certain parts of the jungle in the uplands. This stone resembles the Lapis which is usually found in gold mines.' Gervaise, however, seemed more interested in gold than precious stones, as this is his only reference to gems.

Ulrich Guhler has given us an authoritative account of the history of gems in Thailand, unfortunately in an obscure Thai journal (Guhler, 1947), beginning with Ma Huan. From the fifteenth century onwards, various travellers to Thailand mention in passing that rubies and sapphires are found near Chanthaburi. Also included are the sometimes humorous descriptions of the gems, as well as the local people and their customs. Witness the statements of early French missionaries to Thailand (Cartwright, 1908) from before 1770 on the former Thai province of Laos: 'In the province of Laos from whence the kingdom takes its name, there is a deep mine whence rubies and emeralds are extracted. The King possesses an emerald of the size of an ordinary orange.' Regarding the local males, the rather straight-laced priests said this: 'Unbridled in their desires for the opposite sex, they seem to live merely for reproduction.'

Not only did Laotians receive the scorn of the European clergy of the day, but also that of the Burmese: 'Generally speaking, they [Burmese women] are gentle in their way,

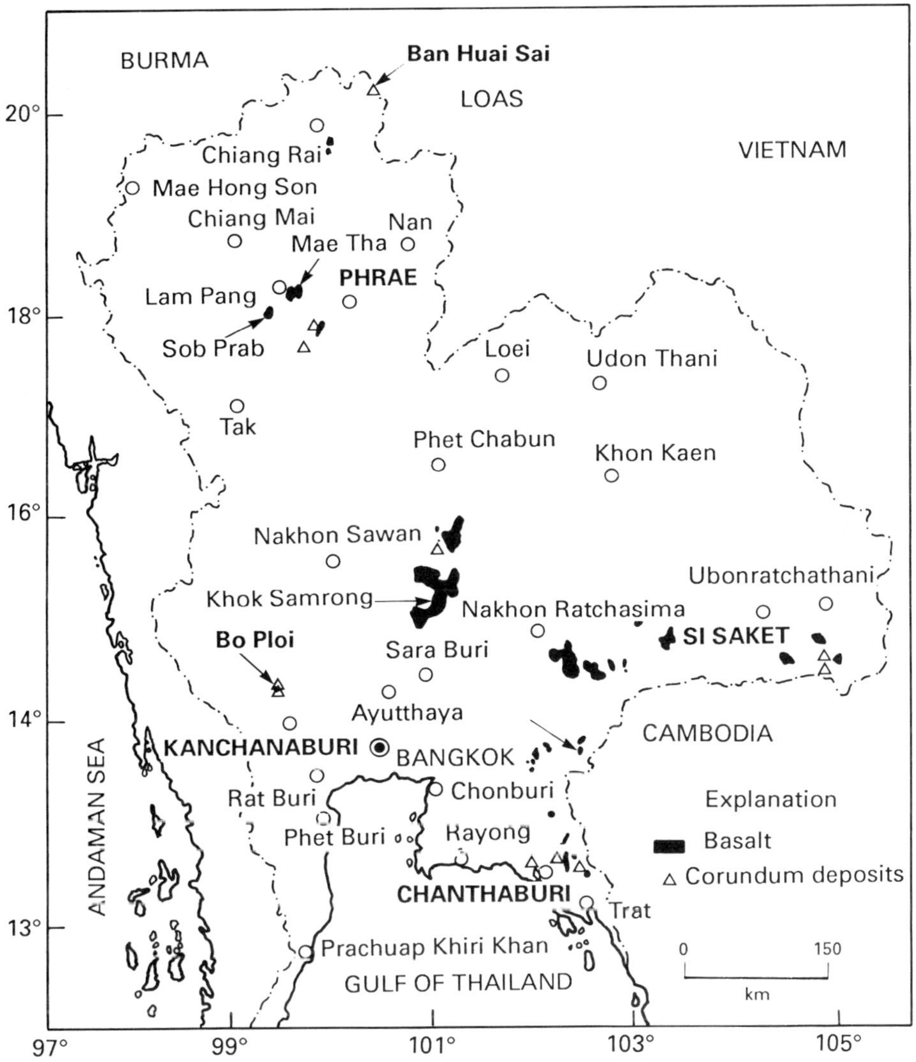

Figure 11.14 The corundum-producing localities of Thailand. Major mining districts include: Chanthaburi/Trat, Kanchanaburi (Bo Ploi), Phrae, Si Saket and Ban Huai Sai (in Laos). (After Vichit, 1978)

but very voluptuous. Lewd and licentious, they have quite abandoned all sense of shame.'

With regard to doing business in south-east Asia, a quaint custom is described that tourists everywhere might wish to revive: 'If someone is convicted of breach of trust, the creditor is allowed by law to seize their wives, children and slaves and to expose them to the glare of the sun at his door.' Certainly many Burmese and Laotians might wish the same upon those French missionaries after reading the previous excerpts. But we have digressed somewhat from the subject at hand.

De la Loubere (1693), French envoy to Siam in 1687, also describes Thailand as 'abounding in mines of rubies and sapphires'. He added that the stones usually found their way into the possession of the monks who were very secretive as to where they came from and who employed them as charms. This is probably because their value as jewels was not fully appreciated in the absence of trained lapidaries. However, for a more detailed description we must wait until the nineteenth century when John Crawfurd, British envoy to Siam and Cochin China (1828), provided greater detail, much of it accurate:

> The only gems which are ascertained to be minerals of Siam, are the sapphire, the Oriental ruby, and the Oriental topaz [yellow sapphire]. These are all found in the hills of Chan-ta-bun [Chanthaburi], about the latitude of 12 degrees, and on the eastern side of the Gulf. The gems, from what we could learn, are obtained by digging up the alluvial soil at the bottom of the hills, and washing it. The gravel obtained after this operation is brought to the capital for examination. Both the ruby and sapphire of Siam are greatly inferior in quality to those of Ava (Burma). Several specimens were shown to us during our stay, but none of them of any value. The mines of Chan-ta-bun, not withstanding this, are a rigidly guarded monopoly on the part of the King.

Pallegoix (1854) also gave a quite descriptive account of the mines:

> Precious stones are found without doubt at several places in the Kingdom of Siam, as, when travelling, I often came across them in the beds of streams and amongst the gravels of rivers, but nowhere are there so many as in the province of Chantaboon. The Chinese, who plant pepper all around the large mountain of Sabab [Khao Sabab, the large mountain which dominates the view from the town of Chanthaburi, sometimes incorrectly referred to as the mountain of precious stones, Khao Ploi Waen] collect them in quantities. In the high mountains, which surround the habitations of the Xong tribe, and in the six hills West of the town these stones are hidden in such quantities that the planters of tobacco and of sugarcane, who have established themselves at the foot of these hills, sell them by the pound: those of the smallest size cost 16 francs per pound, those of medium size 30 francs and those of the largest size 60 francs. The principal stones, which the Governor of Chantaboon has shown me, are topazes [yellow sapphires], hyacinths [zircons], garnets, sapphires of a deep blue and rubies of various tints. One day I went with a number of our Christians for a walk through the hills in the neighborhood of Chantaboon and I found scattered over them black and greenish stones, semi-transparent [corundum], amongst which were garnets and rubies; within one hour we had collected two handfuls of them. As there are no lapidaries in the country, the inhabitants, who have collected some precious stones while planting their tobacco or sugarcane, do not know what to do with them and sell them at low prices to travelling Chinese traders, who forward them on to China. Yet it must be noted that the King of Siam has reserved for himself certain loçalities where the best stones occur in greatest abundance; the Governor of Chantaboon is charged with the exploitation and sees to it that the stones reach the palace, where some second-rate Malay lapidaries polish them and cut them into brilliants.

Pallegoix's description is quite complete and accurately sums up the situation during the first half of the nineteenth century. Sometime about 1857, however, Shan traders from Mogok, Burma, 'rediscovered' the mines of Chanthaburi, starting a gem rush which has continued to this very day. It is said that the rush began after a certain Nai Wong went fishing one day. Instead of prawns and fish, however, he brought up in his net a gravel which contained many rubies. These were exchanged for clothes which he obtained from the Gula (Shan) people and the rush was on (Guhler, 1947). Thousands of Burmese Shans from Mogok, who are actually ethnically related to the Thais of Siam, made their way to Chanthaburi, Trat and Pailin, which was up until 1907 part of Thailand. It was the Shans, or 'Gulas' as they were called, who developed the mines, and the descendants of these original immigrants even today hold prominent positions in the gem trade at Chanthaburi.

The best description of this period in

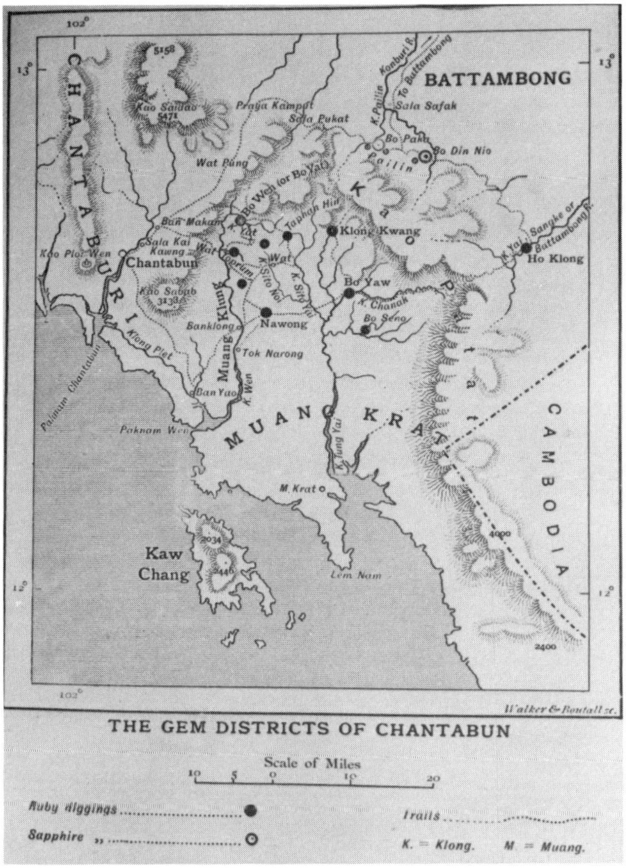

Figure 11.15 Map of the gem districts of Chanthaburi (Chantabun) and Trat (Krat), Thailand. (From Warington Smyth, 1898)

Chanthaburi's history is that of H. Warington Smyth's *Five Years in Siam* (1898). Warington Smyth, an Englishman, was employed by the Department of Mines of Siam in the late nineteenth century and gives an authoritative account of the mining areas in Chanthaburi, Trat ('Krat') and, in what is now Cambodia, at Pailin.

Shan speculators leased the mines from the Thai government and brought in their own men from Burma to work them. However, there was little success until, in 1880, a financial genius named Mong Keng arose amongst them. Gathering all the mines into his hands and gaining control of the opium and gambling monopolies too, he worked the mines profitably for many years, eventually becoming known as the 'king of precious stones'. With his fame also spread the fame of the blue sapphires from Pailin, not only in Siam, but also to India and Europe. In 1895, however, the government made a concession of parts of the district to a British firm, turning the 'Sapphire King' into a pauper almost overnight. Then, in 1907, Battambang Province, containing the valuable Pailin mines, was ceded to the French in Cambodia by the Thai government. Up until this time, Pailin supplied as much as 90% of the world's fine sapphires. Although most of the production continued to pass through Chanthaburi and Bangkok, after 1907 Thailand's production of sapphires dropped dramatically because of the loss of Pailin. The eventual discovery of new sapphire mines in Kanchanaburi and Phrae helped somewhat, but in losing Pailin Thailand lost a major portion of its sapphire-mining potential. The decline of the Burmese ruby and the subsequent rise of the Thai ruby, however, has ensured the importance of Thailand and Chanthaburi as a trading centre for rubies and sapphires.

Figure 11.16 A Gula (Shan) horseman at Pailin, Cambodia. (From Warington Smyth, 1898)

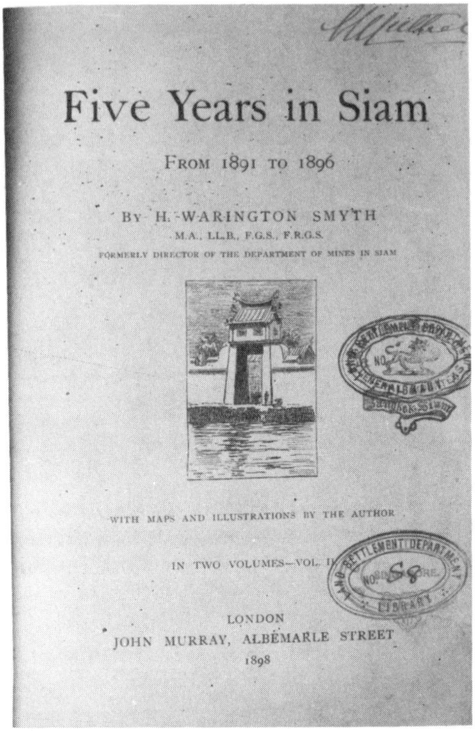

Five Years in Siam

From 1891 to 1896

By H. Warington Smyth
M.A., LL.B., F.G.S., F.R.G.S.
Formerly Director of the Department of Mines in Siam

With maps and illustrations by the author

In two volumes—Vol. I

London
John Murray, Albemarle Street
1898

Figure 11.17 Title page of H. Warington Smyth's *Five Years in Siam* (1898). This two-volume work contains the best nineteenth-century account of the ruby and sapphire mines of Thailand/Cambodia

H. Warington Smyth (1898) has provided us with a fascinating picture of mining in Chanthaburi and Pailin during this period:

> The Shan seems by nature designed for the pursuit of gems. He is bitten with the roving spirit, and in addition he has the true instinct of the miner, to whom the mineral he lives to pursue possesses a subtle charm, which constrains him never to rest or weary of its search against all odds. He detects colour and recognizes quality with a rapidity and accuracy to which few attain. No Siamese, no Lao, no Chinaman can compete with him. The Burman is about his equal, but has not his industry or constitution, and is therefore chiefly found in the capacity of middleman, buying and exporting.
>
> At first whole parties were decimated by the fever. Not one in thirty returned to the sea alive. But there were others to take their place, and gradually the clearing of the jungle, the improvement of communications, the superior shelter, and the comforts which were introduced had their effect; and although the mines still have an evil name, and the opening up of each new district calls for further sacrifices, the rate of mortality among the Shans is now comparatively moderate.
>
> Should the Shans therefore leave for any reason, there is no other force which can be utilized to do the necessary work. The departure of these people would doom the mines. They are necessary, if only to bury the others [other nationalities].
>
> The Gula [Shan] digger is proud and independent. He cherishes the freedom of his life, and he brooks not much official interference. Restraints which may be applied to the African negro will not do for him. When the [British] Company came to Nawong, diggers were notified (inter alia) that if they worked on the company's territory they must sell stones

Figure 11.18 Miners' village at Na Wong, Chanthaburi/Trat, Thailand. (From Warington Smyth, 1898)

to its agents, at prices to be settled by the latter. This was felt to be an infringement of the right to sell in the open market, and was resented. Thereafter the company attempted to enforce the right of search on the persons and belongings of the diggers. Rather than submit they left en masse, some for the Pailin mines, others for their homes in Burma. The result was that, out of two thousand eight hundred diggers whom the company found in and about Nawong, a couple hundred were left when we visited the place; and at Bo Yaw, instead of twelve to fifteen hundred men, there were fifty-four at our visit.

Thus the Gula will be seen to be an individual whose susceptibilities must be given some consideration.

Warington Smyth also discusses the mining methods in some detail, as well as providing a fascinating account of the buying of gemstones:

In buying stones, it is well to remember you are entering the arena to pit your knowledge against the other man's. He regards it as a sporting contest, and he would fleece or ruin his dearest friend as part of the game, if his dearest friend were tool enough to let him do it. The moment you bargain about a stone, a trial of strength (or knowledge) has begun between you, as much as if you were boxing or fencing. Everything is forgotten but the object in view, to protect yourself and get home on the person of your opponent.

Your eye must be in training too, and if you have not looked at a stone for six months, get some one else to do your buying for you, or you will be badly hit.

All in all, the above is one of the best descriptions of the bargaining process known to the present author and is typical of the accurate and informative commentary provided in the writings of H. Warington Smyth. His respect for and love of the peoples of this region shine through clearly, providing a pleasant contrast to the racist diatribes of so many European writers of his day.

As mentioned previously, upon the change-over of Pailin from Thai to Cambodian (French) rule, Thailand lost her most important gem mines. The traditional trading channels, however, remained unchanged and most of Pailin's production continued to pass through Chanthaburi and Bangkok. This was true up until 1975, when the rise of the bloody Khmer Rouge regime in Cambodia (today called Kampuchea) halted all mining activity for several years. With the installation of the Vietnamese-sponsored government in 1979, mining was again started. Today the mines at Pailin are said to be operated by Soviet technicians and it is believed that the entire production (other than that pilferred by the local workers) is being shipped to Moscow to pay for the Vietnamese war debt. Rumour has it that a large French dealer has made an exclusive arrangement with the Soviets to market the production from Pailin, but evidently the

Soviets have decided to keep all of the larger sizes for the time being, only selling the sub-carat stones to the French firm.

The war in Cambodia has inflicted tremendous suffering on civilians in Vietnam, with food and goods of all kinds being in very short supply. This has resulted in numerous jokes making the rounds in Bangkok among the many foreign aid personnel stationed there. The following is a sample, courtesy of Bangkok writer, John Hoskin:

> Vietnamese cable to Moscow: 'Please send aid!'
> Soviet reply: 'Please tighten belts!'
> Vietnamese answer to Moscow: 'Please send belts!'

With Pailin gone, the ruby and sapphire mines of Chanthaburi and Trat declined in importance. This was mainly due to the low quality of the gems being produced, with the sapphires being mostly too dark and the rubies lacking the rich fluorescent red of Burmese stones. However, in the early 1960s a change of government in Burma and subsequent nationalization of the mines caused a drop in production at the famous Mogok Stone Tract. Suddenly the world was starved of ruby and the Thai stones immediately began to look much better in the eyes of the trade. Although other sources, such as Sri Lanka, at times produced rubies of better colour, none could produce large enough quantities of facetable stones except Thailand. In addition, it was learned that Thai rubies could be greatly enhanced in terms of their colour by a relatively simple heat-treatment process. The result was that Thailand soon supplanted Burma as the world's major supplier of faceted ruby, a situation which has continued to the present. Today the ruby market finds itself in what some might consider a very unusual situation. Fully 95% of all faceted rubies originate from a source which was roundly condemned by every major writer on the subject 50 years ago as producing stones of inferior colour! With the prevalence of heat-treated rubies and sapphires, the trade faces a very real danger of losing sight of what the natural stone looks like. Soon it may become impossible to sell the genuine unadulterated article. For those who think this is unlikely, we need only refer to the decline of the natural pearl market to see that it is possible.

Ulrich Guhler (1947) has summed up best the differences between Thai and Burmese rubies:

> The color of the Siamese ruby is inferior to that of the Burmese ruby. The color tends to be purplish or even brownish and generally it is too dark. A Siamese ruby inspected under natural or artificial light will show its color only on those facets which are directly hit by the light. Those facets in the shade [today termed 'extinction'] will appear almost black. A Burmese ruby will appear red throughout, even on those facets which are not directly exposed to the source of light. It goes without saying that on the markets of the world a Siamese ruby is less valued than a Burmese ruby. It should, however, be stated that there are sometimes found at Chanthaboon [Chanthaburi] rubies which are entirely similar to those of the famous Mogok mine in Burma.

Although Thailand does occasionally produce stones of fine colour *before* heat treatment, the vast majority are lacklustre stones with blackish overtones because of the lack of fluorescence. Heat treatment does improve them to a degree, as evidenced by the fact that some sell for as much as \$10 000 per carat or more. At times this author suffers pangs of nostalgia, yearning for those days, long since past, when gemstones were simply dug from the ground, cut and polished, and then set, creating beautiful jewellery.

In the year 1918 blue sapphires were discovered near the town of Kanchanaburi in western Thailand (Graham, 1924). This created a minor gem rush, and the small town of Bo Ploi was founded near the mines. It seems, though, that not much was found in the first few years of mining (Guhler, 1947). At the beginning of 1929, however, rich finds were made in the northern part of the field, with the output exceeding \$3000 during the first months of the year. Over 100 sapphires found, weighing more than 3300 carats, were valued at \$4250. Later the mines were abandoned, believed to be exhausted. They were not started up again until well after World War II. Today, several hundred people are engaged in mining at this locality, which has become Thailand's major source of blue sapphires, especially large stones of 10 carats or more.

Small amounts of deep-blue sapphire are also produced in Chanthaburi and Trat provinces, particularly near the town of Bo I-Ram. The third major sapphire-mining area in Thailand is at Huai Mae Sung and Ban Bo Kaeo in Phrae Province in the north. Mining has taken place here since the 1970s, but the deposits have been known since before 1927.

In 1890 some Shan diggers discovered black spinel, considered to be an accessory mineral of corundum, at Chiang Kong (Ban Huai Sai) which lies just across the Mekong River from Thailand, in Laos. With their typical patience they carefully prospected the area and eventually turned up some sapphires. Mining then proceeded but soon stopped due to the low quality of the stones found.

A Thai official at Chieng Kong (Kawng) discovered some small red fragments of garnet at the mines. Taking them to be rubies, he thought that he would obtain royal support for the mining, so he went out and bought the best ruby he could afford. This was duly sent to the king with a note stating that it was found by the aforementioned official at Chieng Kong. The king dispatched a team of geologists led by H. Warington Smyth to report on the mines. Suspicion was soon aroused, however, as all the miners claimed never to have found even a single ruby in the area. Thus, the report sent back to Bangkok did absolutely nothing to advance the career of the official at Chieng Kong (Warington Smyth, 1898).

Today, mining is proceeding at Ban Huai Sai (Chiang Khong) in Laos, just across the Mekong River from Thailand. Sapphires of a dark-blue colour are found there in a deposit believed to be derived from a small basalt plug (Vichit *et al.*, 1978). The material is said to be somewhat reminiscent of that found in Pailin and Chanthaburi/Trat, with most being overly dark. Yellow, green and black star sapphires are found, in addition to the blue. Generally the material is smuggled across the Mekong River for sale in Thailand.

Gem corundum is also being mined in north-eastern Thailand near the town of Si Sa Ket. The mines are located at Ban Ta Koi, Ban Khok Sa-Ard, Plon Thung Yao and Huai Ta-Ak, all of which lie close to the Cambodian border. Because of their proximity to the border, the area has been declared a restricted zone by the Thai military; this has hampered mining activity somewhat, but not stopped it. Some mines are of the simple pit type while others are mechanized. As with other basalt-derived corundum deposits, the suite of gems found here consists of dark-blue, green and yellow sapphires, as well as dark-red zircons (which can be heat treated to blue or colourless), red garnets and black spinels. Small quantities of ruby have also been found. As the corundum deposits of Chanthaburi and Trat are becoming exhausted, more and more miners are moving up to Si Sa Ket to try their luck there.

Mining areas

Gem-quality corundums are found at three major localities in Thailand and Cambodia. They are as follows:

1. Chanthaburi and Trat Provinces in Thailand and the nearby Pailin District in Cambodia.
2. Bo Ploi District, Kanchanaburi Province, western Thailand.
3. Den Chai District, Phrae Province, northern Thailand.

Chanthaburi, Trat and Pailin

The corundum mines of this region can be divided into four areas based upon the colours found. At Khao Ploi Waen (the 'hill of gems') and Bang Kha Cha, blue, green, yellow and black star sapphires are found in small amounts. Ruby is entirely absent. Khao Ploi Waen is located just 6.5 km west of Chanthaburi and consists of placers derived from an isolated butte some 70 m above the surrounding plains. This is said to have been the first place in Thailand where corundum was found. The hill consists of a volcanic plug and the sapphires are mined from placers derived from the lava. Mines are scattered all around the base of the hill. Most are simple pit mines, but recently mechanized mines have been started, too.

The blue sapphires found here are of a pure blue colour, but suffer from being very

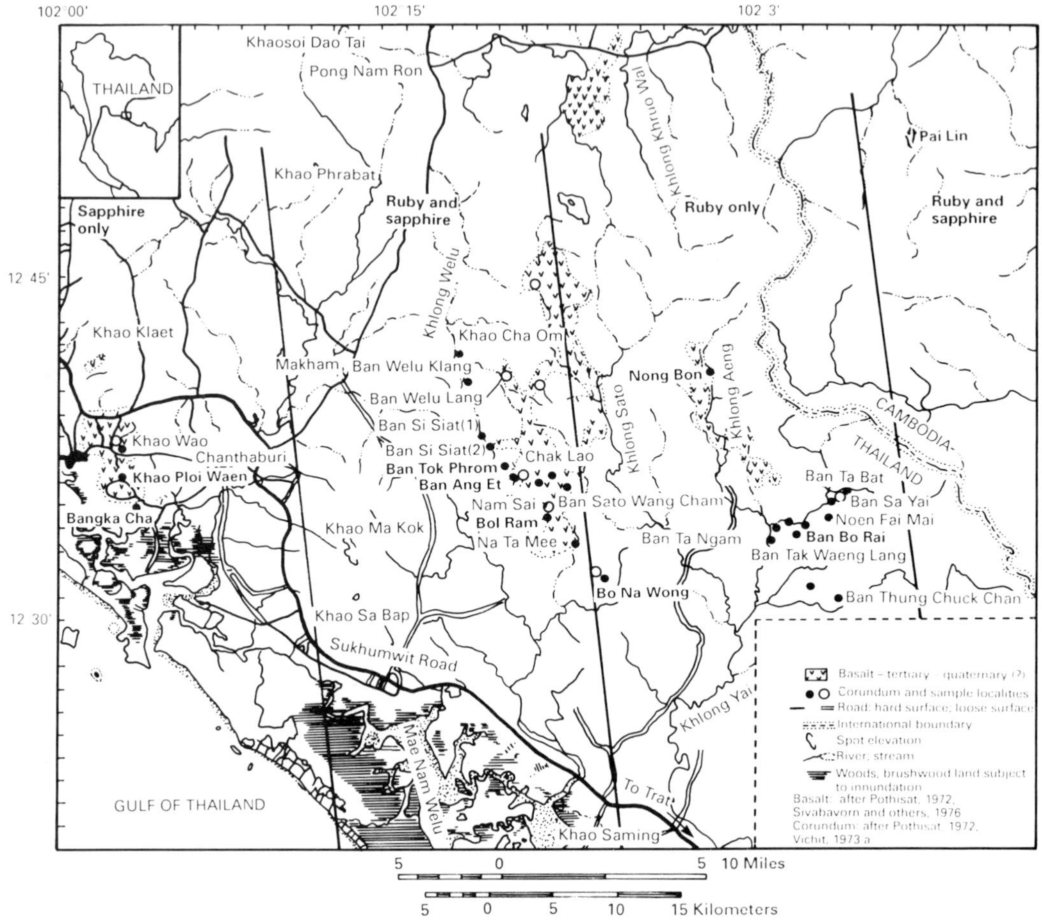

Figure 11.19 The corundum localities in Chanthaburi and Trat Provinces, Thailand and Battambang Province, Cambodia. Major mines are located at: Khao Ploi Waen and Bang Ka Cha (sapphire), Ban Tok Phrom and Bo I Ram (ruby and sapphire), Bo Na Wong (ruby), Nong Bon (ruby), Ban Bo Rai (ruby) and Pailin, Cambodia (ruby and sapphire)

dark. Good black star (many with 12-rayed stars) and green sapphires are found, but the mine is most famous for producing yellow sapphires of a characteristic 'Mekong Whisky' golden-yellow to orange colour. This colour is much in demand in the local market and the stones sell for as much as $500 per carat or more. The rare 'golden-star' black star sapphires are also found here and fetch very good prices locally.

The placers of Bang Kha Cha are located but a few kilometres from Khao Ploi Waen and produce an identical suite of sapphire colours. They are situated on a flat swampy plain on the Khlong Hin River estuary, which is underlain by alluvium. Not only are the swamps mined, but also the mud from the bottom of the river. In addition to sapphires, zircon and pyrope garnets are also found, as at Khao Ploi Waen.

The second major corundum-producing zone is found behind the granite mountain of Khao Sabab, which dominates the view from Chanthaburi. These mines are right on the border between Chanthaburi and Trat Provinces. The following are the major mines of this area:

Bo Waen	ruby only
Bo Na Wong	ruby only

Wat Tok Phrom	ruby only
Ban Bo I-Ram	mainly blue sapphire

Although transportation and roads to this area are poor, a number of large mechanized mines are found. Mostly ruby is found, with the exception of I-Ram, where sapphires dominate. The sapphires are of a very deep but pure blue colour and resemble somewhat those from nearby Pailin. This material is good for cutting small sizes (½ carat or less) because it holds the colour very well. As with the other mines in Thailand, the deposits are alluvial or eluvial in nature. The combined output of ruby from this district forms a significant portion of Thailand's ruby exports.

The third major area is further to the east, in Trat Province, and includes the very large and well-known mechanized mines at Nong Bon and Bo Rai. These are the largest mines in Thailand and are accessible by good roads. Strictly ruby is found here, with larger stones found at Nong Bon, and finer colours coming from Bo Rai. Both Bo Rai and Nong Bon lie less than 10 km from the Cambodian border, and with the war in Cambodia shelling is often clearly audible. It is estimated that, at times, as many as 10 000 people daily make the journey across the border to mine rubies inside Cambodia, as the deposits there have not been so heavily worked as those inside Thailand. This is not without its dangers, however. Roving bandit gangs and Vietnamese army patrols take a steady toll of miners. Those that are not killed outright may be seriously injured by landmines and booby traps set up along the border. Were this not enough to contend with, the area is also thickly jungled and home to the world's worst malaria. The mines at Bo Rai and Nong Bon on the Thai side of the border are mechanized, with bulldozers, earthmovers and other heavy equipment in use; those across the border in Cambodia, however, are primitive pit mines only, due to the dangers just described.

In terms of stone sizes, Thai rubies, like all others, are found in much smaller sizes than sapphires. The largest fine ruby mentioned by Guhler (1947) was a faceted gem of 10.25 carats that was named the 'Star of Chanthaboon'. With the tremendous increase in mining in recent years, larger gems have been unearthed, including several of fine quality in the 10 carat plus range. The biggest ever found in the area was a 150 carat giant (in the rough) unearthed in 1985 in Trat Province. Due to the even coloration of Thai rubies, however, most of the gems above 5 carats (cut) tend to be overly dark.

Further east, across the border in Cambodia, are the famous Pailin gemfields. The lavas here form four separate hills around which the mines are situated. Westernmost are those of Phnum O Tang and Phnum Ko Ngoap, rising about 40–60 m above the surrounding plain (Jobbins and Berrange, 1981). These two areas produce primarily ruby, with small amounts of sapphire. The rubies here are virtually identical to those found nearby in Thailand. To the east, at Phnum Yat near the small town of Pailin, the situation is reversed with sapphires

Figure 11.20 A large-scale ruby-mining operation at Nong Bon, Trat Province, Thailand. (Photograph by the author)

Figure 11.21 The ruby-mining town of Bo Rai in Trat Province, Thailand. This is one of the major ruby-producing mines in Thailand. The mountains in the distance form the border between Thailand and Cambodia. (Photograph by Wimon Manorotkul, AIGS, Bangkok)

dominating. It is from this locality that most of the famous Pailin sapphires have been unearthed. The fourth lava outcrop, located in a coffee plantation, is only 200 m in diameter and is thought to be a volcanic pipe.

One interesting feature of the corundums from the Pailin area is the virtual absence of colourless, yellow and green sapphires. Local diggers only occasionally find a yellow stone; the others are not found at all (Jobbins and Berrange, 1981). Colour-change sapphires are found with some frequency, as in Thailand, and can often be had for a song as they are mixed into parcels of inferior rubies. These stones appear a light- to deep-greenish violet in daylight and change to a purplish pink in incandescent light, similar to those from Umba.

Although the rubies from Pailin are of good quality after heat treatment (similar to those from Thailand), it is the blue sapphires that are its boast. In fact, the word for blue sapphire in the Thai language is *pailin*, its name being taken from this important locality. Ranging in colour from a medium to deep blue, in the best cases the blue sapphires may be amongst the world's finest, with only those from Kashmir and Burma being better. Pailin sapphires strongly resemble those found at Bo I Ram in Thailand, except that the latter tend to be too dark. This is also the major flaw with most Pailin stones; it does mean, however, that small stones (below 0.50 carats) hold their colour very well. Star sapphires have been found at Pailin but are extremely rare. Guhler (1947) says the following about Pailin sapphires:

> Best colors, often with reddish tints. Darker colors are predominant. The gems are inclined to lose color in artificial light and appear rather dark. Flaws are frequent. No star sapphires are known from this location. Large gems are rare. The largest fine stone weighed 18 carats. It was sold for $600.

Since 1975, little mining has taken place in the Pailin area. Recently Soviet technicians have reportedly been brought in to work the mines, with all production not stolen being shipped to Moscow, as mentioned earlier. Zircon and pyrope garnets are found, in addition to the corundums.

Bo Ploi, Kanchanaburi

With the ceding of the Pailin mines to the French (Cambodia) in 1907, Thailand lost its major source of fine-quality blue sapphires. This has been compensated for, to some degree, by the output from Bo Ploi, in Kanchanaburi Province, better known as the site of the famous 'Bridge on the River Kwai'.

Discovered in 1919, the mines of Bo Ploi are today Thailand's major producer of blue

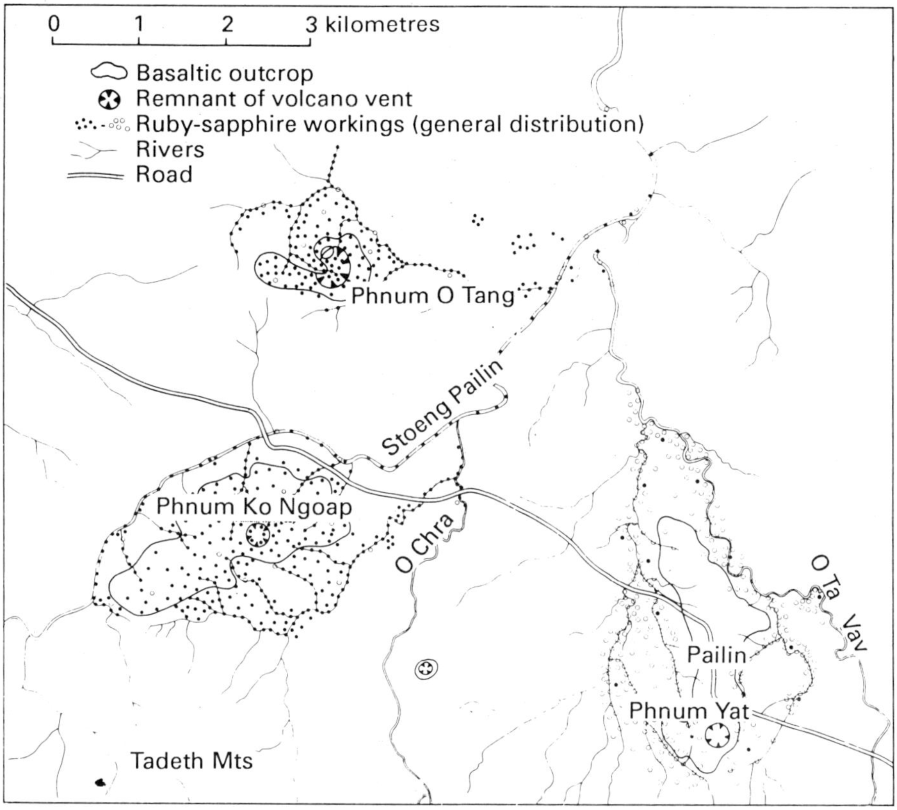

Figure 11.22 The distribution of ruby and sapphire mines at Pailin, Cambodia. (From Jobbins, 1981)

sapphires. They lie at a distance of about 30 km north of Kanchanaburi city and consist of a group of low basaltic hills which are covered by large boulders. Weathered basalt forms a characteristic red soil which has accumulated at the foot of the hills; in this soil are found black spinels (magnetite – strongly magnetic), locally termed 'nin' and faceted into inexpensive gems. Although sapphires have rarely been found attached to the basalt matrix, most are mined from eluvial and alluvial deposits.

In 1978 a new deposit of sapphires was found 6 km north of Bo Ploi, at the village of Ban Chong Dan. As of late 1986, most work was continuing in this area.

The mines of Bo Ploi and its surroundings produce mostly blue sapphire, with the occasional yellow, pink and star sapphire being found. Guhler (1947) describes them as follows:

> Very transparent [actually they are usually somewhat milky] with perfect color; reddish tints are rare. Lighter colors are more frequent. Large specimens are frequent. Rather less flaws than Pailin, except zonar striations which are very common and frequently seen in hexagonal arrangement. Star sapphires of poor quality are sometimes found.

Kanchanaburi sapphires strongly resemble the heat-treated blue sapphires from Sri Lanka, particularly with regard to their strong colour zoning. Because the market demand for Sri Lankan sapphires is greatest, most Kanchanaburi stones end up being mixed into parcels of Sri Lankan stones and sold as the latter. This is indeed a shame, for the stones from Bo Ploi are quite beautiful in their own right. Were it not for the slight milkiness that they always display, these

Thai stones would be the equal of many from Pailin or even Burma. As mentioned, large sizes (20 carats or greater) are frequently found at Bo Ploi. The author has seen several fine pieces larger than 20 carats and Guhler (1947) mentioned one faceted stone of fine colour and fancy shape which weighed 51 carats. Others of top quality cut from a single crystal weighed 28 and 16 carats respectively. The rough which produced these two stones was originally bought for $600, a not inconsequential sum of money 40 years ago.

The Bo Ploi stones are easily separated from those of Sri Lanka and other localities by reference to the concentrations of minute exsolved particles always contained within the stones. These particles follow exactly the hexagonal colour zoning of the stones and are most heavily concentrated where the colour is deepest; colourless areas are generally free of these particles which impart a distinctive 'milky and silky' look to the stones. It is under this 'milky and silky' moniker that they are informally known in Thai gem circles. The particle concentra-

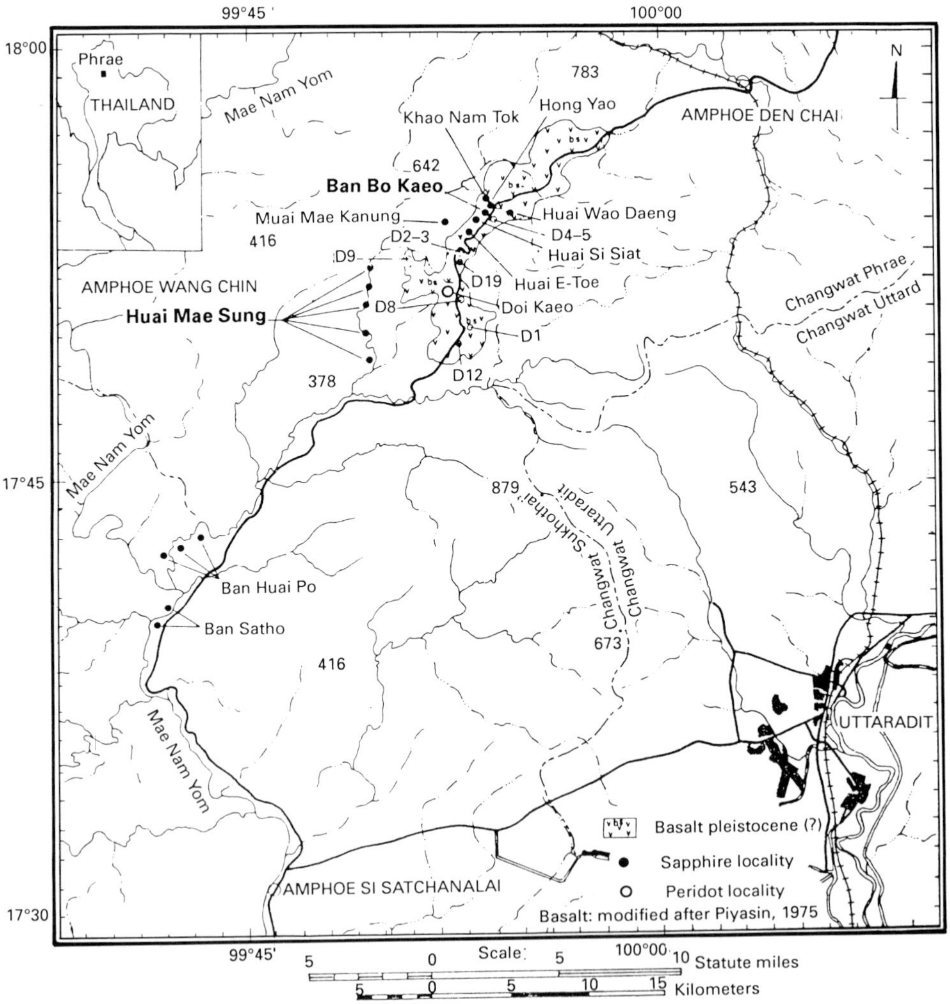

Figure 11.23 Corundum localities in Phrae Province, northern Thailand. Major sapphire mines are located at Ban Bo Kaeo and Huai Mae Sung. (After Vichit, 1979)

tions are not made up of actual rutile silk; thus star sapphires are very rare. Other than corundums, the only additional facetable gem mineral found at Bo Ploi is the black spinel ('nin').

Phrae Province, northern Thailand

The corundum deposits of Phrae in northern Thailand are located in Den Chai District and, although known since the 1920s, have only been worked since the 1970s. Phrae basalt is the largest of the corundum-bearing lavas in Thailand and is bisected by the Nam Yom River. Deposits are found on both banks, with the most important at Ban Bo Kaeo and Mae Sung. During the author's visit to the area in 1984, the mines of Bo Kaeo were found abandoned, with work proceeding only at Mae Sung. All deposits in the area consist of recent alluvium in the streams. Blue and green sapphires are found associated with black spinel, sanidine, augite, olivine and zircon. Only the sapphires are of economic importance.

The blue sapphires of Phrae are of a very deep inky-blue colour which sometimes is slightly violet. Large sizes (5 carats or more) are very rare, but the deep colour does lend itself to cutting melee. Green sapphires are also found, as well as the occasional poor quality ruby and red spinel. Of the various corundum mines in Thailand currently being worked, Phrae is among the least important in terms of production, but the occasional fine blue sapphire turns up every now and then. The best Phrae sapphires have a distinctive deep blue colour that verges on the violet, with blackish overtones. Most, however, are too dark in colour.

Mining methods

There are two basic types of mining utilized for corundum in the region: primitive pit mining and mechanized mining.

The earliest and simplest type of mining is pit mining. These mines can be found to a certain extent at all localities, particularly those producing sapphires. In this type, a narrow circular pit just 1 m in diameter is sunk down to the gem-bearing gravel, which is termed 'kee ploi' and in most localities consists of small round nodules of a hardened clay. The gem-bearing layers are in some places just a metre or so beneath the surface while in others they may be as deep as 20 m or more. They are deepest at Bo Ploi in Kanchanaburi, and so the pits are larger in size and shored up with strips of bamboo to prevent cave-ins. Usually steps are cut into the pits' sides to allow access, but at Bo Ploi bamboo ladders are used.

One worker will remove the earth at the bottom of the pit and another above will man the bamboo lift, a simple type of crane which brings the earth to the surface. After sufficient quantities have been brought up, it will be taken to a small man-made pond for washing.

Figure 11.24 'I've found one!' A sapphire miner at Khao Ploi Waen (Chanthaburi Province), Thailand, holds up the gem she has just found. (Photograph by the author)

Washing is done with a bamboo tray similar to that used for husking rice. Waist deep in water, the washer moves the tray containing the gravel round and round, allowing all of the mud and lighter materials to be washed away. Once the gravel is completely clean, the tray is then tilted upwards and the gravel spread out by hand. Any gem fragments can then be picked out as they catch the eye by reflecting the direct sunlight. In earlier times, pit mining as just described was the only method of mining.

Variations on the above are found depending on local conditions. At Bank Ka Cha in Chanthaburi Province, divers scrape the gem-bearing mud from the bottom of the river and load it onto a small boat. It is then brought ashore for washing. Washing of the alluvial gravels from river beds is also done at Pailin in Cambodia. Open cuts here are also made by hand and the overburden is removed in rectangular areas which descend in steps to the gem gravels. The gravel is then removed for washing. These open cuts are also made at a number of sites in Thailand.

Overall, pit mining offers an inexpensive method suitable for small teams of workers.

Figure 11.25 Washing the sapphire-bearing mud at Bang Ka Cha, Chanthaburi Province, Thailand. Divers scrape the mud from the bottom of the river estuary and place it in shallow boats. It is then brought ashore for washing

It is, however, a very inefficient method and so has been replaced at the larger mines by mechanization.

Mechanized mining allows the processing of much greater quantities of earth in a more systematic manner and has, by and large, replaced the simpler pit mining in the region. The only mines today which still lack mechanization are those in Cambodia where, for obvious reasons, it cannot be done. We can well imagine that the Vietnamese would hardly allow a bulldozer to be driven across the border.

Mechanized mining in Thailand makes use of bulldozers, backhoes and other heavy equipment to move large quantities of earth. The jungle has been stripped of huge tracts of land in an effort to get at the gems below, creating an ecological disaster. Problems with deforestation at one point forced the Thai government to ban the use of all bulldozers and heavy equipment, but this did nothing to stop the destruction. Mine owners merely operated them during the night to avoid the ban and today they work openly during the day.

Bulldozers first strip away all the overburden, exposing the gem-bearing soil. This is then scraped into a big hill, where water cannons gradually force it into a pump. The pump carries the soil to a jig modified from that used in mining tin in southern Thailand. As the mud passes over the jig, the heavier minerals are trapped while the lighter material is carried away. At the end of the day, the pump is turned off and the gems are

Figure 11.26 Mining at Nong Bon, Trat Province, Thailand. Earth movers shift mountains of earth in the search for the elusive ruby. (Photograph by the author)

Figure 11.27 A miner at Bo Rai, Trat Province, Thailand, uses a water cannon to wash the ruby-bearing soil into the separation jig. (Photograph by John Rouse, AIGS, Bangkok)

selected from the gravel trapped by the jig. Small screened troughs are also used for washing the gem gravel, in conjunction with high-pressure water cannons.

One problem with the jigs is that sometimes the water current is too strong and takes away not only the mud but also the smaller gems. Perhaps a reduction in water pressure or the insertion of two or more jigs in line, one after the other, would allow the recovery of more stones, particularly in the smaller sizes. Even under the best of conditions the recovery of gems is small, with a day's work at a typical ruby mine yielding but a tiny handful of gems of mostly inferior qualities and sizes. Witnessing the day's production at one ruby mine proved to be a truly humbling experience for this author. It gives a real appreciation of just how very rare a large and fine ruby really is.

At Pailin, the only modern equipment used up until 1975 was small gasoline pumps for dewatering. All other mining was done by hand (Davies, 1975). With simple pit methods the theft of stones is a very real problem; one mine owner at Pailin estimated that up to 40% of his production was being lost through theft. Davies (1975) prepared a report on potential mechanized mining at Pailin using methods which would largely eliminate the possibility of theft. Water cannons and pumps would send the gem gravel through a series of sorting troughs, with the sorted gravel emptying into a locked box. This would then be taken to an enclosed sorting compound for final hand selection under strictly controlled supervision. Thus, human contact with the gems would be almost entirely eliminated. Unfortunately, the rise of the bloody Khmer Rouge in Cambodia put a rather sudden end to these plans. It seems, however, that there is no reason why his plan could not be effected across the border in Thailand, where the deposits are so similar. After studying the mining in Pailin, Davies remarked that the methods were wasteful and inefficient, with much room for improvement.

The ruby-mining area near Bo Rai and Nong Bon in Trat Province is an amazing place. It looks like a war zone, with

Figure 11.28 Heavy-media separation jig at Bo Rai, Trat Province, Thailand. Ruby-bearing earth is pumped into the jig where the heavy minerals (including ruby) are separated from the lighter waste. (Photograph by the author)

Figure 11.29 Ruby mining in Thailand proceeds at a frantic pace and sometimes even beneath the miners' huts. Here, miners at Nong Bon (Trat Province) are using a water cannon to pump the ruby-bearing soil into a tunnel that runs beneath their homes. On the other side of the tunnel a jig will separate the rubies and heavier minerals from the worthless overburden. (Photograph by the author)

destruction everywhere as the soil is ripped asunder in the quest for the tiny red gems. The towns are little more than collections of lean-tos as more permanent dwellings are shunned, and for good reason. In many cases they lie on ruby-bearing soil and as other areas are worked out the miners stare with little-disguised lust at the soil their houses are built on. At Nong Bon, the author witnessed mining directly on both sides of the town, with drainage tunnels carrying the gem-bearing mud from under one of a small group of houses to the jig on the other side! Also at Nong Bon there was a Chinese temple atop a small hill that had been cut down more than 30 metres on every side, save one. The only access to the temple was by this narrow strip of land with a big drop on either side. On a later visit the author saw that even this road had been cut away and the temple perched at a precarious 45° angle.

From a hill near Bo Rai can be seen the thickly jungled hills of Cambodia in the distance, and sometimes the occasional thump of artillery fire can be heard from the direction of that war-torn land. However, the gently rolling hills surrounding the town of Bo Rai show another kind of activity – that of ruby mining. Once covered in forest and farmland, the hills today have been stripped bare in the quest to find rubies. The green is gone, and only brown earth is left in its place. Although land reclamation laws do exist, enforcement is lax as the ruby mining provides employment and badly needed foreign exchange. The rape of the land therefore continues unabated.

In Sri Lanka, the government forbids the use of heavy equipment in mining gemstones, reasoning that the mines have given employment to many for over 2000 years and hopefully will continue to do so for many years to come if worked slowly. Unfortunately, with world demand and prices for ruby at such high levels, mining in Thailand is proceeding at a frantic pace. We can only hope that foresight and reason will overtake the desire for immediate profits; if not, Thailand may soon exhaust this precious and non-renewable resource. Not only would the rubies be gone but if the land is not properly reclaimed, as at present, the loss of topsoil would render it useless for farming. This is perhaps the greatest danger of all.

The land upon which the mines are situated in Thailand varies in ownership. Much is privately owned and leased out to mining companies, who may then sublease it to smaller groups. In other cases, the land is owned by the government. All miners must possess a mining licence, which is easily obtainable at low cost.

Mining gemstones requires relatively large amounts of water, so the best time for mining is during the rainy season (June–October) when plenty of water is available for washing. Unfortunately the rainy season is also the best time for mosquitoes, and as a result the incidence of malaria amongst the miners is very high. This is doubly true of those that try their luck across the border in Cambodia. Chanthaburi and Trat are among the wettest areas of Thailand and so water is less of a problem than at Bo Ploi in Kanchanaburi, which is very dry for much of the year.

Miners at the big ruby mines earn less than $3 per day. The thousands of peasants from surrounding areas who dig at random in the forest find barely enough to eke out a living. There is always the chance of finding 'the big one', however, and everyone knows of somebody who has. This is what keeps them going.

Poorer people are found at each of the mechanized mines washing gravel by hand. This gravel may at times be dug from nearby pits, or the very tiniest gravel trapped by the jig can also be purchased for approximately $20 per barrel. Of course, only very tiny rubies are obtained from this jig sand. Poor women and children may also be seen washing the mud immediately as it flows out of the jig. Again, it is mostly only small stones that are obtained in this way. The author can testify to this as he tried it one day at Nong Bon. After about an hour's work, a very tiny pile of ruby 'sand' had been found: all the larger red stones were pyrope garnet, of which fairly large sizes are found here.

The biggest problem facing the ruby mines of Chanthaburi and Trat today is exhaustion. The simple fact is that fewer and fewer stones are being found. During the ten-year period that the author has been visiting these mines, he has seen no new deposits opened up. Instead, the same pieces of ground are being worked over and over again, inevitably yielding fewer stones. Although accurate statistics are not available, the general consensus is that production is declining rapidly. Unless new deposits are unearthed, it appears that the already rare ruby will become even more scarce in the future.

Trading and cutting

From the time that the stones are won at the mines to the point of export from Thailand, they may pass through as many as seven or eight different pairs of hands, steadily increasing in value all the way. Once a stone is found, the miner need not venture far to find a buyer as many small dealers wait patiently at the mines for just such an event. Each morning in mining towns such as Bo Rai and Nong Bon a rough stone market is held where miners and dealers exhibit their wares. It is a colourful scene, with buyers and sellers crowding around small tables in a line, examining stones. These markets are mainly for the smaller miners, while the big firms sell direct to cutters in Chanthaburi and Bangkok, or cut the stones themselves 'in house'. Smaller sizes and cheaper qualities are offered in large lots, with the more expensive pieces being sold singly.

A 'wild-west' atmosphere pervades the towns of Bo Rai and Nong Bon at these times; gun-carrying police officers stand nearby, ready to restore order should the need arise. And arise it does. In 1981 over 50 people in Bo Rai were murdered, usually in shootings (Nordland, 1982). Most of the shootings, according to Bo Rai's chief of police, Major Anan Rakmitr, were over conflicting land and mineral rights. There are 140 police officers in Bo Rai for the town's population of 10 000, but Rakmitr claims it is not enough. This attitude is understandable because his predecessor was killed in an ambush by gunmen. Today the police carry M-16 rifles and people do not get so angry at them. 'They hate each other too much', Major Rakmitr explained (Nordland, 1982).

From Chanthaburi, Trat and Cambodia the gems flow to the town of Chanthaburi for cutting, heat treatment and eventual sale. Every Friday, Saturday and Sunday a cut stone market is held and the Bangkok dealers descend on the town to look over the week's finds. Along Treerat Road the business is fast and furious, with little time for contemplation. Brokers, or runners, offer the stones, as they can ask higher prices without feeling embarrassed. Offers are made, the parcels are sealed, and off goes the runner to consult with the owner. In time the runner reappears, usually with a counteroffer, and so it goes until a deal is made. Payment is normally by a three-day post-dated cheque to allow the buyer to test the stones in Bangkok and cancel the cheque if a stone turns out to be synthetic. Most stones, of course, are genuine, but should a newcomer try to buy they must first wade through the many synthetic stones that suddenly appear from the broker's pockets. Passing this 'initiation test' allows them to proceed to examine the real thing. Failure (the purchase of these 'gems') suddenly produces large numbers of brokers, all eager to fleece the *farang* (foreigner). It is definitely not a good place for the inexperienced buyer.

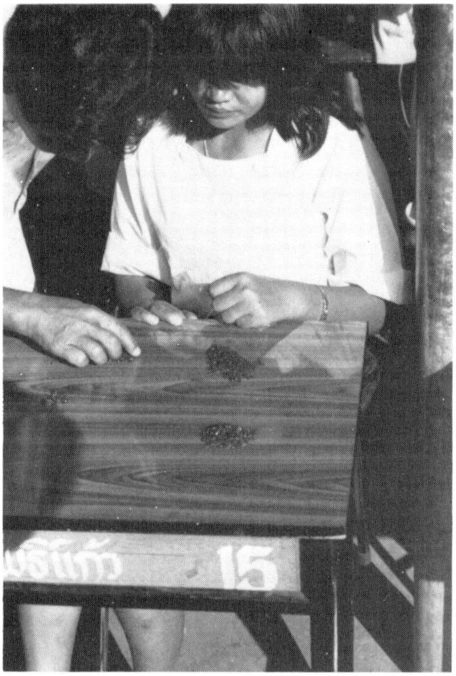

Figure 11.30 Early morning rough market at Bo Rai, Trat Province, Thailand. From 07:00–09:00 every morning rough rubies are traded on the main streets of most major mining towns in Chanthaburi/Trat Provinces. (Photograph by the author)

The Chanthaburi market is almost exclusively devoted to the sale of cut stones; when rough is offered it is inevitably a cleverly beat-up Verneuil synthetic. Not only are Thai rubies and sapphires sold, but also heat-treated Sri Lankan sapphires and other corundums from around the world. This is because many are brought to Chanthaburi for cutting and heating. Even stones from Bangkok may be sent to Chanthaburi for sale, since it is where the buyers are found. From Chanthaburi, the stones proceed to Bangkok for sale to international gem buyers and then pass into the major consumer markets of Japan and the West. The chain of distribution and resulting prices are shown in Table 11.1.

From the above example we can see that a 5 carat rough stone sold by the miner for \$500 (\$100 per carat) could cost the final consumer ten times that amount at retail. This is by no means an extreme case, but represents average mark-ups throughout the chain of distribution.

TABLE 11.1

Distributor	*Purchase price*	*Mark-up (%)*	*Selling price*
Miner	5 carat rough ruby	–	\$100 per carat
Smalltime rough buyer	\$100 per carat rough	30	\$130 per carat rough
Cutter – rough to cut = 20% yield 5 carat rough = 1 carat cut	\$130 per carat	30	\$845 per carat cut
Chanthaburi dealer	\$845 per carat cut	20	\$1014 per carat cut
Bangkok dealer	\$1014 per carat cut	28	\$1300 per carat cut
US buyer	\$1300 per carat cut	40	\$1820 per carat cut
US manufacturer	\$1820 per carat cut	50	\$2730 per carat cut and set
US jeweller	\$2730 per carat cut and set	100	\$5460 per carat cut and set
US consumer	\$5460 per carat cut plus price of setting	–	–

Bibliography

Anonymous (no date) *Corundum deposits in Chanthaburi and Trat Provinces, Thailand*, (in Thai), unpublished report of Department of Mineral Resources, Bangkok

Aranyakanon, P. and Vichit, P. (1979) *Gemstones in Thailand*, unpublished report, Economic Geology Division, Department of Mineral Resources, Ministry of Industry, pp. 1–24

Aranyakanon, P. and Vichit, P. (1983) The gem deposits of Thailand. *Proceedings of the Conference on Geology and Mineral Resources of Thailand*, section B, pp. 1–11

Bauer, M. and Spencer, L. J. (1904) *Precious Stones*, Griffin, London

Bunopas, S. and Bunjitadulya, S. (1975) Geology of Amphoe Bo Phloi, North Kanchanaburi with special notes on the "Kanchanaburi Series". *Journal of the Geological Society of Thailand,* **1**, 51–67

Cartwright, B. O. (1908) *The History of the Kingdom of Siam*, American Presbyterian Mission Press, Bangkok

Crawfurd, J. (1828) *Journal of an Embassy to the Courts of Siam and Cochin China*, Oxford University Press, London (reprint)

Davies, E. B. (1975) *Proposals for mechanized mining of the Pailin gem deposits – Khmer Republic*, unpublished report, Ministry of Overseas Development, London

Finlayson, G. (1826) *The Mission to Siam and Hue, the Capital of Cochin China, in the years 1821–2*, Murray, London

Gervaise, N. (1928) *The Natural and Political History of the Kingdom of Siam A.D. 1688*, Bangkok

Graham, W. A. (1924) *Siam*, Alexander Moring Ltd, London

Guhler, U. (1947) Studies of precious stones in Siam. *Siam Science Bulletin,* **4**, 1–38

Hoskins, J. (1987) *The Siamese Ruby*, World Jewels Trade Center, Bangkok

Jobbins, E. A. and Berrange, J. P. (1976) The geology, gemmology, mining methods and economic potential of the Pailin ruby and sapphire gem-field, Khmer Republic. *Institute of Geological Sciences, Overseas Division, Report no. 35*, pp. 1–31

Jobbins, E. A. and Berrange, J. P. (1981) The Pailin ruby and sapphire gemfield, Cambodia. *Journal of Gemmology,* **17**, 555–567

Keller, P. C. (1982) The Chanthaburi–Trat gem field, Thailand. *Gems and Gemology,* **18**, 186–196

Louis, H. (1894) The ruby and sapphire deposits of Moung Klung, Siam. *The Mineralogical Magazine,* **10**, no. 48, 267–272

Mendenhall, H. E. (1969) Gem opportunities missed in Thailand. *Lapidary Journal,* **22**, no. 10, 1360–1370

Monsignor Pallegoix (1854) *Description de Royaume Thai ou Siam*

Nordland, R. (1982) On the treacherous trail to the rare ruby red. *Asia*, October, 35–42

Pavitt, J. A. L. (1976) Thailand – gem cutting and trading. *The Australian Gemmologist*, May, 307–309

Saurin, E. (1957) Some gem occurrences in Cambodia. *Rocks and Minerals,* **32**, 397–398

Taylor, G. C., Buravas, S. and Charaljavanaphet, J. (1951) Geologic reconnaissance of the mineral deposits of Thailand. *USGS Bulletin*, no. 984, 144–150

Vichit, P., Vudhichativanich, S. and Hansawek, R. (1978) The distribution and some characteristics of corundum bearing basalts in Thailand. *Journal of the Geological Society of Thailand,* **3**, M4-1–M4-38

Vichit, P. and Hansawek, R. (1983) *A field trip to Kanchanaburi gem field, Thailand*, unpublished report, Economic Geology Division, Department of Mineral Resources, Ministry of Industry, pp. 1–7

Warington Smyth, H. (1898) *Five Years in Siam – From 1891 to 1896*, Scribner's, New York

Warington Smyth, H. (1934) *Chase and Chance in Indo-China*, Blackwood, London

Workman, D. R. (1977) Geology of Laos, Cambodia, South Vietnam and the eastern part of Thailand. *Overseas Geology and Mineral Resources*, no. 50, 1–33

Sri Lanka (Ceylon)

The island of Sri Lanka (formerly Ceylon) has for over 2000 years supplied the world with very fine rubies and sapphires. In fact, it is more than likely that Sri Lanka was the original source of the corundum gems. Burma may produce finer qualities, but only Sri Lanka has produced rubies and sapphires in large sizes for such a long a period of time. All varieties are found here, including some such as the 'padparadscha', which are unique to the island.

History

Historical evidence reveals that gemstones were an important item of commerce in Sri Lanka from very early times. Tennent (1859) quotes the *Mahawanso*, a Sinhalese chronicle written in the fifth century AD, as mentioning a gem-encrusted throne owned by the Naga king before the Aryans conquered the island in 543 BC.

Written accounts of the island do not go back earlier than 543 BC. However, with our present knowledge of Sri Lanka it is possible to make a few assumptions regarding the first discovery of gems on the island. The gem deposits lie scattered across the southern two-thirds of the island, in literally thousands of different locations mostly connected with rivers and drainage systems. They appear in such abundance that it is not stretching the truth to say that they were probably discovered shortly after humans began inhabiting the island. As corundum is present at nearly every locality where gems are found, it is entirely possible, and even probable, that corundum was found shortly after the first people arrived from India. In fact, many stone implements have been discovered in the caves and gem pits of Sabaragamuva Province, of which the gem centre of Ratnapura is the capital. These implements and tools date back to the Stone Age (Anonymous, 1968). This means that Sri Lanka is, if not the oldest, then certainly one of the oldest sources of gemstones in the world, having produced continuously since

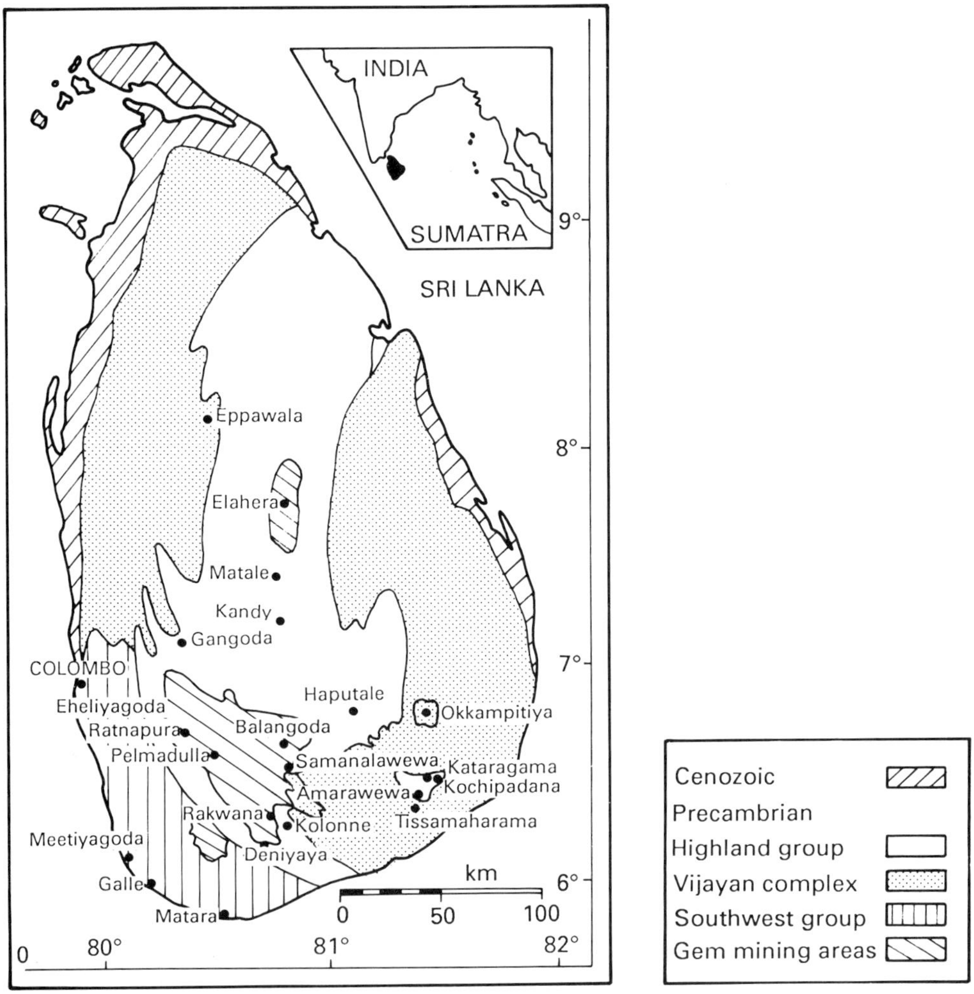

Figure 11.31 Geological map of Sri Lanka showing the major gem-producing areas. (From Zwaan, 1982)

Stone Age times. It also means that corundum has probably been in use since this time as well.

Sapphires and rubies from Sri Lanka first appeared in Western jewellery among the Etruscans (600–275 BC) and were used by the Greeks and Romans from approximately 480 BC onwards (Ball, 1931). C. W. King, in his book *Antique Gems* (1860), describes a sapphire cameo of Hadrian's time (AD 117–138) from Sri Lanka or India with a hole drilled in it, apparently for use as an earring or necklace.

Chinese accounts also tell of the gems of Sri Lanka (Ball, 1931). Fa-Hien, the Chinese Buddhist traveller of the fifth century AD, claimed that Ceylon was visited by gem traders before Buddha's time (624–544 BC). With the arrival of the Aryan conqueror Vijaya in 543 BC, exploitation of the gem deposits undoubtedly increased. Pliny (AD 77) states that in the time of Emperor Claudius (AD 41–54) ambassadors of the Island of Taprobane (Sri Lanka) boasted of the fine precious stones it produced. Ptolemy, the Greek astronomer, refers to beryl, sapphire and gold among the products of the island in the second century AD (Punchiappuhamy, 1985). From this point on, references to the famous gemstones of Serendib (Sri Lanka) are numerous.

Among the more famous references to Sri Lanka are those in the 'Travels of Sinbad the Sailor' from the Arab epic work, *Tales of Arabian Nights*, as well as the travels of Marco Polo (Yule and Cordier, 1921). Marco Polo had this to say about the gems of Serendib.

> Now I will quit these particulars, and tell you of the most precious article that exists in the world. You must know that rubies are found in this Island and in no other country in the world but this. They find there also sapphires and topazes and amethysts, and many other stones of price. And the King of this Island possesses a ruby which is the finest and biggest in the world; I will tell you what it is like. It is about a palm in length, and as thick as a man's arm; to look at, it is the most resplendent object upon earth; it is quite free from flaw and as red as fire. Its value is so great that a price for it in money could hardly be named at all. You must know that the Great Kaan sent an embassy and begged the King as a favor greatly desired by him to sell him this ruby, offering to give for it the ransom of a city, or in fact what the King would. But the King replied that on no account whatever would he sell it, for it had come to him from his ancestor.

There seems to have been circulating since the sixth century AD onwards, among travellers to the region, stories of the great ruby of the King of Sri Lanka (Yule and Cordier, 1921). The Chinese Hiuen Tsang (seventh century) said this about a large gem atop a pagoda in Sri Lanka: 'A hyacinth, they say, of great size and brilliant ruddy colour, as big as a pine cone; and when 'tis seen from a distance flashing, especially if the sun's rays strike upon it, 'tis a glorious and incomparable spectacle.' Marco Polo's contemporary, Hayton, had also heard of the great ruby: 'The King of that Island of Celan hath the largest and finest ruby in existence. When his coronation takes place, this ruby is placed in his hand, and he goes round the city on horseback holding it in his hand, and thenceforth all recognize and obey him as their King.' Tennent (1859), on this subject, quotes a statement from a Chinese work that early in the fourteenth century the Emperor sent an envoy to Ceylon to purchase a carbuncle of unusual lustre. 'This was fitted as a ball to the Emperor's cap; it was upwards of an ounce in weight and cost 100,000 strings of cash. Every time a grand levee was held at night the red lustre filled the palace, and hence it was designated "The Red Palace Illuminator".' Such a story as this may sound ludicrous, but in fact is not as far-fetched as it first seems, since some incredibly large corundums have been unearthed in Sri Lanka. During the author's visit to Sri Lanka in 1986, a stone of fantastic proportions had just been unearthed from the gem gravels. Its weight was said to be a massive 42 kilograms.

From the above accounts, it is now very clear that Sri Lanka has earned a reputation as a vast storehouse of precious stones. Sri Lanka is perhaps unique in that not only are rubies and sapphires found there, but also some 40 other types of gemstones, with many types being found in a single deposit. The list is long and includes spinels, garnets, beryls, tourmalines, quartz, moonstones, topaz, taaffeites, ekanites, sinhalites, etc. With so many fine stones, it would seem correct to assume that the gem industry has always been a major industry on the island; however, such is not the case. Due to government restrictions and the local idiosyncracies of doing business, the industry has not prospered as it should. Sri Lanka has been and continues to be a difficult place to transact business. Government red tape is only part of the problem. Bargaining is the norm in Sri Lanka, as elsewhere in Asia, but the first asking price is usually ridiculously inflated, with the local dealer or miner hoping to make a killing on every sale. Since dealers cannot possibly be experts on every type of stone, if they buy something they are not familiar with in Sri Lanka, they will probably pay much too high a price. As a result, they cannot sell the goods at a profit and so do not return to buy any more. In contrast to this, the dealers of Hong Kong and Thailand show more foresight. By starting with a much more reasonable asking price, they not only save precious time, but also ensure future business. As a result, the business prospers and government regulation is kept to a minimum. This is not the case in Sri Lanka, where the business remains largely a cottage industry. It will only realize its true potential when the government removes all restrictions and, more importantly, when the local dealers and miners reduce their expectations to a reasonable level. By asking such high starting prices, they are killing the goose that lays the golden egg – the customer. Of course, Thai and Hong Kong dealers are only too happy to see things continue as at present in Sri Lanka, for they are the winners. The most important factor in a country's gem industry is not the local production of gemstones – we need only look at Burma and Sri Lanka to see the fallacy of this line of thinking – but the overall business climate in that country. Thailand's success in this regard should serve as an example.

Mining areas

The corundum-producing areas of Sri Lanka are located primarily in the southern two-thirds of the island. Precambrian metamorphic rocks dominate the whole island, with the exception of the north-western coastal belt (Zoysa, 1981). Cooray (1967) describes the division of the Precambrian rocks into three types, known as the Highland, Vijayan and South-west groups. It is within the Highland group that most of the major gem fields are located. Topographically, Sri Lanka is divided into three penoplains as follows, with most of the gem deposits lying in the second penoplain:

First penoplain	coastal plain, up to 30 m above sea level
Second penoplain	average of 480 m above sea level
Third penoplain	average of 720 m above sea level

The corundums of Sri Lanka are mined almost entirely from alluvial gravels. In only a handful of cases have they been found *in situ*. These *in situ* occurrences have been in pegmatites and gneisses, but were of low quality only and so of little economic importance.

Gem mining has traditionally been centred on and around the town of Ratnapura, 'the city of rubies', or as it is more often termed, 'the city of gems'. Mining is not confined only to this area, but is found in scattered locales over the entire lower portion of the island. It seems that every few years a new area is opened up. What makes Sri Lanka perhaps unique is not only the widely scattered mining areas, but also the fact that at nearly every mine a number of different gem species are found, many unique to the island. To date, more than 40 different gem species have been found.

The varieties of corundum found in Sri Lanka are: ruby (including pink sapphire), blue sapphire, yellow sapphire, orange sapphire (padparadscha), violet and purple sapphire, colourless sapphire, green sapphire (rare), colour-change sapphire, and star ruby and star sapphire (all colours).

The rubies of Sri Lanka can be extremely fine, approaching the best Burmese stones in quality. In general, however, they tend to be less intense, more pink than red. In the past, these pink stones were termed 'female rubies', as opposed to the deeper red stones that were called 'male rubies'. A search of the gemmological literature reveals that the term 'pink sapphire' did not appear until the beginning of the twentieth century. Prior to this, all corundums of a red colour (pink is merely a light red) were referred to as rubies. Then someone decided, incorrectly, that pink was not red. As ruby is only red, it was decided that pink must therefore be a sapphire and problems began. Where can the line be drawn between pink sapphire and ruby? This leads to the ridiculous situation of stones being brought to laboratories solely to determine if they are rubies or pink sapphires. We do not have this problem with blue sapphires – light blue or deep blue, they are still blue sapphires. In the author's opinion, all corundums of a red colour, regardless of depth or intensity, should be termed rubies, just as was done before the twentieth century. This would eliminate the above problem entirely.

The derivation of the word *padparadscha* is also very interesting. This is a Sinhalese word which today in the West refers to a Sri Lankan (only Sri Lankan, properly) sapphire of a pinkish-orange colour. As padparadscha sapphires are extremely rare and valuable, we again encounter the question: when is a sapphire a true padparadscha? The term is actually derived from the word *padmaraga* which means 'lotus flower colour'. As anyone who has looked at lotus flowers can confirm, they are not orange at all, but pink. This term used to refer to pink sapphires, but today has come to mean a

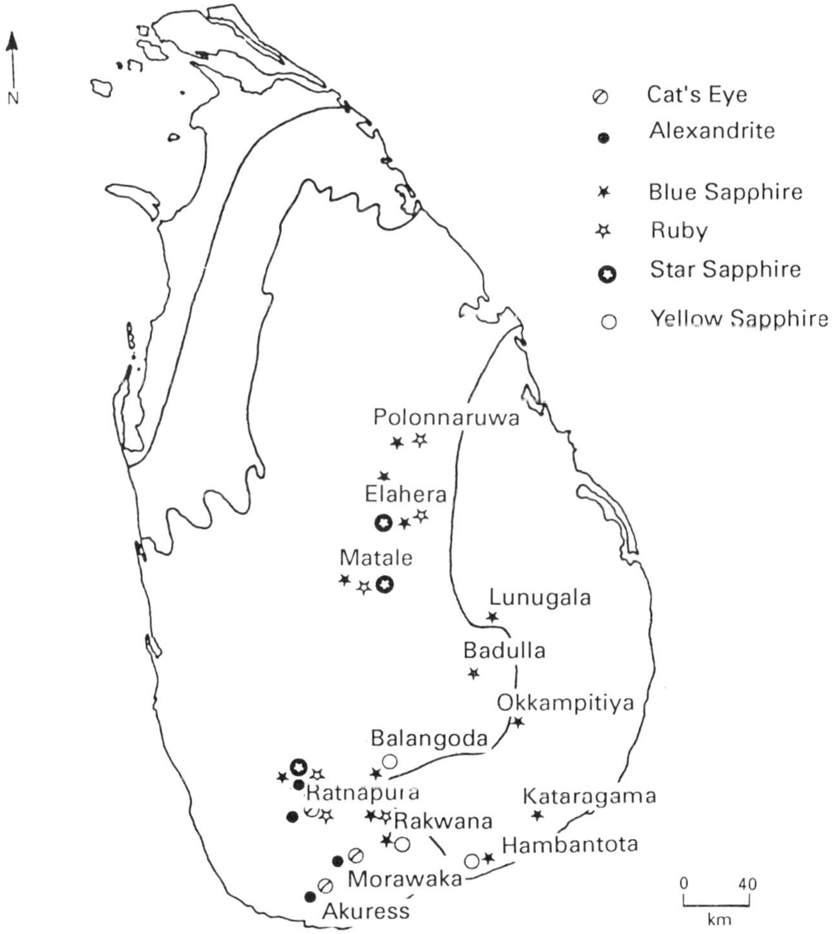

Figure 11.32 Gem localities of Sri Lanka. (From Zoysa, 1982–1983)

pinkish-orange sapphire instead. Which is correct and what to do? In this case, the author does not have a clue as to what is right. Suffice it to say that in most gem circles, a padparadscha is a pinkish-orange sapphire, not a pink sapphire (Crowningshield, 1983). It is probably best, in the case of gem identification reports, for gemmologists to place the padparadscha, the pigeon's-blood ruby and all other such jargon into a capsule and fire it into outer space, for nobody can agree upon just what it is they are supposed to mean.

It is the blue sapphire that is Sri Lanka's pride and joy. Today, with the prevalence of huge quantities of heat-treated Sri Lankan blue sapphires on the market, it may not be appreciated that formerly the island was not a major source of blue sapphires. In olden days, before the rise of the 'geuda', Sri Lanka did not produce many fine blue sapphires in the smaller sizes. The reason is that the rough material is unevenly coloured. Instead, good stones with a rich *and* even blue colour mostly came in larger (above 5 carats) sizes only. With the advent of heat treatment, however, stones of smaller sizes and good colour became available, so much so that Sri Lanka is now the major source of fine blue sapphires in all sizes.

Over the years Sri Lanka has produced some fantastic natural blue sapphires of large size, both faceted and stars. These include top-quality gems in the 100 to 300 carat range. Many of the largest sapphires in museum collections around the world have originated from Sri Lanka.

Mining methods

Mining techniques in Sri Lanka are similar to those of other corundum placer deposits in Burma and Thailand. Gem placers are formed through the weathering action of the natural elements. Weathering releases the gems from their host rock and carries them to streams and rivers. Whenever a river loses velocity suddenly, it deposits its load of coarse sediments and stones carried in suspension. This occurs at the outlet of rivers entering the plains from the mountains, at hollows in the river bed, at the junction of a smaller stream with a large river, and at concave sections in the bends of rivers. Gem placers are also found beneath the surface of rice paddies where rivers once flowed. It is from such alluvial sites as these that gems are mined.

Ratnapura, Fakwana, Balangoda, Okkampitiya and Morawaka are the main alluvial gem beds found in Sri Lanka. Although many different species are found at each location, each site is famous for one in particular, as follows (Zoysa, 1981):

Ratnapura	ruby, blue sapphire
Rakwana	blue sapphire
Balangoda	yellow sapphire
Okkampitiya	blue sapphire, hessonite garnet
Lunugala	blue sapphire
Morawaka	chrysoberyl

The depth of the gem gravel, or 'illam', varies from 3 m to 20 m; at Pelmadulla it reaches 40 m. Some areas contain only one gem-bearing horizon while others contain two or three different layers of illam. The age of the alluvial deposits at Ratnapura has been determined to be middle to late Pleistocene (Zoysa, 1981).

Residual gem deposits occur in the Elahera area, which is one of the major gem-producing areas of Sri Lanka. Blue sapphire, ruby and star stones are found here. The Elahera deposits lie at a depth of between 3 and 10 m. The third major type of gem deposit in Sri Lanka is eluvial. These deposits are found in the Horton Plains area where the depth of the gravel is between 5 and 10 m.

The search for gems is a highly speculative operation and usually carried out by a group of native workmen on a share basis. One-fifth goes to the owner of the paddy field, another fifth to the financier and the rest to the actual workers.

After selection of a likely spot for mining, the ground will be tested from time to time as the pit deepens by using a long steel rod sharpened and tempered at the point (Adams, 1927). By pushing the rod into the soft earth and twisting it round and round, and by tapping the rod and carefully listening to the sound, an experienced operator can tell at what level the illam lies, its thickness and its character. When the point of the rod passes through the illam and strikes the underlying decomposed bedrock,

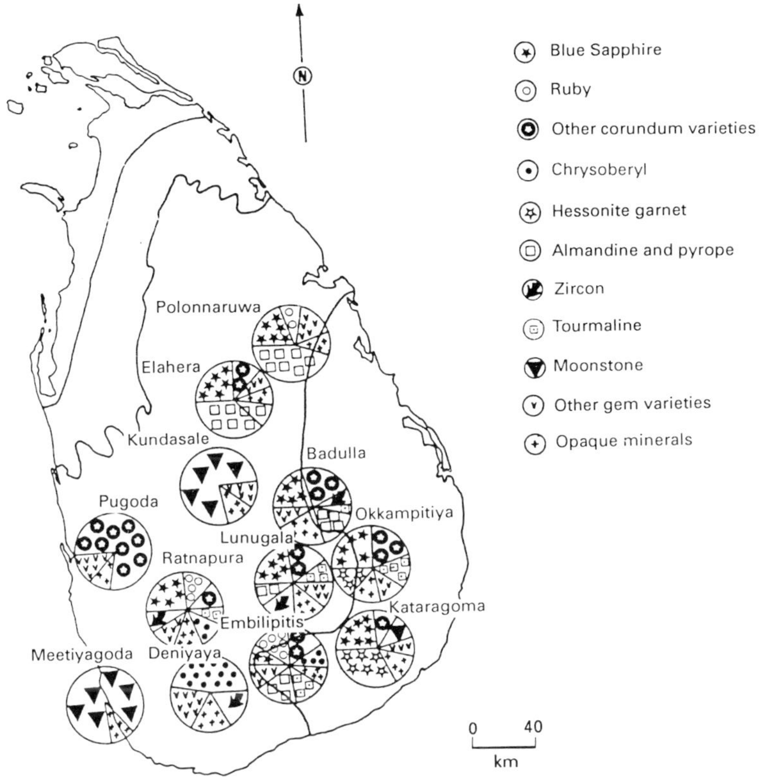

Figure 11.33 Percentage abundance of Sri Lankan gems. (From Zoysa, 1982)

clay will be found adhering to its point; if the rod's surface is scratched this would indicate the presence of quartz or corundum, the two most common constituents of the illam.

To reach the illam, a shaft is sunk. This is round in the case of smaller pits, and square for large pits. The shaft is lined with small branches and vertical poles driven down into the mud to prevent cave-ins. One or two men work at the bottom of the pit, shovelling the clay and mud into a small bamboo basket which, when filled, is adroitly thrown upwards into the waiting arms of another man at the surface. Deeper pits utilize a rope or a bamboo pole lift.

When it is reached, the illam is brought to the surface and placed on a clean flat piece of ground prepared in advance. The work in the gem pits is hot and extremely muddy; workers wear nothing but a loin cloth, fitting attire perhaps for those working in the shadow of Adam's Peak. Adam's Peak is a major site of pilgrimage and is said to have the footprint of Adam (or Buddha, depending on one's faith) at the summit. Even for the atheist a trip to the summit is definitely worthwhile, for when the sun rises the mountain casts a huge shadow in the form of a perfect pyramid on the ground below.

As a result of the accumulation of water in the pits, bailing (or pumping) is required. Once sufficient illam has been accumulated, washing can begin. This is carried out in a nearby stream or pond specifically constructed for the purpose, and is similar to the method employed at Mogok, in Burma. Each man stands waist deep in the water with a basket filled with mud. A circular motion is employed, like panning for gold, to remove all lighter sediments, leaving the 'nambuwa' at the bottom of the basket. When this is done the basket is brought to the shore. After several baskets have been prepared, another man, expert in the

Figure 11.34 Gem pit near Ratnapura, Sri Lanka, 1981. (Photograph by the author)

recognition of rough gems, takes the baskets and examines each one carefully in succession. The basket is tilted up so that the sun's rays shine on the gravel and the man then invokes a prayer to the gods to give him good luck. With a rapid motion he sweeps the gravel back and forth, picking out the large fine precious stones, which are termed 'jathi' (L. J. Punchihewa, personal communication). The residue left behind after all the jathi has been removed is termed 'toura-mali', a word from which our tourmaline is derived. As the better stones are found they will be handed for safe keeping to the man who finances the operation. This financier is always present during washing to ensure that theft is kept to a minimum (it can never be eliminated entirely). When one man finishes sorting a basket it will be handed to another for resorting to make sure that nothing is missed. Once all of the illam is washed, the gems are taken to the market for sale, with the proceeds being divided *pro rata* among the partners in the claim (Adams, 1927).

Figure 11.35 Washing the 'illam' (gem-bearing gravel) in Sri Lanka, 1939

Not all mining takes place in paddy fields; river beds are also mined. A small dam of reeds is constructed across the river below the area marked for mining (Punchihewa, 1983). Special long-handled rakes are then used to scrape the gravel off the river bottom and towards the embankment. Thereafter, washing and sorting are done as usual.

By law, only the simplest of equipment is allowed for gem mining in Sri Lanka. Mr Bruce Davidson of Colombo has built and sold small floating sluices for use by scuba divers. Divers suck up the gravel in what amounts to a big vacuum cleaner; this is then piped to the floating sluice for separation. Several of these are in use today in Sri Lanka. Apart from this, the only modern equipment used is the petrol-operated pump for dewatering the pits. In those areas slated to be submerged due to the construction of large dams for hydroelectric projects, the government has permitted modern mining before the inundation. Overall, however, the gem industry in Sri Lanka remains today a primitive 'poor man's' business. This is not necessarily a bad thing, though, for it has provided, and apparently will continue to provide, employment for thousands of needy villagers for many years to come. Countries such as Thailand, which is going through her gem deposits at breakneck speed, would do well to examine very carefully the potential benefits of such a policy.

The market in Sri Lanka

With so many fine gems being found in Sri Lanka, we might expect that the gem trade would be big business there, but such is not the case. Government regulation and the attitudes of local traders have combined to limit the volume of trade, as discussed previously. To put it bluntly, it is difficult to do business in Sri Lanka, compared with other Asian gem centres such as Thailand and Hong Kong.

The importance of Sri Lanka in the international gem marketplace increased dramatically with the discovery of suitable heat treatments for Sri Lankan geuda sapphires in the mid to late 1970s. By 1980, literally hundreds of Thai buyers were streaming into the country to purchase the once lowly geuda sapphire. Sri Lankan miners profited greatly from this situation, but dealers and other middlemen became increasingly jealous. This jealously reached a peak in 1984, when dealers in Ratnapura and Colombo petitioned the government to restrict the visas of Thais, claiming that they were breaking currency regulations and other laws banning the export of rough gemstones. These claims were true to a large extent, with currency and rough smuggling rampant. The answer to these problems, however, is not to bar Thais from Sri Lanka as local dealers would wish the government to do. In order for the gem business to prosper, local dealers, cutters and treaters must improve themselves to become more competitive.

The Sri Lankan gem-cutting industry provides a perfect example of the lack of competitiveness. In 1939, a government committee published a study of the gem industry in Sri Lanka, listing problems and giving suggestions for improvement. It could have been written yesterday! One of the suggestions was that the local cutting industry must improve, the main complaint being that cutters placed too much emphasis on weight retention, with the result that most stones exported would need recutting to improve their proportions. Nothing has changed. Today Sri Lankan gem cutting is the standard with which all other *poorly* cut stones are compared. The government, in an effort to protect the local cutting industry, bans the export of rough gems. Instead of giving protection, however, it merely perpetuates the low standards of cutting, hurting the industry in the long run.

Steps must be taken to improve, not protect, the local industry. This means removing duties on modern cutting machines, encouraging local schools with foreign technological assistance, and encouraging Sri Lankan businessmen to travel abroad to learn first hand about the methods and needs of foreign markets.

In 1971, the State Gem Corporation of Sri Lanka was set up with the aim of promoting the local gem industry. However, it has merely added to, and become a major cog in, the overwhelming bureaucracy of the trade. By law, all exports must be channelled through the corporation, which uses up precious time – yet another reason for avoiding Sri Lanka.

The gem industry differs from most others in that it is extremely difficult to regulate; efforts to do so merely force it underground. The Dutch learned this lesson the hard way, as their efforts to regulate the diamond trade caused it to move across the border into Belgium. If the gem trade is to prosper in Sri Lanka the government must remove all regulations, starting with the ban on rough exports. This would force the local cutters either to improve or go broke, and local dealers would quickly learn that there is more profit in repeat business with a 20% profit margin than in a single business transaction with a 200% margin. Only by the removal of such protective laws will the local industry be forced into improvements. After all, these protective laws have been on the books for so long, *without any sign of an improvement*. In the short term, losses may result, but over the long term the only direction it can go is up.

Some type of modern gemmological laboratory and school is also sorely needed in Sri Lanka. Currently the State Gem Corporation provides free gem testing to the trade and public, but their facilities are antiquated to say the least. The foundation of a modern school and attached laboratory would kill two birds with one stone by improving local standards and by providing accurate gemmological reports of gem identification. In addition, this would benefit the world gemmological community as a whole, as there is certainly much that Sri Lankan gemmologists could teach the world about Sri Lankan gemstones. A step in the right direction has already been taken with the recent publication of the high-quality *Journal of the Gemmologists Association of Sri*

Lanka. The founding of the Asian Institute of Gemological Sciences in Bangkok has aided the Thai gem trade immeasurably and there is no reason why a similar institute would not have a similar effect in Sri Lanka.

In closing, the author wishes to make it very clear that he bears no ill will towards Sri Lanka and its people. On the contrary, he has, through his visits, developed an abiding love and respect for the island and its inhabitants, unique in so many ways. It is out of this love for the country that the above criticisms are offered, with the sincere hope that greater wealth and prosperity will be realized. Nature has endowed Sri Lanka with mineral and gem wealth greater perhaps than that of any other country of similar size. It is disappointing, however, that to date the inhabitants have been unable to realize the full potential of this gift. With a new day comes new hopes and ideas. Let all the residents of Sri Lanka – Sinhalese, Tamil and Muslim alike – put aside their differences, for it is the very diversity of its peoples which is the island's strongest asset.

Bibliography

Abeyaratne, M. (no date) *Ratnapura as seen by Government Agents*, Department of Government Printing, Sri Lanka

Adams, F. D. (1927) A visit to the gem districts of Ceylon and Burma. *Annual Report, Smithsonian Institution, 1926*, pp. 297–318

Adams, F. D. (1929) The geology of Ceylon. *Canadian Journal of Research,* **1**, 425–511

Anonymous (1939) *Report of the Sub-Committee of the Executive Committee of Labour, Industry and Commerce on the Marketing and Cutting of Ceylon Gems*, Colombo, pp. 1–34

Anonymous (1950) *The Wealth of India*, Council of Scientific & Industrial Research, India

Anonymous (1968) *A Guide to the Ratnapura National Museum*, Department of National Museums, Sri Lanka

Ariyaratna, D. H. (1976) *Gems of Sri Lanka*, Colombo

Ball, S. H. (1931) Historical notes on gem mining. *Economic Geology,* **26**, 681–738

Ball, V. (1881) *A Manual of the Geology of India, Part 3 – Economic Geology*, Geological Survey of India, Calcutta

Ball, V. and Crooke, W. (1925) *Travels in India by Jean-Baptiste Tavernier*, Oxford University Press, London

Barnes, J. (1982) Ratnapura: the heart of gem country. *Asia Magazine*, 5 December, 12–15

Brown, J. C. and Dey, A. K. (1955) *India's Mineral Wealth*, Oxford University Press, Bombay

Chernush, A. (1980) The dazzling jewels of Sri Lanka. *Reader's Digest*, pp. 61–64

Coates, J. S. (1935) Geology of Ceylon. *Spolia Zeylanica (Ceylon Journal of Science),* **19**, part 2, 101–191

Cohen, H. (1934) Mining sapphires in Ceylon. *The Gemmologist,* **3**, no. 32, 231–236

Coomaraswamy, A. K. (1903) Occurrence of corundum in situ near Kandy. *Geological Magazine*, 4th decade, no. 10, 348–350

Coomaraswamy, A. K. (1904) The crystalline rocks of Ceylon. *Spolia Zeylanica,* **1**, part 4, 105–111

Cooray, P. G. (1967) An introduction to the geology of Ceylon. *Spolia Zeylanica,* **31**, part 1, 1–324

Cooray, P. G. and Kumarapeli, P. S. (1960) Corundum in biotite–sillimanite gneiss from near Polgahawela, Ceylon. *Geological Magazine,* **97**, 480–487

Count de Bournon (1823) *Observations sur Quelques-uns des Mineraux, soit de L'Isle de Ceylon, soit de la Cote de Coromandel, Rapportes par M. Leschenault de la Tour*, Paris

Crowningshield, R. (1983) Padparadscha: What's in a name? *Gems and Gemology,* **19**, 30–36

Dahanayake, K. and Ranasinghe, A. P. (1981) Source rocks of gem minerals – a case study from Sri Lanka. *Mineralium Deposita,* **16**, 103–111

Davidson, B. (no date) *Mineral and Geological Study of Areas to be Inundated Under the Accelerated Mahaweli and other Programs*, Corporate Finance and Management Services Ltd, unpublished report

Davy, J. (1818) On the geology and mineralogy of Ceylon. *Geological Transactions,* **5**, 311–327

Ferguson, A. M. (1888) *All about Gold, Gems, and Pearls in Ceylon and Southern India*, Colombo

Gillson, J. L. (1933) A day in a Ceylon gem field. *American Mineralogist,* **18**, 300–308

Gübelin, E. J. (1968) *Der Edelsteine der Insel Ceylon*, Gübelin, Lucerne

Gunaratne, H. S. (1976) On the occurrence of gem corundum in Kolonne. *Journal of Gemmology,* **15**, 29–30

Hakluyt, R. (1904) *The Principal Navigations Voyages Traffiques & Discoveries of the English Nation*, vol. 5, MacLehose, London, pp. 365–505

Heilmann, G. and Henn, U. (1986) Blue sapphire from Elahera, Sri Lanka. *The Australian Gemmologist,* **16**, no. 1, 2–4

Herath, J. W. (1980) Mineral resources of Sri Lanka. *Geological Survey Department of Sri Lanka, Economic Bulletin*, no. 2, 1–70

Hulugalle, H. A. J. (no date) *Gem Stones of*

Ceylon, Department of Information, Colombo
Iyer, L. A. N. and Thiagarajan, R. (1961) Indian precious stones. *Bulletin of Geological Survey of India*, no. 18, 1–105
Katz, M. B. (1986) Review of the geology of the gemstones of Sri Lanka. *The Australian Gemmologist*, **16**, no. 2, 52–56
Moseley, C. W. R. D. (1983) *The Travels of Sir John Mandeville*, Penguin, Harmondsworth
Parkinson, K. (1950) The gem industry is still primitive in Ceylon. *The Gemmologist*, **19**, no. 229, 165–170
Parkinson, K. (1951) A gemmologist in Ceylon. *The Gemmologist*, **20**, nos 235–239
Parkinson, K. (1953) Ceylon, the island of gems. *The Gemmologist*, **22**, no. 258, 5–7
Punchiappuhamy, T. G. (1984) Well known gems of Sri Lanka. *Journal of the Gemmologists Association of Sri Lanka*, no. 1, 17–20
Punchiappuhamy, T. G. (1985) Historical references to gems of Sri Lanka. *Journal of the Gemmologists Association of Sri Lanka*, no. 2, 24–27
Punchihewa, L. J. (1983) *Gem Buying Hints*, Gem Trade Laboratory, Colombo
Robinson, E. L. (1951) From mud to milady's neck. *The Gemmologist*, **20**, no. 245, 251–254
Rupasinghe, M. S. (1985) Coloration of geuda – new treatment methods. *Journal of the Gemmologists Association of Sri Lanka*, no. 2, 1–6
Schrader, H. (1985) Inclusions in corundum from Sri Lanka. *Journal of the Gemmologists Association of Sri Lanka*, no. 2, 37–41
Squires, S. J. (1954) Ceylon – gem shop of the world. *The Gemmologist*, **23**, no. 271, 21–23
Tennent, J. E. (1859) *Ceylon*, reprinted by Tisara Prakasakayo Ltd, Colombo
Wadia, D. N. (1945) Bibliography of geology of Ceylon. *Records of the Department of Mineralogy, Professional Paper*, no. 1, 33–38
Wadia, D. N. and Fernando, L. J. D. (1945) Gems and semi-precious stones of Ceylon. *Records of the Department of Mineralogy of Ceylon, Professional Paper*, no. 2, 13–44
Waidyanatha, W. G. S. (1985) Special report: the gem industry in Sri Lanka. *Economic Review, Colombo*, **11**, 2–18
Waite, G. G. (1966) A visit to the isle of jewels. *The Lapidary Journal*, November, 941–951
Warmington, E. H. (no date) *The Commerce Between the Roman Empire and India*, Curzon Press, London
Watt, G. (1893) *A Dictionary of the Economic Products of India*, Allen, Calcutta
Watt, G. (1908) *The Commercial Products of India*, Murray, London
Wayland, E. J. (1923) Notes on the sources of origin of Ceylon gemstones. *Economic Geology*, **18**, 514–516
Wijesekera, N. (1980) Gemstones of Sri Lanka. *Lapidary Journal*, October, 1616–1618
Wijesuriya, G. (1985) The technique of heat treatment of geuda to blue sapphire. *Journal of the Gemmologists Association of Sri Lanka*, no. 2, 31–36
Williams, H. (1950) *Ceylon – Pearl of the East*, Hale, London
Yule, H. and Cordier, H. (1921) *The Book of Ser Marco Polo*, Murray, London
Zoysa, E. G. G. (1981) Gem occurrence in Sri Lanka. *Journal of the Gemmological Society of Japan*, **8**, 43–49
Zoysa, E. G. G. (1982–1983) Gem occurrence in Sri Lanka. *Journal of the Gemmological Association of Hong Kong*, **4**, 13–28
Zoysa, E. G. G. (1986) Gem deposits in Sri Lanka with special emphasis on recent discoveries. *The Australian Gemmologist*, **16**, no. 3, 110–114
Zwaan, P. C. (1981a) Gem minerals from Sri Lanka. *Journal of the Gemmological Society of Japan*, **8**, 41–42
Zwaan, P. C. (1981b) Gemstones from the Tissamaharama area in Sri Lanka. *Journal of the Gemmological Society of Japan*, **8**, 51–60
Zwaan, P. C. (1982) Sri Lanka: the gem island. *Gems and Gemology*, **18**, 62–71
Zwaan, P. C. (1986) Gem minerals from the Embilipitiya and Kataragama areas in Sri Lanka. *The Australian Gemmologist*, **16**, no. 2, 35–40

India

The history of the use of precious stones in India is an extremely long one, and this includes corundum. In fact, the term 'corundum' is derived from the Sanskrit word *kurivinda*, and ruby and sapphire have since earliest times been considered to be among the 'Maharatnani', or 'Great Gems', in India.

Although both ruby and sapphire have long been venerated in India, the history of *important* corundum production in the country is a relatively short one. Ruby mines do exist, and have existed for perhaps more than 1000 years, but India only became a major corundum-producing country in the 1880s, with the discovery of the famous sapphire deposit of Kashmir. It is to these mines that we will go first in our examination of the corundum deposits of India.

Kashmir sapphires

The famous sapphires of Kashmir are mined in a remote region high in the Zanskar District of the Great Himalayan Mountains in north-western India. Lying at an elevation of approximately 4500 m, they are located in a small valley near the hamlet of Sumjam, in the Paddar region of Zanskar, in Kashmir.

The exact date when sapphires were first discovered in Kashmir is not known with absolute certainty, but it is thought to have been in or just before 1881. Various stories are told of the original discovery. According to one told to T. H. D. La Touche (1890), a certain shikari, having lost the flint of his rifle while out hunting, picked up a likely looking stone as a replacement. He carried it with him for some time, but eventually sold it to a Lahouli trader who then took it to Simla (the former British summer capital in north India), where it was recognized as sapphire. Enquiries made later resulted in the discovery of the spot where it had been found. For some time thereafter, until guards were posted on the spot by the Maharajah of Kashmir, in whose territory the deposits were found, large quantities of the stones were brought to Simla and sold at absurdly low prices, the asking price being only one rupee per seer. Another story states that as some traders were emptying baskets of borax from Rupshu at a merchant's shop in the Simla bazaar, a stone fell out and was thrown into the street. Just at that moment, a well-known jeweller named Mr Jacobs happened to be passing by and, as chance would have it, was struck by the stone. Probably picking it up with the intention of flinging it back from whence it came, he noticed something unusual about it and so bought it off the merchant for a trifling sum. This second story, however, would seem to suggest another locality for sapphires in the region as the borax caravan passed nowhere near the known locality at Padar. Thus it seems that the first story is the more likely account.

The most complete account of the sapphire mines of Kashmir is that of C. S. Middlemiss (1931), from which much of the following information was obtained.

Once the Maharajah learned of the deposits, he dispatched an armed guard to the site to keep an eye on things and prevent looting. The theft of stones was a big problem, and remains so even today, with gangs of hardy smugglers appearing out of nowhere, ever ready to take advantage of the extreme remoteness of the locale to pilfer stones. Because of the altitude, conditions were very difficult at the mines. Even in the best years mining was limited to the three short summer months of July–September. In some years, barely 30 days of mining were possible because the ground was covered in snow.

In the year 1887, on finding a steady decrease in revenues from the mines, the Maharajah approached the Government of India for assistance in assessing and developing the site. T. H. D. La Touche, a trained geologist, was dispatched to the site in September of that year. His account was later published by the Geological Survey of India (La Touche, 1890) and was the first scientific description of the area. He found that material was obtained from two different sites. The first of these, now called the 'Old Mine', was a group of shallow pits sunk into an actinolite–tremolite rock containing small pegmatite lenses high on the north-east wall of Kudi Valley. It was in these pegmatite lenses that the vast majority of all fine stones were found. Sapphires were also mined from the placers 250 m below the Old Mine on the valley floor, but were generally of lower quality. La Touche also traced the pegmatite-bearing rock through the ridge to the other side, but was unable to locate more sapphires. Since that time others have also attempted to locate sapphire-bearing lenses on the opposite side, but without success.

At the time of his visit in 1887, La Touche found the Old Mine to be practically exhausted. Although another site, the 'New Mine', was later found, it produced very little. What this means is that virtually all of the large fine Kashmir sapphires in existence were taken from the site known as the Old Mine during the period from 1881 to 1887! In just six years this mine produced such a quantity of large fine stones that they achieved a reputation second to none among sapphires. So fine were these stones that today they remain the standard with which all others are compared. It is utterly incredible, but absolutely true.

Finding the Old Mine exhausted, La Touche turned his attention to the placers

Figure 11.36 The Kashmir sapphire mines in north-western India. (From La Touche, 1890)

below and worked them with mostly mediocre results. One success, however, was the discovery of a 933 carat parti-coloured giant. In 1888 he returned for another try, but found little.

From the period 1889 to 1906 there was a lull in official mining; the only work done was that by smugglers. In 1906 the Maharajah leased out the mines to private interests. C. P. M. Wright reworked the placers after much study and obtained many fine stones. The year 1907 saw the discovery of what later became known as the 'New Mine', some few hundred metres south-east of the Old Mine. Wright, however, was eventually forced to abandon his efforts due to the many difficulties encountered in mining in such an inhospitable region. Active efforts were not to resume again until 1924.

In 1926, Lala Jagan Nath reopened the New Mine and extracted over 60 kg of corundum. His licence was revoked just one year later, in 1927, for irregularities. This year marked the end of the Kashmir sapphire mines. Over 450 kg were taken from the New Mine in just 15 days, but few fine stones above 10 carats cut resulted. Middlemiss, in his report of 1931, had great hopes for the mines. These were based, in part, on the possibility of discovering the sapphire outcrop on the opposite side of the ridge. Unfortunately for sapphire lovers, his hopes were never realized. Since 1930 the mines have been worked intermittently, but with no real success. Every few years the Kashmir Government considers leasing out the mines, but all attempts end in failure (the last reports on government efforts to lease the mines date from 1979). Thus, the world waits, ever patient, for the most beautiful sapphires of all to rise again from the snows of Kashmir.

Mining methods

The mining methods in Kashmir have always been very primitive as a result of the remoteness of the locality and, therefore, the difficulties in transporting equipment into the area. Even today the mines remain largely inaccessible, with no roads going to them. The nearest road is at Kishtwar, a hard eight-day march from the mines. Much of the journey is over narrow mountain paths fit only for man and small pack animals. In many places large rivers must be crossed, spanned only by narrow bridges not designed for large loads. This, as much as anything, has ensured that mining methods remain primitive.

The sapphires of Kashmir occur in outcrops high upon the wall of the Kudi Valley. Small pegmatite lenses occur within an actinolite–tremolite rock, and it is within these lenses that the sapphires are found. Originally a landslip exposed the sapphires at the surface, allowing discovery. At first, huge quantities were obtained by simple digging. In places they were as thick as 'plums in a pudding', and sometimes of enormous sizes. Many of the finest stones

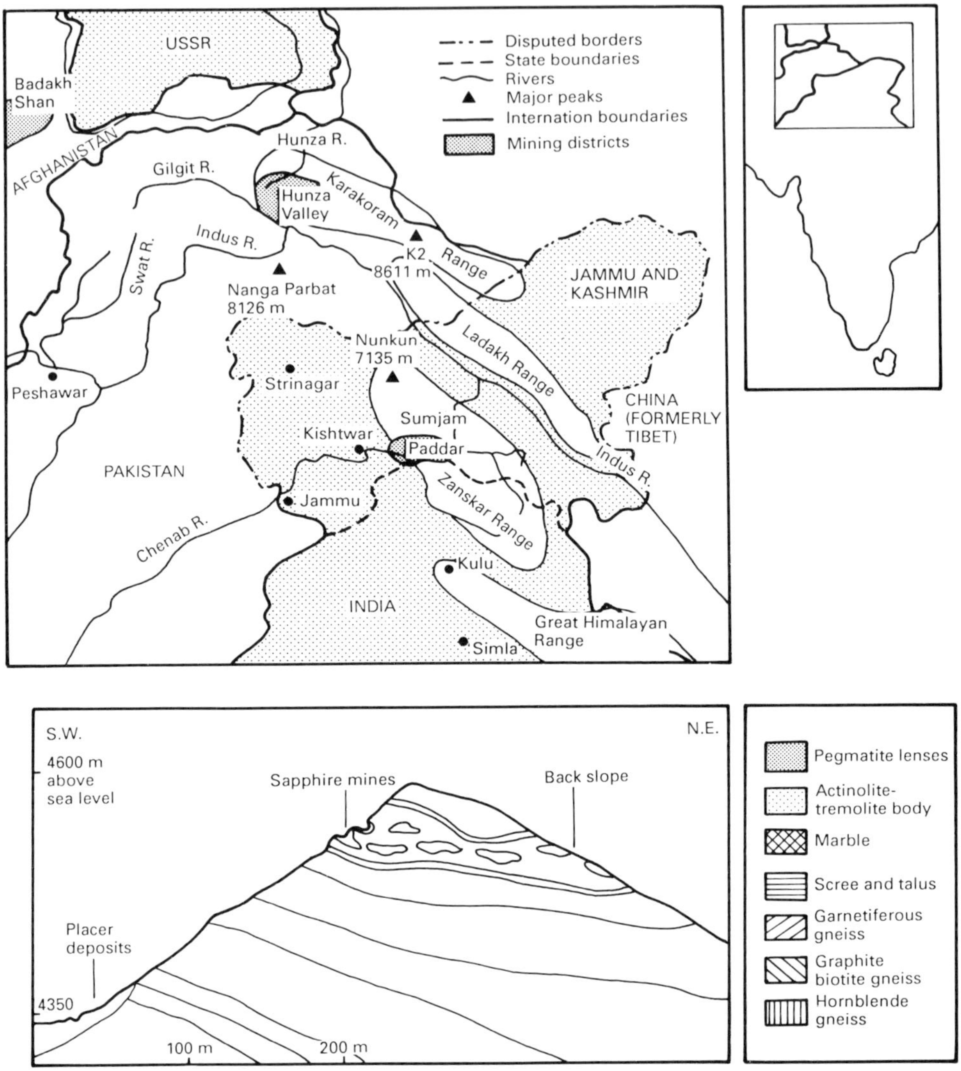

Figure 11.37 The location and geology of the famous Kashmir sapphire mines. (From Kothvala and Atkinson, 1983)

were obtained by the Maharajah and were stored at the Jammu and Kashmir State Treasury. C. S. Middlemiss (1931) described this king's ransom as follows:

> We are aware that one of these outcrops, namely that of the Old Mine, continued yielding gemstones for an appreciable time, and gave an extremely good output of very large stones from about the year 1881 to about 1887. This is a historical fact and is well known to many living people. A few specimens of sapphire then collected are still preserved, jealously guarded by the State, in the toshakhana [treasury], and have been seen by the writer. Of these there is at least one large piece, bigger than a polo or croquet ball, and others smaller all of a rich blue colour. There are also many cases of cut gems of pendant size which are superficially as large as florins.

Incredible! Bigger than a croquet ball! What has become of these stones, you may be thinking? The answer is that we just do not know. In the late 1970s reports from India stated that a caretaker had been caught smuggling pieces, one at a time, out of the palace in Kashmir; since then, however, nothing has been heard of the 'hoard of Kashmir'.

Middlemiss also discussed the failure of all the geologists first on the scene to describe the actual occurrence of sapphires from which these fine stones emanated:

> But of the details of the in situ rock occurrence whence these magnificent trophies were won we unfortunately know hardly anything, nor have we any recent descriptions by the Mineral Survey or Mining Engineer of the nature of the quarry, pits or other openings made by the early pioneers at this place. It is a curious fact that all the geological and mining men who have visited and reported on these mines, from La Touche downwards, though agreeing as to the position of the Old Mine working, have one and all seemed to shirk any description of them. Are they rock-face workings, irregular burrowings, tunnels, pits or what, what is their extent and how deep from the surface do they go? We simply do not know! La Touche, in his published paper simply states 'here the face of the rock has been laid bare by a landslip, and at first the sapphires were taken out of the granite itself: but when I visited the mines this patch of rock had ceased to yield any for some time, nor did the closest search bring any more to light'. Labhu Ram in his report says 'the Old Mine is also located in the same actinolite–tremolit mass that contains the New Mines . . . No trace of pegmatite veins is found near it and the mine has not yielded any stones for very many years since the late eighties' . . . None of the others who visited the mines, including the Mining Engineer, have anything to say at all on this matter.

Evidently, those who did have a chance to observe the workings at the Old Mine were so impressed by what they found that they completely forgot to describe them! This means that we know very little about how these incredible stones were obtained. Today, all that remains of the Old Mine are a few shallow burrows dug into the rock.

About 110 m from the Old Mine are a series of shallow adits distributed over a small area. In recent years much blasting has been done to get at the sapphire-bearing pegmatite (Atkinson and Kothavala, 1983). Today the adits are heavily barred to prevent entry and the entire valley is closely guarded by a small team of police stationed at its mouth all year round.

The sapphires found at the New Mine differ considerably from those obtained from the Old Mine. This difference is important in understanding the Kashmir material. Kashmir rough comes in two basic types, both of which are coated with a tenacious white clay. In all types, the blue colour is found mainly at the outer edges of the crystals, especially at the tips. Virtually all are of a spindle-shaped hexagonal bipyramid habit, as shown in Figure 11.38. Apart from the blue tips and faces, the rest of the crystal is colourless (the New Mine also produces the occasional stone with blue tips and a pink core). What this means is that the fine blue stones were largely cut from the tip of the crystals, similar to the way in which unburned blue sapphires from Sri Lanka are cut.

The major difference between stones from the Old Mine and those from the new (and also those obtained from the placer deposits below) is that in the New Mine stones, the blue faces have been corroded away. Rather than having flat well-formed faces, like the stones from the Old Mine, they have instead deeply pitted faces; thus the fine-coloured areas are, by and large, missing owing to surface corrosion. In stones from the Old Mine these faces (and thus the colour) are found intact, allowing fine stones to be cut from the crystals. This fact alone may account for the great scarcity of fine Kashmir sapphires, as the Old Mine has produced virtually nothing since 1887. Although well-formed crystals are occasionally found in the New Mine and the placer deposits, the vast majority of material is of the corroded type unsuitable for cutting because of its lack of colour.

All is not lost, however. Experiments done by burners in Bangkok on Kashmir material have shown that it is ideal for heat treatment. One large lot examined by the

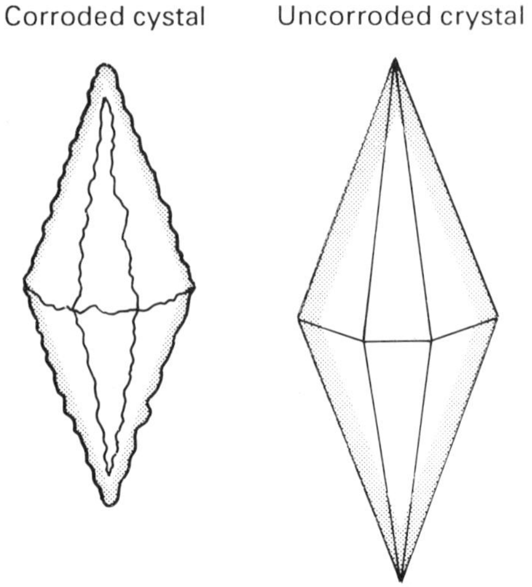

Figure 11.38 The famous sapphire mines of Kashmir in India produce two different types of rough sapphire. Some of the crystals display heavily corroded surfaces, while in others the original crystal surfaces are largely intact. Like the 'ottu' sapphires from Sri Lanka, the colour in Kashmir sapphires lies mainly along the crystal faces, with the core of the crystal being colourless. Unfortunately, in the corroded crystals this vital colour layer is largely absent, making them poorly suited for cutting. Both the corroded and the uncorroded crystals, however, have proven good for heat treatment

author both before and after burning showed a success rate of close to 100%, far better than even the best Sri Lankan geuda material. The only problem was that most were too heavily included to be faceted. Should facet-quality material exist, however, it has tremendous potential for heat treatment.

Kashmir sapphire rough is easily recognized from its distinctive mode of occurrence. The author recently purchased 1 kg of New Mine material and found that all stones are covered by a thin coating of white clay. So tenaciously does it cling to the skin of the sapphires that it requires hydrofluoric acid for its removal. The size of the crystals is small, mostly in the 1–4 carat range. Virtually all consist of tapering hexagonal bipyramids, and some feature small brown tourmaline prisms and mica flakes adhering to their surfaces or intergrown with them. Well-formed crystals have a rich blue colour at the tips and beneath the faces, while corroded crystals show only blue spots here and there. The faceting of these crystals reveals the slightly sleepy or milky appearance for which the Kashmir stones are famed. Tiny tubes or needles (probably rutile) give the stones a distinct haziness, resulting in the velvety-blue texture that is their hallmark. This haziness is present in virtually every piece examined by the author, but is not unique to the Kashmir stone; sapphires from Sri Lanka, Thailand (especially Kanchanaburi) and Pailin may also exhibit a certain milkiness, making confusion a very real possibility. The haziness in Kashmir stones, however, is extremely fine in nature, not enough to degrade the clarity seriously, but just enough to impart the distinctive velvety lustre to the stones. An excellent illustration of this effect is given in the fine article on Kashmir sapphires by Atkinson and Kothavala (1983). Incidentally, this article provides the only available current account of the Kashmir sapphire mines and is essential reading for those interested in learning more about this most famous of sapphire mines.

Other corundum localities in India

Kashmir is not the only locality in India producing corundum. Other sources exist, but are of little importance in world markets as the qualities are low. The other locations in India producing corundum are listed in Table 11.2 (from Viswanatha, 1982).

The major localities produce mainly ruby. Facet-grade ruby occurs in the Kangayam area of the state of Tamil Nadu, of which Madras is the capital. Stones from this area are of a reddish colour with a slight darkish tint, but are generally heavily included. Karnataka State is the other important source of ruby in India. The gems come mainly from the Channa–Patna area, but lack transparency and so are suitable only for cabochons and beads.

India is famous, however, not for faceted and cabochon ruby, but for very low-quality star rubies. These are found at Kangayam (Tamil Nadu) and Mysore (Karnataka), particularly the latter.

TABLE 11.2 Corundum localities in india (ruby)

Geological formations	*Gem tracts*
Contact of pegmatite and ultramafic rocks	Kangayam–Karur–Palni areas in Coimbatore, Tiruchirapalli and Madurai Districts of Tamil Nadu State
Anorthosite–gabbro–ultramafic complexes	Mysore, Bangalore, Tumkur, Coorg, Kolar, Mandya, Hassan, Chikmagalur, Chitradurga, Bellary, Raichur and Shimoga Districts of Karnataka State
Sillimanite–garnet ultramafic complexes	Anatapur, Nellore, Knamman and Visakhapatnam Districts of Andhra Pradesh State
Sillimanite–cordierite–garnet graphite gneisses	Betul area in Madhya Pradesh State
Sillimanite–kyanite–quartz graphite schists	Udaipur District in Rajasthan State
Vanadiferous titanomagnetite in mafic complexes	

The Indian star rubies are generally very heavily included and so of poor colour and transparency. They do possess very sharp stars, but because of the lack of transparency the colour is poor, so they are usually a few dollars per carat or less.

Because of their poor clarity Indian star rubies are often dyed and oiled. Most are filled with cracks, polysynthetic twin lamellae and parting planes which allow penetration of oils and/or dyes. The Indian trade magazine *Journal of Gem Industry* (July–August, 1976) suggested that 'the packing material used in wrapping these stones should never be absorbent, if good customer relations are desired'. Perhaps even better customer relations would result by omitting these dyes and oils altogether.

Bibliography

Anonymous (1970) A sale of Kashmir sapphires. *Australian Gemmologist,* February, 28

Anonymous (1976) Karnataka rubies: some points to remember. *Journal of Gem Industry*, July–August, 56–57

Anonymous (1977a) Hunt on for best quality rubies in India. *Journal of Gem Industry*, September–October, 43–44

Anonymous (1977b) Operation sapphire soon in Jammu–Kashmir. *Journal of Gem Industry*, March–April, 45–46

Anonymous (1978) Blue sapphire mine to be denationalised. *Journal of Gem Industry*, November–December, 63–64

Anonymous (1979a) Fate of world's best sapphires hangs in balance. *Journal of Gem Industry*, July–August, 55–56

Anonymous (1979b) Illegal ruby mining alleged. *Journal of Gem Industry*, May–June, 59

Anonymous (1979c) Indian blue sapphire mine being leased out. *Journal of Gem Industry*, March–April, 55

Anonymous (1979d) The Indian ruby bonanza. *Journal of Gem Industry*, July–August, 56–57

Atkinson, D. and Kothavala, R. Z. (1983) Kashmir sapphire. *Gems and Gemology,* **19**, 64–76

Atkinson, D. and Kothavala, R. Z. (1985) Kaschmir - saphir. *Lapis,* **10**, no. 10, 11–22

Ball, V. (1881) *A Manual of the Geology of India, Part 3, Economic Geology*, Geological Survey of India, Calcutta

Brown, G., Hamid, D. and Kelly, S. M. D. (1985) Some information on the Kashmir sapphire. *Australian Gemmologist,* **15**, no. 12, 448–449

Brown, J. C. (1956) Sapphires in India and Kashmir. *The Gemmologist,* **25**, no. 298, 78–80; no. 299, 97–100; no. 300, 129–132

Brown, J. C. and Dey, A. K. (1955) *India's Mineral Wealth*, Oxford University Press, Bombay

Buchanan, F. H. (1820) Account of the mines or quarry of corundum in Singraula. *Edinburgh Philosophical Journal*, no. 2

Fox, C. S. (1921) Notes on Indian precious stones. *Journal of Indian Industries and Labor,* **1**, part 3, 304–326

Gaines, R. V. (1946) The sapphire mines of Kashmir. *Himalayan Journal,* **13**, 73–77

Gaines, R. V. (1951) The sapphire mines of Kashmir. *Rocks and Minerals,* **39**

Gourjon, M. (1785) *The Indian Connoisseur; or the nature of Precious Stones*, London

Halford-Watkins, J. F. (1935) Kashmir sapphires. *The Gemmologist,* **4**, no. 42, 167–172
Heron, A. M. (1930) The gemstones of the Himalayas. *Himalayan Journal,* **2**, 21–28
Iyer, L. A. N. and Thiagarajan, R. (1961) Indian precious stones. *Bulletins, Geological Survey of India*, no. 18, 1–105
La Touche, T. H. D. (1890) The sapphire mines of Kashmir. *Records, Geological Survey of India,* **23**, part 2, 59–69
Lydekker, R. (1883) The geology of the Kashmir and Chamba Territories, and the British district of Khagan. *Memoirs, Geological Survey of India,* **22**, 336
Mallet, F. G. S. (1882) On sapphires recently discovered in the North-West Himalaya. *Records, Geological Survey of India,* **15**, part 2, 138–141
Mallet, F. R. (1879) On corundum from the Khasi Hills. *Records, Geological Survey of India,* **12**, part 3
Mallet, F. R. (1889) *Manual of the Geology of India, Part 4, Mineralogy*, Geological Survey of India, Calcutta
Middlemiss, C. S. (1896) Preliminary notes on some corundum localities in the Salem and Coimbatore districts, Madras. *Records, Geological Survey of India,* **29**, part 2
Middlemiss, C. S. (1931) Precious and semi-precious gemstones of Jammu and Kashmir. *Reports of the Mineral Survey of Jammu and Kashmir*, no. 9, 1–50
Parimoo, M. L. and Raina, B. K. (1968) Geological mapping in the sapphire mines area, Sumjam and in the parts of Bhutna and upper Chenab valleys, Padar area, Doda district, Jammu and Kashmir state. *Geological Survey of India Report*, 1966–1967, unpublished
Parkinson, K. (1952) Kashmir sapphires. *The Gemmologist,* **21**, no. 248, 39–40
Phukan, S. (1966) Studies on inclusions in some Indian gemstones. *Journal of Gemmology,* **10**, 1–7
Raina, B. K. (1961) Geological report on Sumjam sapphire mines area (Paddar), Doda District, Jammu and Kashmir. *Report of the Directorate of Geology and Mining, Jammu and Kashmir Government*, unpublished
Shepard, C. U. (1833) Corundum gems in India. *American Journal of Science*, series 3, **26**, 339–340
Sherwill, W. S. (1845) Note on corundum mines in Singrowlee. *Journal of the Asiatic Society of Bengal,* **14**
Viswanatha, M. N. (1982) Economic potentiality of gem tracts of southern India and other aspects of gem exploration and marketing. *Records, Geological Survey of India,* **114**, part 5, 71–89
Warmington, H. (no date) *The Commerce between the Roman Empire and India*, Curzon Press, London
Watt, G. (1892–1893) *A Dictionary of the Economic Products of India*, Allen, Calcutta
Zaveri, C. K. (1959) Gem mining in India. *Lapidary Journal,* **13**, no. 4, 546–554; no. 6, 618–623

Australia

It seems that with mother nature things always even out in the end. What Australia lacks in terms of flora or population is more than made up for by mineral wealth. Although the country is largely composed of desert wastelands, beneath the wastes lie some of the world's most economically important mineral deposits. And when sapphires were being doled out, Australia received a generous helping. Unfortunately, most are of relatively low quality, so much so that the term 'Australian sapphire' has become synonymous with the dark inky-blue sapphires found in abundance there. Australia does produce small quantities of better stones, but even the best cannot compare with the better Asian sapphires.

While on the subject of world gem distribution, an important point should be mentioned. It has long been known that gems tend to occur more frequently in the tropics. This was once believed to be due to the closeness of the sun, as gems were thought to 'ripen' more quickly in a warmer climate than a cool one. Today we know this to be due to an entirely different cause. Weathering occurs much more rapidly in the tropics; therefore the sediments (from which many gems are mined) are better classified or sorted in this climate, aiding mining tremendously. Furthermore, the tropics managed to escape the many Ice Ages which have ransacked the sediments of the more temperate zones. As the glaciers moved south they mixed up many of the well-sorted sediments, which means that the classification process of weathering had to begin anew after each Ice Age. Gems are mined much more easily from well-sorted sediments, as mother nature has already done much of the work.

The occurrence of sapphire in Australia closely resembles the deposits of Thailand and Cambodia, with iron-rich sapphires found associated with basalt flows. Recent

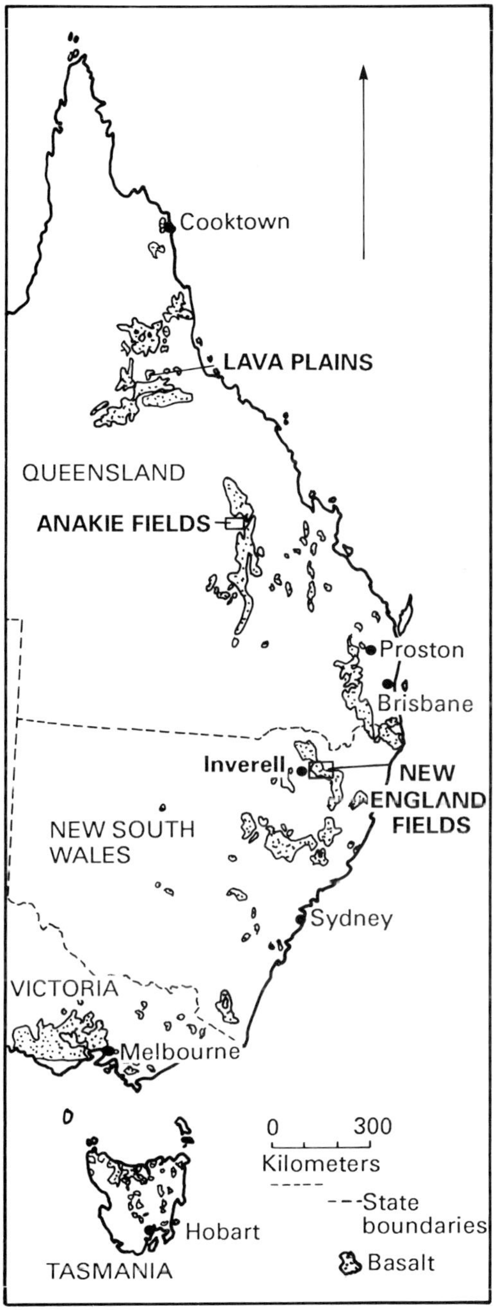

Figure 11.39 The sapphire mines of Australia. Most production is centred around the Anakie and New England fields, with the latter producing the best-quality stones. At present little is being produced from Lava Plains. (After Coldham, 1985)

studies, however, suggest that 'basaltic' tuffs may instead be the source of the corundums (Pecover, 1987). If true, this fact would have an important bearing not only in Australia, but also in south-east Asia (Thailand/Cambodia).

Two sources in Australia produce the majority of the sapphires: the Anakie fields in central Queensland and the New England fields in New South Wales. Of these two sites, New England produces the finest blue gems while Anakie is noted for fine yellow and green sapphires.

History

New South Wales

Sapphires were first discovered in Australia in 1851, when a Mr Strutchburry observed their presence in a gold sluice on the Cudgegong and Macquarie Rivers in New South Wales (Broughton, 1979). A few years later, in 1854, the Rev. Mr W. B. Clark noted sapphires in the tin sluicing concentrates of the Inverell mining district. This is the site of today's mining operations in New South Wales, and the discovery predated that in Queensland by over 20 years.

The crystals initially discovered were proven to be sapphire, but they were at first thought to be of no value. Later, in 1869 and 1870, larger amounts were discovered through gold-mining operations in several localities. Nothing came of this until much later, however.

Commercial mining operations for sapphire in New South Wales began for the first time in 1919, on Fraser Creek. This was at the site of the small present-day village of Sapphire, an appropriate name if there ever was one. Only 8 ounces were recovered that year, but evidently this was promising enough to spur further development. Mining soon progressed in the area of Inverell and Glen Innes, with the gemmiferous gravels of Fraser Creek, Horse Gully and Mary Anne Creek near Inverell and Sapphire, and Reddistone Creek north of Glen Innes, being exploited. The sapphire-mining industry was relatively stable until the Great Depression, after which production virtually ceased until the early 1960s (Broughton, 1980).

Queensland

The discovery of sapphire in Queensland is attributed to a Mr Richardson, who worked for the railroad. He discovered some small red zircons on Retreat Creek near the town of Anakie in 1875. Believing them to be rubies, he sent the red pebbles and their gravel concentrate off for testing. They proved to be zircons, but some of the other pebbles were sapphires. Mr Richardson eventually became one of the partners in a company set up to mine the field for sapphires. The company extended its prospecting to nearby areas, eventually expanding the gem area to 28 square miles. Production began in the 1890s, but was sporadic, with most of the stones going to tsarist Russia via German buyers (Coldham, 1985).

It is said that initial nineteenth-century production at Anakie was minimal due to a lack of incentive caused by low prices (Broughton, 1979). Many mines closed down at the end of the century, but were later reopened as new markets were developed for the dark-blue gems. A number of the early immigrant prospectors came from Germany, and Germans tightly controlled the selling and marketing of the stones in the years up until World War I. The stones were exported to Germany for sorting, cutting and eventual sale to the wealthy Imperial Russian Court (Broughton, 1979). It is said that the deep inky-blue stones, which today are considered inferior, held a fascination for the Russian nobility unmatched by sapphires from other sites. To the present author it seems that their low price may have held the greatest fascination, but this fact is not mentioned by other authors. Anakie produced mainly blue stones, but also fine-quality yellow and green stones. These fancy colours were not appreciated by gem buyers at the time and so the stones were used mainly by the watchmaking industry (Broughton, 1979).

World War I brought about the end of German domination of the Anakie sapphire industry. Unfortunately, it also brought an end to the lucrative Imperial Russian market. Despite the loss of this important market, sales rose after the end of the war, a boom that continued until the Great Depression as a result of the demand from Europe and the United States.

The popularity of Anakie sapphires descended to an all-time low in the period from 1930 to 1960. During the year 1957, a paltry $100 worth of sapphires was mined. It was only in the mid 1960s that the flagging market revived, primarily because of the discovery in Thailand that heat treatment could improve a stone's appearance. Like the Thai ruby, which was considered inferior by nearly all the experts of the late nineteenth and early twentieth centuries, the Australian sapphire today largely owes its market share to the skill of Thai burners in improving its appearance. According to Australian gem dealer Terrence Coldham (1985), 'without heat treatment to clear (remove the silk from) the stone, the mining of Australian sapphire would not be commercially viable'.

Mining methods and stone types

Like the deposits themselves, the mining methods used in Australia somewhat resemble those of Thailand.

The last 10–20 years have seen the development of modern commercial processing techniques in Australia. However, primitive hand sieving is also to be found. In the New England district of New South Wales, mechanized mines cover an area of some 4000 square kilometres, with each miner working a creek or river often hundreds of hectares in area (Coldham, 1985). After samples are taken, mining then proceeds in a systematic manner. In the Anakie district claims are much smaller because state mining laws limit their size. This results in miners working almost on top of each other in an area only a few square kilometres in size. Anakie miners are also more concentrated because the gem gravel tends to occur in thicker and richer deposits (Coldham, 1985).

Both districts feature gem gravel covered by about 1 to 25 m of barren overburden, which is typically removed by either backhoes or bulldozers. As single claims do not have enough work to keep this heavy equipment busy at all times, miners usually band together to hire the machines. After removal of the overburden, the gem-bearing wash is scraped up and loaded onto trucks for removal to the processing plants. Some parts of the Anakie fields are reserved for single miners, or fossickers, who mine by sinking small shafts, as is done in some parts of Thailand.

Gravel brought to the processing plants is tipped into dump boxes where it is washed into a trommel with high-pressure water hoses. The trommel consists of a revolving screen similar to those used at Nong Bon in Thailand, where a coarse gravel is found. This normally has two different sized screens: the first sieves out fine sand-sized particles while the second removes 1¾ to 2½ cm material. Larger gravel passes into rock bins for return to the excavations. The middle sizes of 0.15 to 2.5 cm pass into a pulsator jig for separation. Again this is similar to that of the corundum mines in Thailand, and it is from this pulsator that most of the gems are recovered. Recovery rates are said to be as high as 90–95% (Coldham, 1985).

Unlike Thailand, Australia has strict conservation laws requiring reclamation of the mined-out areas, particularly in New South Wales. All gravels that have been processed are returned to the sites for infilling and are then covered with the original overburden and top soil. This prevents the destruction of the land as occurs in Thailand, and Thailand would be wise to adopt similar measures.

At the end of the day the concentrate is removed from the pulsator for selection. In addition to corundum, zircon, pleonaste and other minerals are found, but only the sapphire and zircon are of use in gems.

Separation of the sapphire involves drying the concentrate and passing it through a magnetic separator to remove iron-rich minerals, such as black spinel. This leaves a concentrate containing the corundum which represents only 10% of the whole before magnetic separation. Women then carefully work the concentrate by hand over mirrors and remove the gems, the mirrors allowing light to reflect back through the gems and expose them (Coldham, 1985). Perhaps light boxes might be even more suitable for this work.

Blue sapphires are typically placed into a separate 'mine run' parcel consisting of all qualities except the very largest and finest stones, which are sold separately. Approximately 95% of the total corundum production consists of blue stones (Coldham, 1985).

Sometimes the mine run is sold 'as is', but at other times it is divided into firsts, seconds and thirds, with each category being further sieved into separate size categories. These are then sold, typically, by the ounce or kilogram. The exact range of qualities will of course depend on the individual mine's grading, which varies from place to place and person to person. Other colours like the green, yellow and parti-colour stones are sold separately in small lots, as they represent but a few per cent of the entire output.

According to Coldham (1985), who has provided the most extensive modern description of the Australian corundum mines, a typical Inverell mine run parcel contains about 10–20% of stones above 2 carats, 40–60% of 0.5 to 2 carat sizes, and 30–40% of 0.20–0.50 carat sizes. Mine runs from Anakie are of a slightly larger stone size overall (20–40% are above 2 carats). Most cuttable Australian sapphires are under 2 carats in the rough, with clean stones above 10 carats being quite rare today. Very large pieces (up to 1000 carats) were once recovered, but are rare now because the large gravel is removed by the trommel. At Anakie, large stones are still recovered occasionally from the areas restricted for individual pit miners without heavy equipment. This is particularly true at the Willows field, which is noted for fine golden-yellow stones (Broughton, 1979).

The blue sapphires make up over 90% of the total production. Generally, those of Anakie tend to be very dark, almost black at times, while finer blues are found in New South Wales. Most material from both localities, however, tends towards the inky-blue shades. This dark material may be deliberately cut slightly or far off the pure blue *c* axis. Although this causes dilution of the pure o-ray, making the stone more greenish, it is preferable to a black stone in cases when the o-ray is too intense. Such an orientation, where the table is parallel to the *c* axis, is termed 'cross-table' in Australia.

There does exist one further sapphire-bearing locality in Australia, in northern Queensland at Lava Plains. Material from this site given to the author by Mr Barry O'Leary displayed a rich deep-blue colour somewhat reminiscent of stones from I-Ram in Thailand or Pailin in Cambodia. The only question regarding development of this site is whether or not material clean enough for faceting can be found. Mr O'Leary, for one, believes the site has great potential.

Other than blue sapphires, the most common varieties in Australia are those of a blue-green hue. Two other colours that are often found are yellow and green. While New South Wales is acknowledged as the source of Australia's finest blue sapphires, Anakie in Queensland is said to produce the finest green and yellow sapphires. Green stones occur in a wide range of hues, from bluish green to yellowish green, but all suffer from the typical green sapphire malady – a lack of purity and intensity.

Australian yellow sapphires may be of very high quality on occasion but, unfortunately, large quantities are not available. These stones range from greenish yellow through pure yellow to the highly prized golden yellow. Occasionally these latter stones may resemble the prized whisky-yellow stones of Chanthaburi in Thailand. Like the blue sapphires, these stones are also heat treated to remove the silk, but the changes are generally slight. In addition to green and yellow sapphires, an unusual variety is found which is of a yellow-green to green-yellow colour. It is termed 'wattle' because of its resemblance to Australia's national flower and is popular locally. Pink, purple and mauve stones are very rarely encountered in the Australian gem gravels. Some of these may even show a colour change from greyish green in daylight to purplish in incandescent light. Black star sapphires featuring both six and 12 rays are also found, including the rare 'golden-star' black star sapphires. These are quite similar to those mined in Thailand.

In terms of marketing the Australian sapphire production, the majority is brought by Thai buyers in Australia. This situation originally developed because of the Thais' ability to heat treat the material. Generally the rough is shipped to Bangkok for treatment and then cutting. From here it is bought by international buyers who then sell the stones worldwide. Unfortunately, because the Australian sapphire has a reputation for being too dark and inky and because much of the Thai production strongly resembles that from Australia, the inferior dark stones of both countries often end up being sold as 'Australian' sapphire while the better stones from both countries are sold as 'Thai' sapphire. This situation could change, however, if a larger percentage of the Australian rough were to be treated and sold in Australia. It is unlikely to happen, though, so Australia will probably retain its reputation for supplying the world with large quantities of dark commercial-grade sapphire.

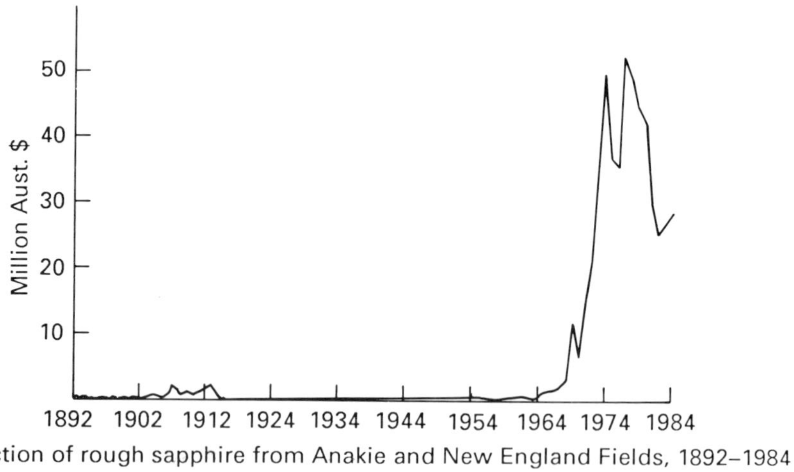

Figure 11.40 Production of rough Australian sapphire. (From Coldham, 1985)

Ruby in Australia

In 1978, the first commercially significant discovery of ruby in Australia was made in the Harts Range, Northern Territory (McColl and Warren, 1980). According to McColl and Warren, the host rock is an amphibolitic complex, interpreted as a development from a terra rossa soil profile formed on a limestone contaminated with pyroclastics and interlaminated with lavas and tuffs.

The ruby occurs as well-formed crystals, but practically no material suitable for faceting has been found. In the early 1980s a company was formed to mine the deposit and a lapidary workshop was set up in Bangkok specifically to cut the material. However, the people involved had little experience in the gem business and so did not recognize the important fact that red corundum alone does not a ruby make. Because of the lack of clarity of the material it was suitable only for the lowest cabochon grades and the company soon floundered.

Bibliography

Anderson, O. (1971) A century of sapphire mining. *The Australian Gemmologist*, August, 11–20

Anonymous (1976) *Economic Geology of Australia and Papua New Guinea*, Australasian Institute of Mining and Metallurgy

Anonymous (1983) *Sapphires in New South Wales* (brochure), Department of Mineral Resources, New South Wales

Anonymous (1985) A new type of sapphire deposit – implications for the sapphire industry in New South Wales. *Minfo*, no. 7, 14–18

Batchelor, H. H. (1938) Queensland sapphires. *The Gemmologist,* **8**, no. 37, 42–43

Batchelor, H. H. (1959) Mining in Australia. *The Gemmologist,* **28**, no. 333, 66–68

Broughton, P. L. (1979) Economic geology of the Anakie sapphire mining district, Queensland. *Journal of Gemmology,* **16**, 318–337

Broughton, P. L. (1980) The gemmiferous gravels of the Fraser and Reddistone Creeks, Inverell–Glen Innes District, New South Wales. *Journal of Gemmology,* **17**, 95–118

Chalmers, R. O. (1952) Australian gemstones. *The Gemmologist,* **21**, no. 250, 82–84

Coldham, T. (1973) Sapphire mining in Northern N.S.W. *The Australian Gemmologist*, May, 14–19

Coldham, T. (1985) Sapphires from Australia. *Gems and Gemology,* **21**, 130–146

Coleman, D. C. (1968) Digging sapphires in Australia. *Lapidary Journal*, November, 1070–1075

Dunstan, B. (1902) The sapphire fields of Anakie. *Geological Survey of Queensland Publications*, 1–26

Gartrell, B. (1960) Sapphire mining in Australia. *The Gemmologist,* **29**, June, 103–108

Jack, R. L. (1982) Sapphire, gold, and silver mines near Witherfield. *Queensland Geological Survey Publications*, no. 81, 1–3

McColl, D. H. and Warren, R. G. (1980) First discovery of ruby in Australia. *Mineralogical Record*, November–December, 371–375

MacDonald, R. (1930) The sapphire-fields of central Queensland. *Chamber's Journal,* **6**, 483–487

MacNevin, A. A. (1972) Sapphires in the New England district, New South Wales. *New South Wales Geological Survey – Records,* **14**, 19–35

MacNevin, A. A. (1973) Sapphire mining in New South Wales. *The Australian Gemmologist*, August, 14–16

Males, P. A. (1976) Ruby corundum from the Harts Range, N. T. *The Australian Gemmologist*, May, 310–312

Mumme, I. A. (1978) The Little River sapphire lease. *The Australian Gemmologist*, May, 177–183

Mumme, I. A. and Ball, R. A. (1978) Notes on sapphires from stream gravels at Fraser's Creek in Glen Innes area, New South Wales. *The Australian Gemmologist*, May, 173–175

Pecover, S. R. (1987) New concepts on the origin of sapphire in Northeastern New South Wales. *The Australian Gemmologist,* **16**, 221

Stratton, V. (1951) Australian sapphire fields not fully exploited. *Gems and Gemology,* **7**, 125–128

East Africa (Kenya and Tanzania)

Many gemmologists and traders believe that Africa is the source of the future. A huge continent with large areas which still remain unexplored, it certainly has great potential in terms of mineral wealth. Since the early 1960s we have begun to develop this potential, with new discoveries of great importance being unearthed. Nowhere on the continent is this more true than in East Africa, particularly Kenya and Tanzania.

Garnets may be East Africa's greatest boast, but fortunately the area is also well endowed with corundums. In this section we will explore the African deposits of greatest economic importance – those of Kenya and Tanzania.

Kenya

Of the corundum gems found in Kenya, it is the ruby which takes first place. Kenya is a land of fabulous game reserves and a pro-Western government; it is also a land of untold gem wealth, much of it just beginning to be tapped. Included in the gem cornucopia of Kenya is the ruby. With ruby being in extremely short supply and the world's important ruby deposits numbering less than the fingers of one hand, the discovery of gem qualities of the famed red stone in Kenya was important in its own right. However, it was for entirely different reasons that the discovery made world headlines in 1974.

In the year 1973 two American geologists, John Saul and Elliott 'Tim' Miller, were prospecting in Kenya's Tsavo West National Park for chromium- and vanadium-bearing gemstones when they found what has become one of the world's richest deposits of ruby. Saul, a long-term Kenya resident, did what any miner would do – he obtained a fully legal permit to mine the deposit. In Kenya, however, things were not quite the same as in other countries. Thinking that local participation would smooth the way, they offered 51% of the deal to a group of high-ranking Kenyans, including the then Vice President, Daniel Arap Moi (*Time Magazine*, 14 October 1974).

Unfortunately for Saul and Miller, others soon learned of the find. One of these was Beth Mugo, President Kenyatta's niece and unofficial lady-in-waiting to his wife, Mama Ngina; another was a wealthy Greek resident, George Criticos. Criticos was a friend of Kenyatta's and involved with his wife in running the Kenya Trade and Development Corporation. According to the *Time Magazine* article from which this account is largely taken, Saul and Miller charged that Mugo and Criticos encouraged other leading Kenyans, including Mama Ngina, to demand a larger share of the take. This the Americans complied with, increasing the Kenyans' share to 72%. The old adage of 'give them an inch and they'll take a yard' was certainly true in this case, for the Kenyans continued to make demands. Within a matter of days, the total 'requests' by influential Kenyans exceeded 150% of the ownership, with no end in sight. In June 1974, John Saul was suddenly declared a 'prohibited immigrant' and given 2½ hours to leave the country. Miller, who had been out of the country, returned but found it advisable to go into hiding; after a month he left for London. Kenyatta, in an oblique reference to the ruby mine, publicly declared after Saul's expulsion that no foreigner should be allowed to exploit Kenya's resources for his own private benefit. Laudable as this stand was, in this instance it rang hollow; the wealth of the mine was apparently intended for the pockets of Kenyatta and his cronies exclusively.

Perhaps Kenyatta should have supplied paperweights to the Ministry of Natural Resources, for it appears that the record of Saul and Miller's original claim mysteriously disappeared, blown away by a strong wind no doubt. Just as mysterious was the appearance of a different claim for the mine, this time in the name of one George Criticos, and probably blown in on the same wind that took the Americans' claim out. This was after a local court temporarily forbade Criticos from working the deposit. The court decision, however, apparently did not even cause Criticos to pause to catch his breath, for he continued to work the claim. After all, why worry about such trivial matters as court orders when one is in partnership with the President and his family? Not George Criticos, evidently.

Despite protests by the then American ambassador, Anthony D. Marshall (in whose safe Saul had deposited duplicate copies of the claim documentation), and a lawsuit by Saul and Miller in the Kenyan courts, the mines were never returned to their rightful owners. The legacy of the original prospectors still lives on in the minds of the people of the region, however. Today the famous ruby mine is known as 'The John Saul Mine' (Bridges, 1982) and its production's colour, the 'johnsaul' colour.

The ruby mines of Kenya lie in the area of Mangari, just within the north-eastern tip of the southern section of the Tsavo West

National Park. Here, at the junction of the seasonal Mwatate and Bura Rivers, are the ruby mines that initially generated such controversy. There are actually two mines, consisting of altered ultrabasic intrusive bodies lying just over 3 kilometres apart. Although they are very close together and of similar geologic occurrence, the quality of rubies found at each is distinctly different.

The more easterly deposit is known as the 'Penny Lane Mine', while the westerly pipe is called the 'John Saul Mine'. Dr W. G. Pohl (1977) has stated that the corundum occurs

1. in desilicified veins cutting through altered and serpentinized ultramafic rocks; or
2. in or near the contact zone (of the serpentinized pipe) associated with desilicified pegmatoidal segregations. Apparently the former refers to the Penny Lane Mine while the latter refers to the John Saul Mine.

At Penny Lane, large open-cast pit diggings have been made into a yellowish-green tremolite–actinolite talc chlorite rock, following narrow mica-rich (sericite, vermiculite?) veins, with which kyanite and crystalline limestone segregations are sometimes found (Bridges, 1982). The rubies occur in pocketed clusters along these veins, especially where mica is concentrated. Mica adheres to the surfaces of the rubies and when removed reveals euhedral to subhedral ruby crystals of a dark-red to dark-pinkish-red colour. These stones fluoresce a very strong crimson under ultraviolet light and like most rubies improve greatly in appearance when illuminated by incandescent light. However, almost all are heavily included and so are suitable for cabochons only. Dark-green tourmalines are also found in the pegmatites of the area (Bridges, 1982).

Rubies found at the John Saul Mine occur within the contact zone of a rhomb-shaped pipe measuring about 200 m across. Plumastic pegmatites occur in this zone and are sometimes 'riddled with rubies' (Bridges, 1982). The pipe itself consists of a greenish 'rotten' rock, similar in appearance to the blue ground of a diamond pipe, and where chlorite is found ruby is also encountered.

Considerable eluvial reserves of ruby are said to exist in the area due to its flatness and the millions of years of weathering that have occurred.

Tim Miller is quoted by Cambell Bridges (1982) as describing the plant at the John Saul Mine as probably the most modern, expensive and efficient of any gem mine in East Africa. Like most, it works on a heavy media separation principle and was chosen because it can process and sort large quantities of material and because, with ore fed in at one end and cleaned ruby returned at the other, there is very little chance for theft of stones, a major concern at any gem-mining operation.

The rubies of the John Saul Mine tend to be lighter and brighter than those of Penny Lane, displaying a rather strong pink to purple-pink element (Bridges, 1982). Minute exsolved particles are often present, but actual rutile silk is absent; therefore star stones cannot be obtained from this mine. The vast majority of the stones are of fine cabochon to sub-cabochon grade, but occasionally a fine small clear piece is found. The stones are all strongly fluorescent under ultraviolet light and at times somewhat resemble the rubies of Burma in terms of their colour. However, Kenyan stones are generally filled with thick wispy liquid fingerprints and feathers, unlike stones from Burma. These thick fingerprints have caused many Kenyan rubies to be misidentified as flux-grown synthetic stones, which is indeed unfortunate as they also contain included crystals and long white boehmite needles along polysynthetic twin lamellae. In addition, they contain zoned white clouds of minute exsolved particles of an unknown nature and very strong hexagonal colour banding. The above inclusions, as well as the occasional euhedral pyrite crystal, together allow for positive identification.

Today many Kenyan rubies are brought to Bangkok for heat treatment, cutting and eventual sale. It has been found, as with most rubies, that their appearance can be greatly improved by heat treatment, and many a stone would be virtually unsaleable without it. The better quality (i.e. cleaner) Kenyan stones often end up being passed off as Burmese because of their strong crimson fluorescence. The author has encountered faceted Kenyan rubies up to 20 carats in size selling as Burmese. Separation is usually straightforward, as the Burmese stones contain actual rutile silk and generally lack liquid fingerprints. Recently, large numbers of Burmese rubies have been heat treated to remove this silk. This also produces numerous fingerprints and feathers which were previously rare in Burmese stones. This makes separation of heat-treated Burmese and Kenyan rubies much more difficult. Both show the obvious signs of heat treatment, such as exploded or melted crystals with tension haloes, but the heat treatment obscures the origin. Perhaps this could be a blessing in disguise. It might finally push the gem trade into concentrating its efforts on separating natural stones from treated stones rather than worrying about source of origin. In the opinion of the author this is one area in which the coloured stone trade could learn from the diamond trade. After all, how many customers ask for, say, a Botswanan diamond?

Rubies and sapphires occur at several other localities in Kenya and, according to Bridges (1982), there appears to be a definite north-north-west striking belt that traverses both Kenya and Tanzania. About 80 miles north-east of Mt Kenya, at Garba Tula, dark-blue sapphires of an inky colour are found that strongly resemble those from Australia. It does not appear that these are being exploited, however, presumably because large enough quantities of facetable stones have not been found. Some fine yellow sapphires have also been recovered from this locality.

Tanzania

Although rubies are mined in Tanzania, it is for sapphires that this country is chiefly famous. These are mined from the Umba River Valley in north-eastern Tanzania, very close to the Kenyan border. From the Umba region come sapphires of virtually every colour: oranges, yellows, violets, blues, colour-change stones and colours for which no word can be found. John Saul once put together a reference collection of 108 distinct sapphire colours from Umba, excluding greys, parti-coloured stones and colour-change stones. Other gems, such as garnet, are also found, making Umba one of the richest sources of gems in the world.

Rubies and sapphires occur in and around a greyish-green serpentinite pipe measuring about 2½ km long in a north-north-east direction by 1 km wide (Bridges, 1982). Pegmatitic veins striking generally north-north-east and dipping almost exclusively eastwards, cut and lie within the serpentinite pipe. Associated with the pegmatite are vermiculite, chlorite, actinolite and sometimes corundum. The country rock consists of amphibolite–garnet gneisses, biotite gneisses, crystalline limestones carrying small flakes of phlogophite and graphite, and siliceous granulites. Corundum occurs mainly in the limestones, but fine gem-quality stones are found almost exclusively in the pegmatitic areas within or bordering the pipe where it is associated with vermiculite and chlorite. Gem material is also mined from alluvial gravels and this has been the main source since mining first began (Bridges, 1982).

Corundum was first discovered at Umba in 1960. It is said that a young Tanzanian of Greek origin made the original find (Sarofim, 1970). Claims were subsequently pegged by Umba Ventures, Ltd, a company formed to work the deposit (Solesbury, 1967).

The sapphire occurrence at Umba is somewhat unique (Sarofim, 1970). Stones are found around the aforementioned pipe, being brought up in shafts scattered in a perimeter of over 3 km. According to Sarofim (1970), ten different shafts were found scattered around the pipe, each producing only one colour of sapphire. The matrix collected from the shafts is processed by knocking it to reduce its size. It is then milled and extensively treated in tumblers. This causes disintegration of the agglomerate and thus frees the sapphires.

Production from the mine in 1969 was said to be 100 000 carats per month. Of this amount, over two-thirds was produced by the company while small miners found less than 20 000 carats. It is interesting to note that a very large proportion of the sapphire mined at Umba is large rough of a facetable grade (Sarofim, 1970).

True ruby-red colours, although very scarce, do occur on occasion at Umba. However, most of the reddish stones found tend towards orangy-red and magenta shades which traditionally have not been considered under the varietal name of ruby.

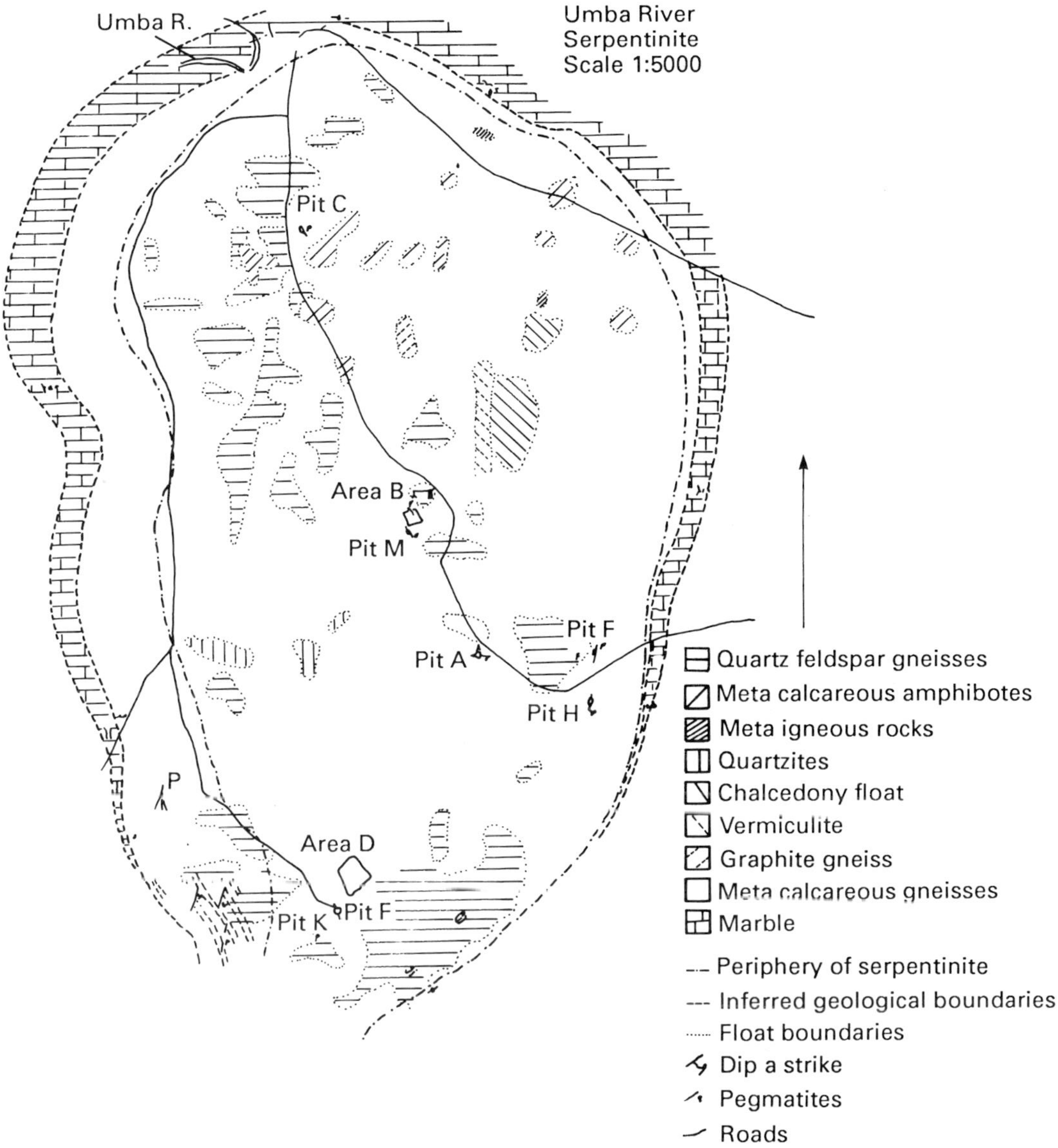

Figure 11.41 Corundum deposits of Umba River, Tanzania. (From Solesbury, 1967)

Instead, it is the multi-hued sapphires which are Umba's forte. These range from light to deep blue (most with a violet tinge), pink, orange, yellow, purple, green, peach, lavender, lilac, violet and even colour-change varieties. Parti-coloured stones are also found. One such specimen described by Cambell Bridges (1982) had a diameter of 4 inches (over 10 cm), passing in colour from deep pink to orange, yellow, green and blue! It is said that at least six of the major colour-producing ions (chromium, iron, manganese, nickel, titanium, vanadium) are present in the pipe at Umba and its environs. This appears to be one factor that accounts for the tremendous variation encountered in the colour of the sapphires from the Umba mine.

One of the more interesting varieties of corundum found is the colour-change sapphire. It changes from a greenish-grey in daylight to a purplish colour under incandescent light and it appears that vanadium may play an important role in causing this shift of colour. Another unusual variety is the orange sapphire. The Umba orange sapphire tends to be a highly saturated brownish-orange colour. It is often incorrectly sold under the name of padparadscha, the term that refers to a very delicate pastel pinkish-orange sapphire from Sri Lanka. Like the term 'heliodor', the author feels that this term should be reserved solely for stones coming from a specific locality. In the case of the padparadscha, that locality is Sri Lanka. So far as the author is aware, no other locality has yet produced sapphires of this unique and valuable colour; Sri Lanka remains the only source.

The first discovery of corundum in Tanzania dates back as far as the early 1900s. A German officer stationed on the Kenyan border at Mt Longido found red stones in the area which subsequently proved to be rubies (Sarofim, 1970). The outbreak of World War I, with the Battle of Merkerstein, fought close to the deposit, put an end to his exploration, but at the end of the war he returned to East Africa to continue the search. In 1924 he formed the Tanganyika Corundum Exploration Corp. Ltd, which, according to Sarofim (1970), was still operating in 1970.

The company's concession covers nearly 300 hectares, consisting of two hills at the foot of Mt Longido. Here the ruby displays a unique origin, being found as hexagonal red crystals of sometimes enormous size (for ruby) encased in 'anyolite' (derived from a Masai word for green), a rock consisting largely of a chrome-green zoisite. The rock is brought to the surface from galleries over 40 m deep sunk into the hills. Anyolite occurs in veins which are not more than 1¼ m wide, but which may measure a few hundred metres in length (Sarofim, 1970). Invariably it contains black hornblende concentrations which appear to affect the colour not only of the rock, but also of the rubies. The best rubies are said to be found in the light-coloured anyolite, while the darker rock yields darker rubies.

After the anyolite is brought to the surface, the rock is hammered to free the ruby crystals, which vary in diameter from very small sizes to more than 50 cm (Sarofim, 1970). Cobbing of the rough is performed, as all the rubies are heavily included. Each stone is broken down until it begins to transmit some light. Those which upon cobbing to 2 mm size still fail to transmit light are then discarded. Prior to export, the rough rubies are commonly tumble-polished to remove cracked sections. According to Sarofim (1970) the amount of usable gem ruby rough in anyolite is said to be of the order of 80 carats per tonne. Of this, only 25% is of facet grade, while the rest is cabochon quality (this appears to be a highly optimistic estimate). Most rough is less than 5 mm in size. Recently it has been found that this rough can be greatly improved in both colour and clarity by suitable heat treatment. Fair quantities of the material find their way to Bangkok for burning and cutting. The stones from Longido tend to be of a deep-brownish-red colour, somewhat similar to the better grades of Indian ruby, and after treatment they are strongly fluorescent. Some of the mine's output goes to gem carvers. The carved material often incorporates the original chrome-green zoisite matrix. Cabochons are also fashioned out of the matrix, and when rubies are present they are sold as 'ruby-in-zoisite'.

Longido is but one of several areas in Tanzania known to produce ruby. One small Longido-type deposit at Lossogonoi on the Masai Steppe produced several thousand carats of superb darkish rubies during a few weeks of digging in the late 1960s. The Morogoro region, 120 km west of the capital Dar es Salaam, also produces the red corundums. In this area the rubies are found in either alluvial diggings or associated with agglomerates of mica and kyanite (Sarofim, 1970). Gravel containing the rubies is usually collected from the surface or extracted from shallow pits. It is then washed in the typical manner of such workings and the rubies selected by hand. Small amounts of spinel are also found. When the ruby occurs in the mica/kyanite agglomerate, the matrix is either crushed or tumbled to free the gems. Morogoro rubies are said to be of a pinkish colour somewhat reminiscent of some Burmese rubies. In 1970, the entire production of ruby in Tanzania was said to be less than 35 000 carats per month (Sarofim, 1970).

Bibliography

Anonymous (1974) Kenya: the ruby rip-off. *Time Magazine*, 14 October, 55–56

Bank, H. (1971) Uber einige Edelsteine aus Tansania und ihre Vorkommen. *Abhandlungen Hess-L.-Amt Bodenforschung,* **60**, 203–215

Bridges, C. R. (1982) Gemstones of East Africa. *International Gemological Symposium Proceedings* (1982), pp. 266–275

Game, P. M. (1954) Zoisite-amphibolite with corundum from Tanganyika. *Mineralogical Magazine,* **30**, 458–466

Gunawardene, M. (1984) Reddish-brown sapphires from Umba Valley, Tanzania. *Journal of Gemmology,* **19**, 139–143

Hänni, H. A. (1986) Korunde aus dem Umba-Tal, Tansania. *Zeitschrift der Deutschen Gemmologischen Gesellschaft,* **35**, 1–13

Hänni, H. A. (1987) On corundums from Umba Valley, Tanzania. *Journal of Gemmology,* **20**, 278–284

Pohl, W. G. (1977) Geology of the Mangari ruby mines. *Austrominerals*, Nairobi

Sarofim, E. (1970) Gem rich Tanzania. *Lapidary Journal*, June, 434–439

Schmetzer, K., Bosshart, G. and Hänni, H. A. (1983) Naturally-colored and treated yellow and orange-brown sapphires. *Journal of Gemmology,* **18**, 607–622

Solesbury, F. W. (1967) Gem corundum pegmatites in N.E. Tanganyika. *Economic Geology,* **62**, 983–991

Webster, R. (1961) Corundum in Tanganyika. *Gems and Gemology*, Fall, 202–205

Zwaan, P. C. (1974) Garnet, corundum, and other gem minerals from Umba, Tanzania. *Scripta Geologica,* **20**, 19–30

Montana, USA

The United States is not generally noted as an important producer of gemstones. None the less, the state of Montana is well endowed with a variety of corundum gems, most notably the blue sapphire. Fine blue gems have been mined at Yogo Gulch, while further west fancy sapphires are found associated with the placers of the Missouri River and at Rock Creek. It is the blue sapphires of Yogo, however, that are Montana's major boast, so it is to this site that we will turn first in our exploration of the Montana sapphire deposits.

Yogo Gulch

Yogo Gulch is situated in eastern Montana, in Judith Basin County. Sapphires were first discovered at Yogo in 1894 (Voynick, 1985b). It was not sapphires that first brought miners to the area, however, but a much different mineral – gold. Prospectors along Yogo Creek found little gold, but noted small blue wafers in the bottom of their sluices. Jake Hoover, a hunter and prospector of the type which the American West seemed to breed like flies, is credited with the original discovery. Although it is likely that the sapphires were actually found by prospectors some 15 years earlier, they did not recognize their actual nature. Hoover, too, at first simply dropped them back into the creek. Only after the gold was played out did he begin to collect the little blue pebbles. Eventually a box of them was sent off to New York, to Tiffany & Co., where the famous gem expert, George F. Kunz, immediately recognized them for what they were – fine blue sapphires. Unlike the off-colour stones already being mined along the Missouri River, these were of a rich blue colour, making the discovery at Yogo the most important occurrence of gemstones in the United States. The box of sapphires was never returned to Hoover. Instead, in its place came a cheque for \$3750, together with a letter stating that the stones were 'sapphires of unusual quality' (Voynick, 1985b). At the little town of Yogo, the rush was on. Kunz, writing in his reminiscences in 1928, stated that the mine was later to produce more gem wealth 'than all the other sapphire mines in America put together'.

The arrival of the Tiffany & Co. cheque brought attention to bear on the sapphires rather than gold. Being found only below the point where the dike cut across the creek, it was not long before the actual primary dike rock was uncovered. In 1896 a partnership was formed, bringing into being the New Mine Sapphire Syndicate. Although the mining, which made use of high-pressure water jets and wooden sluices, was successful, the marketing of the rough stones was not. It was this difficulty that caused the Syndicate to approach a respected London firm, Johnson, Walker & Tolhurst.

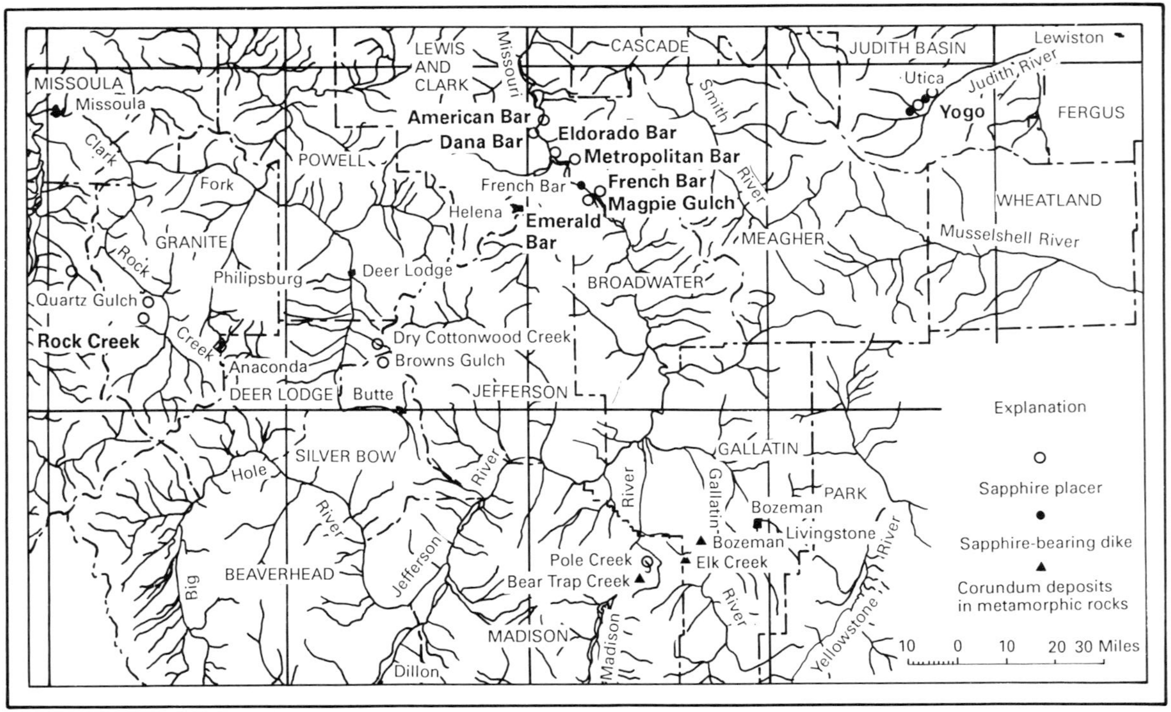

Figure 11.42 The corundum deposits of Montana. Major mines are located at Yogo Gulch, along the Missouri River, and at Rock Creek. (After Clabaugh, 1952)

Johnson, Walker & Tolhurst was already involved in mining sapphires in Sri Lanka and Burma and envisioned setting up a De Beers-style sapphire monopoly. The offer to become sole agents for the sale of Yogo sapphires was thus readily accepted. This began the most profitable period for the Yogo sapphire mine.

In the year 1899 the company dispatched an English mining engineer, Charles Gadson, to oversee operations at Yogo. Gadson arrived with his wife in tow, beginning a relationship with the Yogo mine which was to be severed only by his death in 1954. By 1902 the British firm had acquired full ownership of the mine and Gadson was given complete authority. Yearly production soon topped 1 million carats and the 10% which was of gem quality accounted for 75% of the \$100 000 annual income (Voynick, 1985b).

Initial mining at Yogo consisted entirely of working the accumulated placer deposits and surface workings on the dike itself. Gadson, however, soon undertook underground mining. Hardrock miners recruited from nearby Butte brought the dike rock to the surface where nature was allowed to lend a hand. Although this rock was hard at first, after a year's exposure to the weathering elements at the surface it became soft and crumbly and ready for washing. (This is a technique that has been used to great effect up to the present day.) Sluices were then used to separate the sapphires from the worthless dike rock. As recovery efficiency was vital, Gadson decreed that the tailings be rewashed a second and even a third time. This paid off when in 1910 the largest Yogo sapphire ever found, a stone of 19 carats in the rough, was discovered.

The British company was not the only one to mine the Yogo dike. While the British worked the eastern end, the steep western end was worked, starting in 1910, by the American Sapphire Company of New York (Voynick, 1985b). Although much money was spent developing the American mine, including underground workings and a modern processing plant, the lack of effective marketing of the stones doomed it to failure. Eventually, in 1914, the mine was bought by the British, giving them control of the entire dike. It was never reopened, but the \$80 000 purchase price was soon recovered by the shrewd Charles Gadson, by reworking the old tailings.

Figure 11.43 Excavations in the sapphire-bearing dike at Yogo Gulch, Montana. (From Claremont, 1906)

During World War I production dropped, but soon returned to high levels. The year of 1921 was a record year for the British mine, with 300 000 carats of gem sapphire being mined. This was its last gasp, however. In 1923 a severe cloudburst flooded the underground workings and destroyed the surface facilities. The British mine was not to recover and in 1929 closed permanently. In

Figure 11.44 Washing for sapphires at the New Mine, Yogo Gulch, Montana. (From Claremont, 1906)

39 years the Yogo mine had produced 2.5 million carats of gem sapphire said to be worth as much as $25 million (Voynick, 1985b).

After the closure of the British mine there followed a long period of inactivity when the only mining taking place was that of local residents working illegally with simple tools. The world, it seemed, had forgotten Yogo. Commercial mining resumed in 1958 but quickly failed, and in the period up to 1979 no less than six separate mining ventures had a go, each ending in failure. By far the most unusual of these was an idea to sell small plots of land adjoining the dike to holidaymakers who would, according to the plan, be able to mine the sapphires during their leisure hours. Of course, this idea never really caught on and the venture soon folded. Finally, in 1980, Harry C. Bulloch formed Intergem, Ltd with the idea of reviving the Yogo mine. Lack of adequate operating capital had killed many of the earlier ventures at Yogo, but Intergem managed to raise a large amount of funds and finally it seemed that the Yogo sapphire might again grace the necks and fingers of the world's rich. Delmer Brown, a mining geologist, was brought in to study the deposit. The plan was to begin working the dike surface and eventually to extend mining underground. Studies by Brown indicated that the dike extended deep into the earth's crust and it was believed that all of it was sapphire bearing. This would make the Yogo dike the largest proven sapphire deposit in the world.

Figure 11.45 Mining for sapphires at Yogo Gulch, Montana, around the turn of the century. (From Claremont, 1906)

Intergem brought a very new and innovative approach to the project. Rather than simply selling the rough material to dealers and cutters, as is done at most gem mines around the world, Intergem planned towards true vertical integration, including mining, cutting and, finally, jewellery manufacture. They developed a very novel approach to marketing the finished product, taking out large advertisements in magazines and other publications billing the Yogo sapphire as the world's only guaranteed non-heat-treated corundums. Only with Yogo sapphires could this claim be made, as the entire production was controlled by Intergem from the mine. Sapphires from other localities, of course, are commonly heated to improve the colour and clarity. In contrast, Yogo stones did not need such treatment because they came out of the ground in a fine blue colour and with a very high clarity.

Unfortunately, Intergem's advertising won it few friends in the gem trade. With stocks full of treated gems, jewellers and dealers did not want anyone telling the public the truth about the frequency of gem treatments. Thus, sales were not as strong as expected, creating cash shortages. Eventually this brought about the downfall of Intergem and an end to the experiment. In 1986 the mines at Yogo closed for good, with the underground mining having barely gotten started. Thus, another chapter in the Yogo sapphire story has come to an end. Only time will tell whether or not the famed blue stones of Yogo will ever again rise from the earth.

Mineralogy of the Yogo sapphire

The Yogo sapphires differ from other sapphires around the world in a number of important respects. Chief among these differences is the fact that the Yogo stones are recovered from the original rock matrix; this has given scientists an unparalleled opportunity to study their genesis.

Yogo sapphires as recovered from the jig concentrate sometimes possess a hard outer coating which is believed to originate from a late magmatic stage reaction of the magma with the previously formed sapphires. Great care must be used in the sorting stages if this coating is not to obscure the crystals themselves.

Another unusual feature of Yogo sapphires is their rich blue colour. A deeper blue than the sapphires from Sri Lanka (untreated), it is lighter than those from Australia. In stones of 1 carat and larger the colour is very close to ideal. The only problem is that stones of this size are extremely scarce. More amazing than the colour itself is the great consistency of colour from one stone to another. Virtually all are of the same even-blue hue, with less than 1% being of a lilac colour. Delmer Brown (1982), chief geologist for Intergem, discovered just two rubies and ten green sapphires out of 300 000 carats of Yogo rough, and these odd colours were all of insufficient size to be faceted!

The clarity of Yogo sapphires is also outstanding. Although liquid, gaseous and solid inclusions do occur, for the most part they are entirely absent. When present, the inclusions tend to resemble those found in Thai rubies. Colour zoning, which disfigures many a sapphire from other localities, is also entirely absent. What is it, then, that has kept the Yogo sapphire from being the world's most desirable sapphire? The answer is simple – their size. Field investigations have shown that less than 10% of the stones recovered exceed 1 carat weight in the rough. The largest cut Yogo sapphire on record is a mere 8 carats. Were it not for this fact, the Yogo sapphire might indeed be the most renowned of blue corundums.

In addition to their small size, Yogo sapphires exhibit a very flat tabular crystal habit, making it doubly difficult to recover large stones in cutting. Crystals are typically combinations of large basal pinacoid and smaller rhombohedron faces. As a result of crystallographic symmetry, the rhombohedron faces often develop such that the pinacoid faces take on a triangular shape. In some cases, this triangular pinacoid is raised slightly above the surface. When equally developed triangles are present on each end of the crystal, observing the stone from above gives the appearance of a six-pointed 'star of David'. This is due to the fact that each triangle is reversed with respect to the other; when they are superimposed, a star is seen.

In many cases crystals exhibit deep etching on their faces, providing further evidence of a reaction with the magma as the sapphires, which formed at depth, were brought to the surface. In most cases, however, this etching has not completely destroyed the crystals' symmetry, but merely rounded their edges. Occasionally, pieces are found which show no trace of their original euhedral outlines. Many would take this rounding as evidence of water transport, but this has not been the case. Again, it has resulted merely from a partial dissolution of the crystal due to attack by the magma.

Other Montana corundum localities

In addition to the famous mine at Yogo Gulch, gem sapphires are also found at other localities in the state of Montana. These, however, are of relatively little importance.

The other localities are to the west of Yogo, along the gravel bars of the upper Missouri River near Helena, and further west, along Rock Creek. These mines are worked primarily by amateurs or mainly for gold, with sapphires recovered as a by-product of the gold mining.

Missouri River

Sapphires were first found in Montana in the gravel deposits along the Missouri River. The original discovery was made by a gold prospector, Mr Ed R. Collins, in the year 1865, somewhere along the Missouri River. This did not create a great deal of excitement, however, as the stones were mostly of a greenish hue. Eventually, in 1889, the Montana Sapphire and Ruby Co., Ltd was floated in London to mine the stones along the Eldorado Bar and adjacent areas. Unfortunately little interest was shown in the sapphires, because of their lacklustre colour, and the company soon folded. Other companies also tried, but none met with any success. In addition to the Eldorado Bar, other areas where mining has taken place along the Missouri River include the French Bar, American Bar, Dana Bar, Metropolitan Bar, Emerald Bar, Ruby Bar and Magpie Gulch.

In 1890 George Kunz (Clabaugh, 1952) described the Missouri River sapphires as follows:

> The Montana specimens rarely exceed 1/4 inch to 1/2 inch in length. They are brilliant but usually of pale tints. The gems are usually of a light-green, greenish-blue, light-blue, bluish-red, light-red, and intermediate shades. They are usually dichroitic [*sic*] and often blue in one direction and red in another, or when viewed through the length of the crystal, and frequently all of the colors mentioned will assume a red or reddish tinge by artificial light. A fine piece of 9 carats was found of a rich steel blue. A very beautiful piece of jewelry, in the form of a crescent, was made of these stones by Tiffany and Co., in 1883; at one end the stones were red, shaded to bluish-red in the center, and blue at the other end; by artificial light the color of all turned red. Perfect gems of from 4 to 6 carats each are frequently met with. Occasionally crystals are found which would afford ruby and sapphire asterias of a poor quality.

According to Clabaugh (1952), most of the stones are pale in colour, with deeper hues being rare. Pale blue-green, blue and green stones are most common and pale purplish-blue sapphires are not all that rare. Very attractive amethyst and green stones have been cut from the darker pieces. Deep-blue, pink, reddish-purple and yellow gems are rare. Colour-change gems have also been found on occasion.

The sapphire crystals from the Missouri River commonly exhibit a faint zonal distribution of colour with the lightest zone in the centre along the *c* axis. Some have more brightly coloured central spots and others show distinct bands of colour. Many exhibit a silky sheen due to exsolved mineral particles.

Most of the stones are slightly rounded and worn, but crystal faces can generally be recognized. The crystals themselves show development of the basal pinacoid, hexagonal prism, bipyramid and various rhombohedron faces. Some are of the thin wafer shape so common at Yogo Gulch, but others are barrel shapes or hexagonal prisms.

In February 1988 the author was shown several parcels of blue sapphires obtained from the Missouri River by the American gem dealer, Mr Tom Palmer. The stones originated from a gold mine located on a ranch opposite the Eldorado Bar on the Missouri River, the sapphires being merely a by-product of the gold mining. When they were taken from the ground their colours were said to be transparent light blues and greens, with no cloudiness or silk present at all. However, after heat treatment and cutting in Bangkok the colour changed to a much richer and more intense blue. One piece shown to the author displayed a beautiful and intense blue hue and was priced at $400 per carat (wholesale price in Bangkok).

Most of the material from this locality is below 1 carat in size after faceting. Good stones of up to 5 carats cut, however, have been found. Although the material is of little value before treatment, proper heating has yielded quantities of stones of definite commercial value. Apparently the sapphire mining alone is not profitable, but fortunately the stones are found alongside commercial quantities of gold which is the major product of the mine. This discovery that heat treatment can produce some very fine blue stones may result in the expansion of sapphire mining along the Missouri River.

Rock Creek

Sapphires of various colours are also obtained in the Rock Creek area, near the town of Phillipsburg in Granite County. Mining is centred chiefly along two tributary gulches of Rock Creek, namely Anaconda Gulch and Sapphire Gulch.

According to Kunz (Clabaugh, 1952), sapphires were first discovered on Rock Creek about 1892, being found in a gravel bed 4 feet (1.2 m) thick overlain by 3 feet (0.9 m) of loam. At first the sapphires were extremely plentiful, with ten to 20 pieces being found in every pan of gravel. As with the Missouri River deposits, various companies were floated to work the claims, but all failed eventually because of the poor colours of the stones found. Today, small-scale commercial mining is proceeding, as well as fee-basis amateur diggings.

Rock Creek sapphires are similar to those found along the Missouri River except that a somewhat wider range of colours is common, with many stones of deeper hues being obtained as well. During the years 1899–1901, it was reported that about 6% of the sapphires mined were suitable for cutting, but Clabaugh (1952) doubted that such a large percentage was so classified in later years.

Most of the sapphires found at Rock Creek are pale in colour, some are dull and translucent, and many show irregular or zonal distribution of colour with a brightly coloured spot near their centre. Bicolour or parti-coloured sapphires are found on occasion and are termed 'pinto' sapphires locally.

The most common colours are pale blue, blue green, green, pink and yellow. Blue or blue-green sapphires with a central orange or yellow spot are also found. In many ways the large assortment of odd colours at Rock Creek brings to mind the well-known fancy sapphires from the Umba Valley region of Tanzania.

Both the Rock Creek and Missouri River sapphires are sometimes heat treated to improve their colour, but the changes are not great. Some of the stones contain minute exsolved mineral particles which are removed by heating. In certain cases, small crystals of illmenite, an iron–titanium oxide, are partly or completely dissolved by the heat, sending these elements into solid solution and consequently giving a blue colour to that part of the stone. The 'cannibalizing' of these inclusions for their colouring agents sometimes forms perfect blue spheres around these crystals, producing a most spectacular effect under the microscope.

The Rock Creek sapphires are, according to Clabaugh (1952), similar in size and shape to their cousins from the Missouri River. Some crystals are sharp and distinct, with pitted surfaces, while others are abraded and rounded. Most of the crystals are hexagonal plates, many are roughly equidimensional, and a few are prismatic crystals with greater length than width. Crystal forms present are the same as those from the Missouri River, namely basal pinacoid, rhombohedron, hexagonal prism, and hexagonal bipyramid.

Bibliography

Brown, D. L. (1982) *Geology of the Yogo sapphire deposit, Judith Basin County, Montana*, unpublished report

Christensen, S. F. (1979) Cornish sapphire. *Gems and Minerals*, August, 46–47

Clabaugh, S. E. (1950) Pegmatites of Montana. *Economic Geology*, **45**, 254–257

Clabaugh, S. E. (1952) Corundum deposits of Montana. *USGS Bulletin*, no. 983

Claremont, L. (1906) *The Gem-Cutter's Craft*, Bell, London

Crocker, R. C. (1956) The Yogo sapphire mine. *Gems and Gemology*, **8**, 323–330

Droullard, S. (1978) Yogo, Montana sapphire. *Gems and Minerals*, August, 40–43

Heinrich, E. W. (1950) Syenitic corundum pegmatites near Bozeman, Montana. *Economic Geology*, **45**, 378–380

Kunz, G. F. (1892) *Gems and Precious Stones of North America*, Dover, New York (reprint)

Leiper, H. (1969) Five miles of sapphires. *Lapidary Journal*, **22**, 1278–1282

Mineralogy, Inc. (1983) The pot of gold at the end of a sapphire dike. *The Goldsmith*, September, 42–45

Pratt, J. H. (1906) Corundum and its occurrence and distribution in the United States. *USGS Bulletin*, no. 269

Themelis, T. (1985) Montana sapphires (Sapphire Mts.). *AGA Gemological Report*, unpublished

Voynick, S. M. (1985a) *The Great American Sapphire*, Mountain Press, Missoula, MT

Voynick, S. (1985b) The sapphires of Yogo Gulch. *Rock and Gem*, April, 52–56

Willard, H. M. (1981) The Yogo sapphire, Montana's elusive treasure. *Lapidary Journal*, July, 868–872

Young, M. W. (1981) Sapphires at Gem Mountain. *Gems and Minerals*, August, 82–84

Index